CW00793251

A Natural History of Ladybird Beetles

The Coccinellidae are a family of beetles, known variously as ladybirds or ladybugs. In Britain alone, some 46 species belong to the Coccinellidae family, although only 26 of these are recognisably ladybirds. Composed largely of Professor Michael Majerus' lifetime work, and updated by two leading experts in the field, this book reveals intriguing insights into ladybird biology from a global perspective. The popularity of this insect group has been captured through societal and cultural considerations coupled with detailed descriptions of complex scientific processes, to provide a comprehensive and accessible overview of these charismatic insects.

Bringing together many studies on ladybirds, this book has been organised into themes ranging from anatomy and physiology to ecology and evolution. This book is suitable for interested amateur enthusiasts and researchers involved with ladybirds, entomology and biological control.

Michael E.N. Majerus (1954–2009) was Professor of Genetics in the Department of Genetics at the University of Cambridge and Fellow of Clare College, Cambridge. He was a world authority in his field, a tireless advocate of evolution and an enthusiastic educator of graduate and undergraduate students.

Helen Roy is a Group Head and Principal Scientist at the NERC Centre for Ecology & Hydrology, where she leads zoological research within the Biological Records Centre (UK focus for terrestrial and freshwater species recording). She is an ecologist with a particular interest in the effects of environmental change on insect communities.

Peter Brown is an ecologist and Senior Lecturer in Zoology at Anglia Ruskin University, Cambridge, where he is also Course Leader for MSc Applied Wildlife Conservation. His research for over ten years has focused on ladybird ecology and, with Helen Roy, he has co-authored two recent books on ladybirds.

Michael, Helen and Peter collaborated for many years and produced numerous peer-reviewed papers on various aspects of the biology of ladybirds.

A Natural History of Ladybird Beetles

M.E.N. MAJERUS
Formerly of Cambridge University, UK

Editors

H.E. ROY
Centre for Ecology & Hydrology, UK

P.M.J. BROWN
Anglia Ruskin University, UK

CAMBRIDGE
UNIVERSITY PRESS

University Printing House, Cambridge CB2 8BS, United Kingdom

Cambridge University Press is part of the University of Cambridge.

It furthers the University's mission by disseminating knowledge in the pursuit of education, learning and research at the highest international levels of excellence.

www.cambridge.org
Information on this title: www.cambridge.org/9781107116078

First published 2016

Printed in the United Kingdom by Clays, St Ives plc

A catalog record for this publication is available from the British Library

Library of Congress Cataloging in Publication data
Names: Majerus, M. E. N.
Title: A natural history of ladybird beetles / M.E.N. Majerus, formerly of Cambridge University.
Description: New York : Cambridge University Press, 2016. | Includes bibliographical references and index.
Identifiers: LCCN 2016016322 | ISBN 9781107116078 (Hardback)
Subjects: LCSH: Ladybugs.
Classification: LCC QL596.C65 M25 2016 | DDC 595.76/9–dc23 LC record available at https://lccn.loc.gov/2016016322

ISBN 978-1-107-11607-8 Hardback

Contents

Editors' Foreword and Acknowledgements

It was with great pleasure and gratitude that we accepted the invitation from Cambridge University Press and Christina Majerus to work on this book. We both had the privilege of working with Professor Michael Majerus (Mike) and over the years had learnt so much from him, not only about ladybirds but also about his approaches to science in general. Mike was a mentor to many and the fondness with which he is remembered is testament to his inclusive approach, leading to collaborations far and wide. So it is particularly poignant that his last book has a global perspective building on his previous books on ladybirds (Majerus and Kearns, 1989; Majerus, 1994a).

Mike was a prolific and eloquent writer. He published many books throughout his lifetime and discussed the text for this book with us before his untimely death. He was inspired by the progress that had been made in the understanding of ladybird biology since the publication of the New Naturalist *Ladybirds* (1994a). The resulting book was beautifully written and represented an overview of ladybird biology organised in themes ranging from anatomy and physiology to ecology and evolution. In editing the book, we have endeavoured to ensure that changes to Mike's original text are minimal and that we have simply provided more recent research perspectives. We should also mention that we have not been alone in making these edits. A number of reviewers provided insightful comments and suggestions. We have endeavoured to address these.

So what have we added? Chapter 1, 'Ladybird, Ladybird...', provides cultural perspectives and is both richly detailed and beautifully written. We simply added a few references; perhaps the most significant was to note that the Naturalist Handbook that Mike had written with Peter Kearns in 1989 has now been revised and published (Roy et al., 2013). There have been a number of changes to the classification of ladybirds and throughout Chapter 2, 'The Structure of Ladybirds' (and subsequently the entire book), we have recognised the new taxonomy, which is outlined in detail by Nedvěd and Kovár (2012).

We worked with Mike to plan *Ladybirds (Coccinellidae) of Britain and Ireland* (Roy et al., 2011) and information gathered by volunteers and submitted to the UK Ladybird Survey (www.ladybird-survey.org) has enhanced our understanding of the associations of ladybirds with other species and their habitat associations. Therefore, we were able to add to Chapter 3, 'Where Ladybirds Live'. Indeed, new technology, vertical-looking radar, has been used to examine flight patterns of ladybirds (Jeffries et al., 2013). Mike published a number of studies on intraguild predation, mainly in response to the arrival of *Harmonia axyridis*, an alien species that Mike described as the 'most invasive ladybird on Earth'. Many people have prolifically continued this research theme and therefore we have included an overview of intraguild predation within Chapter 4, 'What Ladybirds Eat'.

Mike was an evolutionary ecologist and his fascination for this topic is revealed through Chapter 5, 'Sex and Reproduction'. This extensive chapter highlights Mike's deep understanding and represents colloquial reflections coupled with detailed description of complex scientific processes. There have been a number of advances in the field, particularly through molecular studies on paternity (Haddrill et al., 2002, 2008), that Mike would have been aware of but that have only been recently published. There is no doubt that this was the most difficult chapter for us to edit, because it represents the focus of Mike's incredible research profile. There was little we could add.

There had been few studies on the overwintering of ladybirds but over the last few years a number of publications have addressed this gap. There have been a number of recent publications on aggregation pheromones and winter survival of ladybirds that were worthy of addition to Chapter 6, 'Ladybird Dormancy'. Field studies on overwintering are particularly lacking but recent research such as that of Ceryngier and Godeau (2013) assessing the negative relationship between soil humidity and abundance of overwintering *Vibidia 12-guttata* are advancing understanding.

Chapter 7, 'Ladybird Death', has been a source of fascination for many researchers. Mike was no exception and his breadth of understanding from defensive chemistry to parasite taxonomy was impressive. The cocktail of chemicals used by ladybirds for defence continue to be a source of intrigue. Laurent et al. (2005) noted that more than 50 alkaloids have been identified across the coccinellid family. Mike had already noted that *H. axyridis* was more resistant to parasites compared to other ladybirds. There has been further work in this regard and evidence suggests that the species has arrived in enemy-free space within the invaded range (Comont et al., 2014) but may be host to hitchhikers (Vilcinskas et al., 2013). Research on the role of natural enemies in shaping ladybird assemblages is set to continue and the male-killing symbionts that captivated Mike are particularly appealing study organisms.

It is enjoyable to think back to discussions with Mike on the conundrum of colour pattern polymorphism in aposematic ladybirds. The evolutionary links between the defensive chemistry and elytral patterning were described by Mike in Chapter 8, 'Ladybird Colouration', in which he poses three questions. Firstly, how does true warning colouration evolve? Secondly, why do all species of ladybird not look more or less the same? Thirdly, why is there so much colour pattern variation among individuals in some species of ladybird? Throughout the book Mike explores these evolutionary perspectives. A recent study by one of Mike's students examines the effects of temperature during pupation on the spot size of *H. axyridis* (Michie et al., 2010). There is still much to reveal about the evolutionary and ecological relevance of colour pattern variation, but Mike's overview throughout Chapter 9, 'Variation and Evolution in Ladybirds', will certainly provide inspiration for further studies in this area.

The final chapter, 'Ladybirds and People', provides a detailed cultural perspective alongside reflections on the beneficial role played by ladybirds through control of pest insects. Mike celebrates ladybirds as biological control agents while recognising that his concerns over the threat posed to biodiversity by *H. axyridis* are symptomatic of wider problems of increasing arrivals of invasive alien species globally. He states: 'many accidentally introduced species have had little or no effect on native ecosystems, and many agricultural and biological control species have been beneficial, but the impacts of some alien species have been highly undesirable' and that 'Biotic homogenisation now has widely recognised ecological and evolutionary consequences and is considered among the greatest threats to biodiversity (Olden et al., 2006).'

We have added a substantial number of new images throughout the book and are grateful to all of the photographers who kindly gave permission for their images to be reproduced. We are especially grateful to Gilles San Martin, who provided many superb photographs, including that featured on the front cover. We would also like to thank Matt Tinsley (University of Stirling), a PhD student with Mike in the 1990s, who commented on the male-killing section within Chapter 7 Olda Nedvěd kindly checked the taxonomy throughout.

We hope that you will enjoy this book as much as we have done. It will be a pleasure to read it again and again but for now we leave this editorial with Mike's words:

> I also enjoy speculating. In this book I will do so liberally. If others, who are caught up in a fascination of ladybirds as much as I have been in the last quarter of a century, subsequently put my ideas and theories to the test, I will be delighted, whether my ideas are verified or refuted.

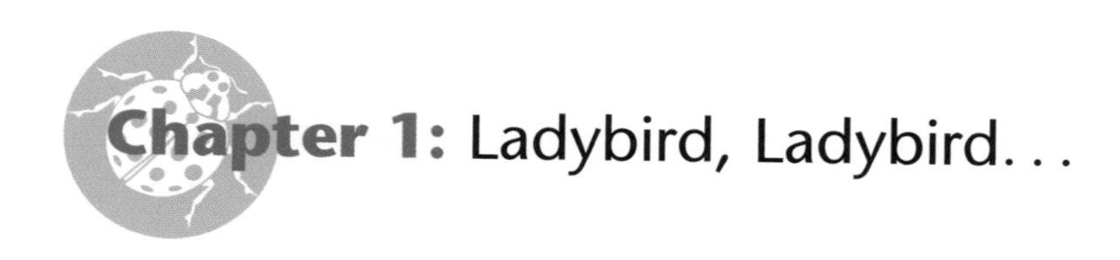

Chapter 1: Ladybird, Ladybird...

Everyone Loves a Ladybird!

> Ladybird, ladybird, fly away home,
> Your house is on fire and your children all gone,
> All except one and that's little Anne,
> And she's crept under the warming pan.

Children throughout the English-speaking world sing this endearing little nursery rhyme. It shows the caring way in which we regard these colourful creatures, which until recently were unique among the insects in being regarded benignly everywhere. Ladybirds are the first insects that most children learn to recognise. In Europe, I have yet to meet anyone who actively disliked ladybirds, let alone had a phobia about them. This is surprising because many people show a strong dislike of beetles, and ladybirds, belonging to the insect order Coleoptera, are beetles. So why do we view ladybirds in this way?

The essence of our rosy view of ladybirds is that they are fun. Their wide popularity is manifest in the commercial and charitable organisations that use them as a motif (Fig. 1.1). In Britain, the shopping chain Woolworth's used the ladybird brand name for its range of children's clothes. Ladybird Books have helped many youngsters in their early reading. In Germany, Coccinelle became a nickname for the early models of the Volkswagen, which, in America, was nicknamed the beetle. The nickname obviously did the car no harm, as the beetle-shaped Volkswagen has been one of the most successful models of car ever produced.

The popularity of ladybirds owes much to their bright colours, most often red or orange and black. They are easily spotted, being active mainly by day, and doing little to hide away. The habit some species have of overwintering in our dwellings, around window frames, or in corners of cool rooms, also brings them to our attention.

The use of ladybirds as pest controllers also contributes to their popularity. Most ladybirds are predators, eating sap-sucking plant pests, particularly

Fig. 1.1. The popularity of ladybirds is evident from their wide use in merchandise and marketing as seen here alongside the Majerus children: Kai, Kara and Nic.

aphids, and they are thus well regarded by farmers and gardeners. Indeed, the introduction of *Rodolia cardinalis*, from Australia into California, in 1888, to control the cottony cushion scale insect, *Icerya purchasi*, which threatened the citrus industry, is widely regarded as the first and still one of the most spectacularly successful instances of biological pest control. In Britain, Kirby and Spence, in their *An Introduction to Entomology*, noted the usefulness of ladybirds as early as 1815, where their importance in controlling hop aphids is noted. More recently, various supermarket chains have fronted campaigns promoting organic food with images of ladybirds and slogans such as 'an insect is better than an insecticide'.

Ladybird Names and Myths

Their bright colours and use in pest control place ladybirds among the most widely venerated of animals. An examination of the vernacular names given to ladybirds in different languages shows strong religious connections. In English, the name ladybird is a dedication to Our Lady, The Virgin Mary. They are 'Our Lady's birds'. This name is thought to derive from one of the most common species across Europe and Asia, *Coccinella 7-punctata* (Fig. 1.2). This species was introduced into North America in the twentieth century and has

Fig. 1.2. *Coccinella 7-punctata* mating pair. (Helen Roy)

become widely abundant. The red colour is said to represent the red cloak that Mary was usually depicted wearing in old paintings and statues, and the seven spots are for the seven joys and seven sorrows of The Virgin. In some parts of Scotland, ladybirds are still called ladyclocks, the clock being a corruption of cloak.

The dedication to The Virgin Mary is not confined to English. It is common in European languages and, due to the colonising endeavours of the Europeans, has spread around the world. Exell (1991) made a life-long study of the common names given to ladybirds in different tongues. In his *The History of the Ladybird*, he cites some 329 names from 55 languages (Exell, 1991). Of these, about a quarter refer to The Virgin Mary, and another 50 are dedications to God. Other names involve Saint Catherine, Saint James, Saint John, Saint Nicolas, Saint Martin, Heaven, Jesus and the Pope. Strangely, one of the Italian names is *Galineta del Diablo* – The Devil's chicken, from Verona, which is applied to a black ladybird and is unique in its dedication to The Devil.

Although it is difficult to date the origin of many of the names, it is likely that some of the European names pre-date Christianity. Indeed, it is notable that The Virgin Mary is not featured in any Celtic language names. In Swedish, *Frejhöna*, *Gullfrigga* and *Gullbagge*, all meaning golden beetle, probably pre-date Christianity (Exell, 1991).

Among the Swedish names is *Himmelska nyckla* – the Keys of Heaven. According to Blackman, The Virgin Mary takes care of the Keys of Heaven. As spring and summer are symbols of Paradise, it is the ladybirds that open the gates of Heaven. Professor Blackman's interpretation of the Swedish equivalent of the English 'Ladybird, Ladybird' nursery rhyme:

> *Maria Nyckelpiga, flyg hem till ditt land*
> *dina barn är i fara, hela hunet stär i brand*
>
> (Mary's Key Maid, fly to your home country,
> Your children are in danger, your whole house is on fire.)

is that it is a warning to beware the threat of purgatory and the fires of hell.

Dr Exell (1991) notes that the idea of the Keys of Heaven being attributed to The Virgin is contrary to the Bible. In St. Matthew's gospel, chapter 16, verse 19, they are entrusted to Peter:

> And I will give unto thee the keys of the kingdom of heaven: and whatsoever thou shalt bind on earth shall be bound in heaven: and whatsoever thou shalt loose on earth shall be loosed in heaven.

Connections with religion are not confined to the Christian world. In Hebrew, the name *Pârat Noshe Rabbênu* – Cow of Moses our Teacher – is used. In Sanskrit, Marathi, Pali and Hindi, dedication is to Indra, the most common name being *Indra Gopa* – Indra's cowherd.

According to Mr Jerry L. King, a gentleman of Cherokee descent, ladybirds enjoy a place of reverence in the mythology of that tribe. Two names are given to them. One, *A Weh Sa*, means creature living in water. This name has an unlikely origin. Because ladybirds are carnivorous and patterned orange with black spots, they are associated with the jaguar, and the jaguar is a cat that is very fond of the water. This name is a common term for ladybird, but is considered too familiar to be used in the visible presence of a ladybird. Then, the name *A Giga U E*, meaning Great Beloved Woman would have been used for these most venerated of insects. This was the title of the highest office a woman could hold in the Traditionalist Native Cherokee Government. On ceremonial occasions, the woman holding this office would have her face painted orange with dyes made from bloodroot, bear grease and white kaolin clay, and spotted with 15 black dots using a paste of charcoal, pine tar, bear grease and blue kaolin clay. These markings are possibly based on one of the 15-spotted ladybirds, probably *Hippodamia convergens, Hippodamia 15-signata* or *Anatis 15-punctata*. The association between the Great Beloved Woman, the ladybird and the jaguar is of ancient origin, and extends to a fourth creature, the owl. It may pre-date the Cherokee language, which evolved about 4000 years ago, and is one of the oldest living languages.

Not all American Indians show such reverence for ladybirds. Their name in Cheyenne is *Mhoynshimo*, meaning playing-card beetle, and is surely a result of the similarity of the spotty patterns on cards and ladybirds. The name is probably post-colonial (Exell, 1991).

Luck-bringers, Match-makers and Weather-forecasters

There are many old wives' tales associated with ladybirds, some of which may be true, but at least one of which is false. The number of spots on a ladybird does not indicate its age. Once a ladybird becomes adult, it develops its basic colour pattern within about 24 hours, and the number of spots it has does not change thereafter.

Ladybirds are considered both fortune-tellers and lucky omens in many parts of the world. One Italian name, from the Genoa region, is *Porta Fortuna*, meaning luck-bringer. Most commonly they are considered portents of romance. The use of ladybirds as foretellers of love may result from their own reproductive behaviour (Fig. 1.3).

Ladybirds are highly promiscuous. *Adalia 2-punctata* (Fig. 1.4) mate on average 20 times in their main reproductive period. They have strong powers of endurance; a successful mating rarely takes less than an hour and a pair may stay *in copula* for over nine hours, although two to three hours is the norm. Certainly, on a sunny morning in spring or early summer, when one's fancy may turn to thoughts of love, it would not be unusual to find mating ladybirds, for up to 50% of a population may be mating at the same time. Indeed, the word ladybird has been used as a term of endearment, synonymous to

Fig. 1.3. *Psyllobora 22-punctata* mating pair. (© Gilles San Martin.)

Fig. 1.4. *Adalia 2-punctata* mating pair.

sweetheart. In Shakespeare's *Romeo and Juliet* (Act 1, scene 3), on the instruction of Lady Capulet, the Nurse, calls for Juliet thus:

> 'What, lamb! What, ladybird! God forbid! Where's this girl? What Juliet!'

Similarly, in *Cynthia's Revels* (Act 2, scene 1) (Jonson, 1599) comes the line:

> 'Is that your new ruff, sweet ladybird?'

The connection between ladybirds and fertility or love is a strong one. The Scandinavian association with the gods of fertility is now reflected when a young girl finds a ladybird and allows it to run around her hand. The maiden should then observe 'She measures me for my wedding gloves'. This alleged ability of ladybirds to foretell romance is seen in many poems. For example:

> This ladyfly I take from off the grass
> Whose spotted back might scarlet red surpass.
> Fly, ladybird, north, south or east or west,

Fly where the man is found that I love best.
He leaves my hand, see to the west he's flown
To call my true-love from the faithless town.

While from Eastern England, where some still call ladybirds Bishie Barnie-bees, comes:

Bishie, Bishie Barnabee,

Tell me when my wedding be,
If it be tomorrow day,
Take your wings and fly away!
Fly to east, fly to west,
Fly to him that I love best.

It is not certain whether ladybirds were dedicated to Bishop Barnabas or to Saint Barnabas. The latter is possible because ladybirds are often associated with fine weather, and Saint Barnabas' Day, in the old calendar, used to fall around Midsummer's Day. However, I think the former derivation more likely, as the name is centred on Norfolk and Suffolk, and Bishop Barnabas hailed from that region. I have been offered two reasons for the dedication to Bishop Barnabas. One is simply that Bishop Barnabas wore a red cloak and a black hat (Forster, pers. comm.). The other notes that Bishop Barnabas was burnt at the stake. As one of the names cited by Exell (1991) is Bishop is Burning, this explanation may have some credibility. However, doubt is again cast by Southey's poem *The Burnie-Bee* (in Newell, 1845) for the Burnie part of the name appears here to be derived from the shiny or 'burnished' appearance of ladybirds.

The Burnie-Bee
Back o'er thy shoulders throw thy ruby shards,
With many a coal-black freckle deck'd;
My watchful eye thy loitering saunter guards,
My ready hand thy footsteps shall protect.

So shall the fairy train, by glowworm light,
With rainbow tints thy folding pennons fret,
Thy scaly brest is deep azure dight,
Thy burnish'd armour deck with glossier jet.

As tellers of the future, ladybirds are also alleged to have the ability to predict the weather and the quality of harvest. In some cases, there appears to be a belief that ladybirds can go further than predicting the weather; they can influence it. From Austria, courtesy of Mrs Gerlinde Southey, comes:

Marienkäfer, Marienkäfer
Flieg' mach Mariabrunn
Bring' uns hent' und morgen
Eine schöne Sunn'!

The translation is:

Ladybird, ladybird,
Fly to Mariabrunn,
Bring us today and tomorrow
A beautiful sun!

Mariabrunn lies about 12 miles from the Austrian capital and features a miracle-working image of the Virgin Mary, who sends good weather to the Viennese.

In northern Germany, it is thought that if most ladybirds (Maerspart) seen have more than seven spots, the harvest will be poor; if ladybirds with few spots are more abundant the harvest will be good. This belief leads to the request:

Maerspart, fly to heaven,
Bring me a sack of biscuits,
One for me, one for thee,
One for all the little angels.

Ladybirds have even been rumoured to have medicinal properties. Newell (1845) notes ladybirds to be a cure for measles and colic. Others have noted that mashing up a ladybird and putting the pulp into the hollow tooth can treat toothache. Jaeger (1859) reports that he tried this treatment twice, and on both occasions his toothache was immediately relieved.

Ladybirds have a unique position among insects in literature and art. In poetry, ladybirds are well featured, particularly in children's nursery rhymes. They are commonly featured on greeting cards of all sorts, and *C. 7-punctata* has been featured on 43 different postage stamps from around the world; more than any other insect species. I have assembled a collection of over two hundred poems, songs and rhymes featuring ladybirds from all parts of the world. Here I have selected just a smattering. I began this chapter with the best known of all ladybird rhymes. The first two lines are relatively invariant but the second couplet has a number of variations. Gordon (1985) cites two versions, the first from Yorkshire and Lancashire, the second from Scotland.

Ladybird, ladybird, eigh thy way home;
Thy house is on fire, thy children all roam,
Except little Nan, who sits in her pan,
Weaving gold laces as fast as she can.

Ladybird, ladybird, fly away home,
Your house is on fire, your children's at home
All but one that ligs under the stone –
Ply thee home, ladybird, ere it be gone.

The derivatives of this nursery rhyme have found their way into other literature. For example, Beatrix Potter's Mrs Thomasina Tittlemouse shoos out ladybirds from her house with the words 'Your house is on fire, Mother Ladybird. Fly away home to your children.'

Poems and rhymes tend to feature our own enchanted perception of ladybirds. Few reflect much about the ladybirds themselves. One song, however, written by Canada's well-known television naturalist, John Acorn, for his show *The Nature Nut*, is an exception, emphasising the aphid-eating abilities and defensive qualities of ladybirds.

Munching on Aphids

Chorus
Ladybug, ladybug, munching on aphids,
Your life is so simple, so plain and so vapid.
Ladybug, ladybug, munching on aphids,
Your life is so simple, so plain and so vapid.

There's birds in the bushes and there's spiders at wait,
For them you could soon be the last thing they ate,
Remember to think when you're plucked from the tree,
The poison to spew as you bleed from the knee.

Chorus
Hold your head high, step light on those legs,
Go fearlessly forth – be not in the dregs,
For you like a skunk with your obvious warning,
In red and black safety you greet each new morning.

Chorus

John Acorn is an accomplished Blue Grass player and a compulsive composer. I recall that one day, while searching for rare, high alpine species in the Canadian Rockies, we had some discussion of the different pronunciations that he and I use for ladybirds/ladybugs. On my return to England, I found this composition in my e-mail.

Divided by a Common Language
You say guttAHta and I say guttAYta,
You say sinuAHta, I say sinuAYta,

You say MyZEa, I say, MIZeeah,
Its hard to know which one of us is keeping the other busier.

You say maculAHta and I say maculAYta,
You say punctAHta and I say punctAYta,
You say HippodAHmia I say HippodAYmia,
You say Propylea, and who am I to blame ya?

Chorus
An Adalia is a Two-spot, by any other name,
Septempunctata is C7, or words that mean the same,
Calvia is the Polkadot, Coccidula is the Snow,
When it comes to naming ladybugs there's so much more to know...

You say ChiloKORus and I say ChiLOCKorus,
You say rhinoSERos and I say rhinOSSerus,
You say PsylloBORa and I say PsylloBORa...too,
Get out your Robert Gordon, cause that's what it's for (woo woo!).

You say reH-GEE-nah, And I say reE-JYE-nah,
Your way is cleaner
But mine rhymes with Va- [Oh! Never mind!].

Chorus
You say AnEEso and I say AnEYEso,
You say conVERgens and I say CONvergens,
You say transversoguttata and I say transversoguttatatatatata,
It's nice to know for all of us the confusion never ends.

Chorus

John Acorn, 2000 (previously unpublished).

Contemporary poets have also not ignored ladybirds. Following publication of results showing that genetically modified potatoes containing an anti-aphid gene from snowdrops can have adverse effects on ladybirds (Birch et al., 1999), this poem appeared on the internet.

Lacewings and ladybirds, mind where you roam.
The plants are all poisoned that once were your home.
They've spliced in a toxin to kill off all pests.

Now friendly bug-eaters will die like the rest.
Ladybird, ladybird, have your children all flown?
The food chain is poisoned – we're left here alone.

Anon

The Cambridge Ladybird Survey

Given all this fascination with ladybirds, their religious, mythological and literary associations, it amazed me to find, in 1984, four years into a study of mating behaviour in *A. 2-punctata*, that no book devoted to British ladybirds existed. Indeed, although many texts had been written on ladybirds in other parts of the world, these were, without exception, taxonomic or reference tomes, catering for scientific researchers. In October 1984, I presented some of my work on *A. 2-punctata* at an exhibition organised by the British Amateur Entomologists' Society, in London, concluding the display with a request for help in obtaining information on all aspects of the natural history of ladybirds in Britain.

At this point, the inherent popularity of ladybirds came into play, with the aid of a man from the British Broadcasting Corporation (BBC). Having seen my appeal for help, he directed a reporter to find out more about our research and the survey. Within a month, more by luck than judgement, a nationwide survey on ladybirds, the Cambridge Ladybird Survey, was born. We had slots on a wide variety of national and local television and radio programmes. By mid-November we had over 500 ladybird recorders across Britain. Over the years of the Cambridge Ladybird Survey, we had further free advertising from all sections of the media, both in Britain and around the world. A full episode of Sir David Attenborough's *Wildlife on One*, entitled simply *Ladybird, Ladybird*, based on information from the survey, was produced. In 1990, the Survey became embedded in British culture when, for a week or two, it became part of the story line of *The Archers*, when Linda Snell, a character in this daily radio drama that has run for over 50 years, became our first fictitious recorder.

In its life (1984–1994), The Cambridge Ladybird Survey had over 5000 individual recorders and thousands of others from the numerous schools and colleges that took part. Much of the information collected during the survey was published in a final report (Majerus, 1995), in a series of scientific papers and in my book *Ladybirds* (Majerus, 1994a). The success of The Cambridge Ladybird Survey in Britain led to the format being used for similar surveys in Canada, South Africa and various parts of Europe, some of which are still ongoing.

The Cambridge Ladybird Survey was reinvigorated in 2005 as the online UK Ladybird Survey, partly in response to the first record of the harlequin ladybird, *Harmonia axyridis*, in England the previous year (Roy et al., 2005). Ladybirds have continued to attract considerable media attention through the UK Ladybird Survey and have featured on popular television programmes, such as BBC *Springwatch* and BBC *Autumnwatch*, and radio programmes, such as BBC *Living World* and BBC *Saving Species*. There have also been many features within national newspapers and magazines. The attention received by ladybirds has undoubtedly contributed to the growing interest in recording of these charismatic beetles.

The data collected during the Cambridge Ladybird Survey and the UK Ladybird Survey, alongside historic data from the Coccinellidae Recording Scheme, represent a rich resource for many investigations including the effect of global warming on insects and on the impact that the recent arrival of *H. axyridis* in England is having on native ladybirds (see www.ladybird-survey.org). Additionally, the data have contributed to a major review of the status and trends of UK wildlife, *The State of Nature Report* (State of Nature partnership, 2013).

The intention of all these surveys was to answer the question posed in the introduction to Sir David Attenborough's *Ladybird, Ladybird* programme: 'But what of the real ladybird?'

A Natural History of Ladybirds

In Britain, the first book to be devoted to the natural history of British ladybirds was that written by Peter Kearns and myself, for the Naturalists' Handbooks Series, in the late 1980s (Majerus and Kearns, 1989). This landmark publication was subsequently revised by Helen Roy, Peter Brown, Richard Comont, Remy Poland and John Sloggett (Roy et al., 2013). In the early 1990s, I wrote a more comprehensive book on the natural history, ecology, evolution and genetics of British ladybirds for the New Naturalist series, published by HarperCollins (Majerus, 1994a).

Since 1994, I have travelled widely, observing ladybirds on all the world's continents with the exception of Antarctica. I have seen and studied ladybirds from the alpine slopes of the Canadian Rockies (Fig. 1.5) and the arctic wastes of Swedish Lapland, through the expanses of the Siberian steppes, to the tropics of Queensland, Uganda, Panama and Malaysia (Fig. 1.6). While *A Natural History of Ladybirds* is modelled on my New Naturalist book, it is a complete rewrite, with a global perspective and many new emphases and sections reflecting our increasing knowledge of the natural history of these fascinating beetles. Although many of the unanswered questions that I highlighted in 1994 have now been answered, many still remain to be addressed, and new puzzles and conundrums on the natural history of ladybirds have arisen as our understanding of these beetles has grown.

As an evolutionary biologist, who believes in Darwinian natural and sexual selection, I acknowledge that my writing has a strong evolutionary perspective.

Fig. 1.5. The alpine slopes of the Canadian Rockies.

Fig. 1.6. Tropical forest, Malaysia.

In his autobiography, Charles Darwin (1887) wrote:

> I have steadily endeavoured to keep my mind free, so as to give up any hypothesis, however much beloved (and I cannot resist forming one on every subject), as soon as facts are shown to be opposed to it.

I also enjoy speculating. In this book I will do so liberally. If others, who are caught up in a fascination of ladybirds as much as I have been in the last quarter of a century, subsequently put my ideas and theories to the test, I will be delighted, whether my ideas are verified or refuted.

Chapter 2: The Structure of Ladybirds

Two of a Kind

On a pine tree, at the base of a cone, rest two beetles (Fig. 2.1). One is very much larger than the other. This larger beetle is bright red with seven black spots, and anyone seeing it would immediately recognise it as a ladybird. The smaller beetle is brown and unpatterned; just a small nondescript brown beetle. Only a specialist is likely to recognise that this is also a ladybird. So what is a ladybird?

What is a Ladybird?

At a basic level, ladybirds are beetles, and so belong to the largest order of organisms on Earth, the Coleoptera (from *koleos* = sheath, and *pteron* = wing). A combination of two features separate beetles from other insects. Firstly, they have biting mouthparts, and secondly, the front pair of wings are modified into hardened wing cases, or elytra, which cover the delicate membranous hind wings and meet along the centre line when closed, rather than overlapping as do the forewings of the true bugs (Hemiptera) (Fig. 2.2).

Ladybird is the common name for beetles of the family Coccinellidae, from the Latin *coccinatus*, meaning clad in scarlet, although many coccinellids are not red at all. The family Coccinellidae comprises almost 6000 described species worldwide, in 360 genera (Vandenberg, 2002; Nedvěd and Kovár, 2012). These are classified into a number of subfamilies (ending in 'inae', e.g. Coccinellinae), and further split into tribes (ending in 'ini', e.g. Psylloborini) (Sasaji, 1968; Lawrence and Newton, 1995; Kovár, 1996; Nedvěd and Kovár, 2012). The classification of these beetles has been revised (Magro et al., 2010; Nedvěd and Kovár, 2012) and here we refer to the subfamilies proposed within this recent phylogeny (see Fig 2.31 for reference).

The family Coccinellidae is one of eight families in the cerylonid series of the Clavicornia section of the superfamily Cucujoidea (Nedvěd and Kovár, 2012). Authorities differ as to the exact position of the Coccinellidae, with Slipinski and Pakaluk (1991) placing them closest to the Alexiidae and

Fig. 2.1. *Scymnus suturalis* and *Coccinella 7-punctata* on a pine tree.

Fig. 2.2. Overlapping forewings of the true bug *Eurydema dominulus* (Hemiptera).

Endomychidae, while Crowson (1955) and Sasaji (1971) saw closest affinity with the Corylophidae and Endomychidae.

The diagnostic features of beetles of the family Coccinellidae are as follows. They are small to medium-sized beetles, 1.3–10 mm in length, with an oval, oblong oval or hemispherical body shape. The upper or dorsal surface is convex, and the under or ventral surface is flat. The eyes are compound and large. The antennae are usually 11-segmented, although this figure may be reduced to as few as seven; are more or less clubbed, and are socketed close to the inner front margin of the eyes or below the eyes. The mouthparts comprise large, strong mandibles, usually with two inner-facing teeth on each; four-segmented maxillary palps lie behind the mandibles, with the terminal segment axe-shaped (Fig. 2.3); a labium, or lower lip, divided into the prelabium and postlabium,

Fig. 2.3. Axe-shaped terminal segment of the four-segmented maxillary palps.

which are connected by a membrane; three-segmented labial palps; and the labrum, or upper lip. The head can be partly withdrawn under the pronotum, the hard plate covering the thorax. The pronotum is wider than the head, is broader than long, and has slight anterior extensions at the margins. The legs are short and may be retracted into depressions under the body. The tarsi (end section of the legs) are usually four-segmented (three-segmented in the genus *Nephus*), but the third segment is small and almost hidden in the end of the deeply lobed second segment, such that the tarsi appear to be three-segmented. Each tarsus bears two claws, which may be simple, divided into two more or less equal pointed claws, or divided into a pointed claw and a blunt claw. The elytra cover the abdomen dorsally, the flight wings being folded away completely between the elytra and dorsal surface of the abdomen when not in flight. The abdomen has ten segments, but from below only five, six or seven are obvious. The first abdominal segment often has a humped central area ventrally. The two ladybirds on the pine cone (Fig. 2.1) share all of these features.

Although most natural history enthusiasts may feel that they could recognise a ladybird if they saw one, two groups of insects often prove problematic. The first are those insects, which, although not ladybirds, are brightly coloured and strongly patterned with spots or stripes. These include other species of beetles (Fig. 2.4) and both nymph and adult true bugs (Hemiptera) (Fig. 2.2). Many of these species mimic ladybirds for defensive purposes. The second group includes coccinellids, such as *Scymnus suturalis* (Fig. 2.1) that are not strikingly patterned with a brightly coloured livery. This group does include some larger species, such as *Cycloneda polita* (Fig. 2.5), some forms of *Harmonia axyridis* and *Hippodamia convergens*, and very old individuals of *Anatis labiculata*. Yet these species are rarely a problem; they still have 'the look' of ladybird beetles about them.

Fig. 2.4. *Endomychus coccineus.*

Fig. 2.5. *Cycloneda polita.*

Rather, it is with the smaller species that problems are more acute. Many species of the subfamily Scyminae are less than 3 mm in length, are black or brown, and unpatterned, or patterned with dull colours. The same is true of a few species from other tribes. Thus, members of the genus *Coccidula* are not only dull, but have a much more elongate shape (Fig. 2.6) than do most coccinellids. Together, these smaller, less obviously patterned coccinellids comprise a significant proportion (about 35%) of the whole family. For those field entomologists who wish to check whether a small, nondescript beetle is a coccinellid or not, examination of the tarsal segments and antennae, with a good hand lens, is usually sufficient for diagnosis.

As beetles, the Coccinellidae belong to the group of insects known as the Endopterygota, that is to say they undergo complete metamorphosis,

Fig. 2.6. *Coccidula rufa*. (© Gilles San Martin.)

consisting of the egg, larval, pupal and imaginal (adult) stages, just as the Lepidoptera, Diptera and Hymenoptera do.

Until it is pointed out to them, many people do not realise that ladybirds are beetles. This is, no doubt, partly a consequence of early learnt prejudices: ladybirds are brightly coloured and 'nice'; beetles are black or brown and 'nasty'. Yet, apart from their bright colours, morphologically and anatomically, ladybirds are fairly typical beetles. The diagnostic features of the Coccinellidae have already been given. However, these are just the bare taxonomic essentials, and give little guidance to the biology of the group.

When considering the design and construction of coccinellids, it is worth keeping in mind the phenomenal success of this design. The beetles are extremely diverse, showing a great range of adaptations to different habitats and conditions. They are economically and ecologically important, and, as Crowson wrote in his magnificent book *The Biology of the Coleoptera* (1981): 'Coleoptera provide excellent illustrations and test cases for almost every general evolutionary principle, and future study of the group may well lead to the formulation of new generalisations.'

It is not surprising that Charles Darwin, who first synthesised the basic tenets of modern evolutionary theory, had a life-long interest in beetles.

Indeed, he once wrote: 'Whenever I hear of the capture of rare beetles, I feel like an old warhorse at the sound of a trumpet' (Darwin, 1887).

The extraordinary number of species of beetle has been noted by at least two evolutionary biologists of renown. Haldane, when asked by a group of theologians what his studies of biology had taught him about the nature of The Creator and His Creation, is reported to have remarked that The Creator must have had an inordinate fondness for beetles.

In a rather less tongue-in-cheek vein, Gould, in *The Panda's Thumb* (1980), marvels at the diversity of life on Earth with the lines: 'And then of course, there are all those organisms, more than a million described species, from bacterium to blue whale, with one hell of a lot of beetles in between – each with its own beauty, and each with a story to tell.'

The coccinellids are only one small part of the Coleoptera, yet even within this single family of beetles there is considerable diversity in form and function. Consideration of this diversity reveals many general biological and evolutionary principles. These will be discussed in ensuing chapters, which will, I hope, be more accessible against a background of the basic morphology and anatomy of the family.

In giving a detailed description of the morphology and anatomy of the coccinellids, it is convenient to consider each of these rather different phases of the life cycle separately. As the adult stage is the most familiar, I shall start with this phase of the life cycle, before considering the immature stages.

Adults

A ladybird's adult life starts with its emergence from the pupa. This process is known as eclosion. The pupal skin splits transversely and longitudinally across the dorsal surface of the thoracic region, and the adult ladybird slowly clambers out. Muscular contractions of the abdomen help to free the adult. The beetle takes several minutes to emerge completely. Once free, it moves around onto its pupal case, where it remains for some time, while its exoskeleton stiffens and its elytra and flight wings expand, harden and dry. At this stage the elytra are soft, have a distinctly matt appearance, and are quite unlike the bright, strongly patterned wing cases so characteristic of many ladybirds (Fig. 2.7). They are usually some shade of off-white, pale yellow or pale orange, and have a washed-out appearance. This contrasts with the pronotum, which is fully coloured and patterned at eclosion. To expand the wings, haemolymph is pumped into them (Fig. 2.8). The flight wings poke out from the back of the elytra during this phase (Fig. 2.9). Once dry, they are folded away by means of a set of folds, mediated by hinge-like breaks in the wing veins. The expansion

Fig. 2.7. Newly emerged *Chilocorus renipustulatus* on its pupal case, noting soft elytra with unpatterned matt appearance. This ladybird will end up black with two red spots. (© Ken Dolbear.)

Fig. 2.8. Haemolymph is pumped into the wings to enable expansion.

and drying of the elytra and flight wings takes about an hour. The deposition of pigments in the elytra takes much longer; it is several hours before the beetle approaches its normal colour pattern, and pigments continue to be laid down

Fig. 2.9. *Halyzia 16-guttata* recently emerged from its pupal case. (© Derek Binns.)

for several days or, in some cases, much longer. Thus, in most red species the shade of red gradually deepens throughout the ladybirds' active lives, so that relatively young adults can be readily distinguished from any survivors of their parents' generation. The ground colour of the handsome *A. labiculata* changes gradually from pale grey through shades of tan and brown to deepest purple during adult life (Fig. 2.10A–C). In *Adalia 10-punctata*, the rate of pigment deposition is variable from one individual to another, this variability being partly heritable and partly due to the temperatures to which pupae are exposed. In other species, such as *Anisosticta 19-punctata*, pigment is deposited in the elytra at a specific stage in the adult's life.

The body of an adult ladybird is made up of the three main sections: head, thorax and abdomen, that are found in all insects. The skeleton is external rather than internal. The main components of this exoskeleton are the cuticular plates, which are joined together by thinner flexible areas, allowing movement. The cuticle is non-cellular, being secreted by the epidermal cell layer below. The cuticle is composed of three basic layers. The outermost is an extremely thin layer, consisting mainly of a hardened protein, cuticulin, and waxes that help reduce water loss through the exoskeleton. Under this is a thicker layer of chitin and proteins, these proteins being chemically 'tanned' to produce sclerotin, a hard, generally brown material, which gives the cuticle its rigidity. This layer is absent from the thin flexible parts of the cuticle. Finally, the third and thickest layer also contains chitin and proteins, but here the proteins are not sclerotised and thus remain flexible.

It is because of the rigidity of the exoskeleton that once ladybirds have eclosed from their pupae and hardened, they do not change in size (length

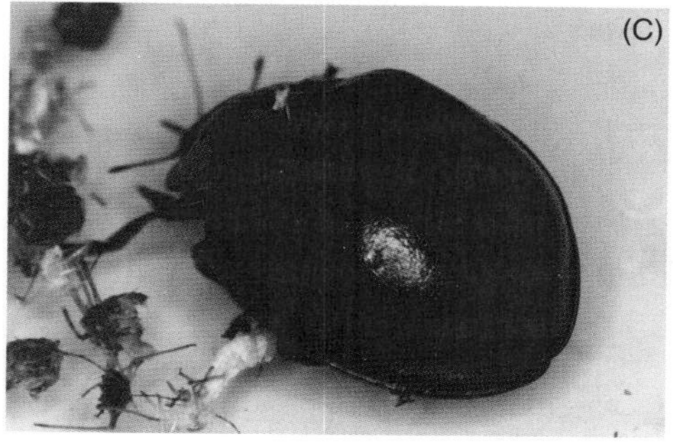

Fig. 2.10. Ground colour of *Anatis labiculata* changes gradually from pale grey (A) through shades of tan and brown (B) to deepest purple (C) during adult life.

and breadth), although they may change in weight. Thus, the common belief that little ladybirds grow into big ladybirds is untrue.

The exoskeleton has two major functions. Firstly, it serves to anchor the muscles and other internal organs and tissues. Secondly, it is protective; adult ladybirds being remarkably resistant to squashing.

Head

The various mouthparts of ladybirds are designed for biting. The mandibles are large and adapted to specific foods. They have two recognisable portions on their inner biting sides: the incisor region and the molar region. In most predatory species, the incisor region bears two large teeth and the molar region carries two more teeth, positioned side by side and orientated obliquely to one another.

In the herbivorous subfamily Epilachninae, the mandibles are specialised for biting off and masticating plant tissues. The incisor region consists of three to five blunt-ended teeth, with jagged inner edges. The molar teeth are absent, being replaced by a long row of small, coarse teeth. Other adaptations are shown by mildew-eating and pollen-eating species.

The antennae are important sensory tools, receiving scent, sound and tactile stimuli. As mentioned, they are most commonly 11-segmented, although this number is reduced in some species, *Chilocorus renipustulatus*, for example, having only seven antennal segments. The size of segments also varies between species. The eyes are positioned to the front and sides of the head and are large and compound, most having a finely faceted surface. Although the eyes appear well developed, studies of the behaviour of ladybirds in respect of mate recognition, detection of prey, and avoidance of predators, suggest that coccinellids do not possess keen sight.

Thorax

From above, the most obvious feature of the thorax is the pronotum, the hard plate that covers and protects the first thoracic segment, or prothorax. The other thoracic segments, the meso- and metathoracic segments, are covered by the elytra and are not obviously visible dorsally, unless the elytra are lifted. In fact, a small section of the mesothorax is visible in the form of the scutellum, a small triangular area bounded by the hind margin of the pronotum and the inner basal margins of the elytra.

It is from the thorax that the wings of ladybirds arise. Most ladybirds have two pairs of wings, the forewings attached to the mesothorax and the hind wings to the metathorax. In a few genera, the hind (flight) wings are absent.

The fore and hind pairs of wings are very different from each other. The elytra are hard, and often brightly coloured and boldly patterned. Their prime function appears to be to provide a protective covering to the delicate flight wings, the meso- and metathorax and the abdomen. However, they also have secondary roles, playing a part in temperature regulation, reducing water loss, and flight. When flying, coccinellids raise their elytra out and forwards, turning them so that the lateral margins point forwards, and are angled to reduce air resistance. In this position, the elytra may help to provide lift, but they are probably more important in steering and in controlling flight pitch, yaw and roll. The exact role ladybird elytra play in flight has not been investigated, but tests have shown that a ladybird that has had the elytra surgically removed is unable to fly.

The hind wings, or flight wings, are thin membranes stretched between veins (Fig. 2.11) and are attached to the metathorax by the axillae, or wing hinges. The flight wings are remarkable structures. As Wootton (1992) pointed out, from an engineering point of view, they act simultaneously as levers, oscillating airfoils and cantilevered beams. They function as levers by transferring energy from the metathoracic muscles and that stored elastically in the cuticle to the air. As airfoils they move air to regulate efficiently the reactive

Fig. 2.11. *Hippodamia 7-maculata* taking off; the hind wings, or flight wings, are thin membranes stretched between veins. (© Gilles San Martin.)

force of the surrounding air against the insect's body, providing lift and movement. During a full wing beat cycle, the aerodynamic forces are constantly changing in strength and direction. The flight wings have to accommodate the continual shifts in these forces as they bend and twist, without structural failure, and as such they may be considered cantilevered beams. As the wings rapidly accelerate and decelerate twice in each wing beat cycle, the forces of stress include a significant inertial component, as well as the aerodynamic component. The wings must also be able to stand up to impact forces resulting from collisions with solid or liquid objects, such as raindrops. The structure of the wing is designed to accommodate such impacts. The basal area, particularly the front vein, is reinforced to aid power transfer from the axilla. This strong, somewhat inflexible region is almost as long as the elytron, and will be partly protected by it in flight. The portion of the hind wing that extends beyond the elytron, known as the deformable tip, is much less rigid and crumples on impact, regaining its normal shape immediately afterwards.

The wing beat cycle is made up of an upstroke and a downstroke. In the upstroke, the wings are orientated to cut through the air with minimum downpush, and to push air backwards, promoting forward motion. The upstroke culminates with the wings clapping together above the insect and peeling apart from the leading to the trailing edge into the downstroke. In the downstroke, the wings are angled to push against the air, to produce lift and movement. The wings are subjected to particularly strong forces during the downstroke, with the aerodynamic forces at first augmented by inertial forces as the wing accelerates, and then opposed by inertial forces as it decelerates into the upstroke.

The main components of the wing – the axilla, the veins, the membrane and the fold and flexion lines – play various roles in flight. The axillae are the points of attachment of the thoracic muscles to the wings. In contrast to the wings of birds and bats, the wings of ladybirds do not contain muscles. The energy generated by thoracic muscles is transmitted to the wings via the axillae. The more rigid parts of the wings, the longitudinal veins and areas of thickened membranes, work as levers to amplify small axillary movements, whether these are primarily linear movements in the upstroke or downstroke, or rotatory movements as the angle of the wing is changed, particularly during stroke transition.

The wing veins are of two main types, longitudinal veins radiating from the base, and cross-veins between the longitudinal veins (Wootton, 1981). Typically, veins are tubes of cuticle that contain haemolymph, and tracheae and nerves may also supply the larger longitudinal veins. The veins are both supporting structures for the wing, and transportation conduits for haemolymph, oxygen and sensory information. Haemolymph is necessary to prevent the wing cuticle desiccating, which would weaken the wing and reduce its flexibility. The longitudinal veins are the supporting beams of the wings. They taper from the base towards the tip, which reduces the energy required, and the stresses generated in the wing during flight. The stresses on these veins are primarily twisting and bending forces. This contrasts with the compression and stretch stresses on the cross-veins. These different stresses are reflected in the structure of the two types of vein, the longitudinal veins being relatively large-diameter, thin-walled tubes, the haemolymph in the vein helping to resist buckling. The cross-veins are corrugated, much as a flexible drinking straw is. This allows the cross-veins to be stretched or compressed while maintaining their flexibility. The veins also help to prevent excessive ripping in the thin, fragile, wing membranes, for they often stop the spread of a tear. The flight wings of ladybirds must be able to change shape, both during flight and when folded away under the elytra. The manner of wing shape changes must be predictable, hence the wings are armed with an array of weak lines. Wootton (1979, 1981) drew a distinction between flexion lines that allow the wing to alter and buckle during flight, and fold lines that are involved in folding the wing away when not in use. The distinction between the two is not absolute, some lines serving both purposes.

The folding away of the hind wings under their protective elytra poses launching problems for ladybirds. Brackenbury (1992) noted that ladybirds are not particularly well equipped for jumping into the air, so it is important that the leg movements are accurately synchronised with the first power stroke of the wings. To allow this, the elytra must be lifted and the hind wings meticulously unfolded first. Watching ladybirds take off, a definite pause is apparent after the wings open and before the ladybird becomes airborne. It is as though the ladybird, like an airline pilot, is going through a checklist to guard against mishap, before applying the power. However, unlike airline pilots, ladybirds do not appear able to shut their systems down quickly during this checking process, for, when things go wrong, the ladybird, rather than furling its wings in an orderly manner, usually falls off the substrate in disorder.

The structure and control of the wings of beetles in general, and the coccinellids in particular, have been subjected to little close scrutiny. This is a shame as study of them may have beneficial consequences for humans. As Wootton (1992) writes: 'As flexible airfoils, much of whose three-dimensional shape is controlled automatically and dynamically by local differences in rigidity, insect wings are unique in nature and have few parallels in technology. Their engineering has great theoretical and potentially practical interest.'

The case for specific study of beetle wings and flight is strengthened by both the phenomenal success and by the unique features of the beetle order. The modification of the forewings into protective coverings, which play only a limited role in flight, must have posed unique and complex problems during the evolution of the ancestral beetle model; problems that have undoubtedly been solved through the modification of the hind wings.

The other main appendages of the thorax are best seen from below. The three pairs of legs, one pair arising from each thoracic segment, are designed for running. They are attached to the thorax by a ball-and-socket joint, the 'ball', or coxa, being lodged and rotating in the socket, or coxal cavity. The coxae are well developed in coccinellids, those of the fore and hind legs being transversely oval, and those of the middle legs almost round. To the coxa is attached the trochanter, a small, somewhat triangular segment, to which is attached the femur. In most ladybirds, the femur is elongate and slender, as is the tibia, which leads to the tarsus. This is usually four-segmented, although, because only three segments are easily visible, it is usually referred to as being cryptotetramerous or pseudotrimerous. Within a species, each leg has the same number of tarsal segments. A claw arises from the final tarsal segment. Claws vary in form from species to species. The legs of most ladybirds are covered in small hairs, or setae. However, in the Chilocorinae these are confined to brush-like pads on the first and second tarsal segments.

The jointed legs of insects provide considerable flexibility of movement. In general, the muscles that allow movement of a particular segment are housed in the previous segment. However, movement of the tarsal segments and claws are served both by the tarsal depressor muscle in the tibia, and by a set of tarsal claw retractor muscles, some of which are housed in the femur. These muscles are attached to the base of the tarsal claw by a long tendon running the length of the tibia and tarsus.

Viewing the ventral surface of the thorax allows one other important thoracic feature to be seen: a soft flexible area of cuticle between the pro- and mesothorax. This thin area of cuticle allows the head and prothorax to move laterally, and vertically, independently of the hind parts of the insect. It also allows the head and prothorax to stretch forwards quite rapidly when foraging, and to be retracted quickly when attacked.

Abdomen

The abdomen of a coccinellid is ten-segmented, but only eight segments are visible dorsally (when the elytra and flight wings are moved aside), and only five, six or rarely seven are visible when viewed ventrally. Appendages of the hind-most two segments are modified to form the genitalia.

The female has no true ovipositor, but the ninth and tenth abdominal segments form a rather slender telescopic tube, which acts as a functional ovipositor. The male has two genital structures, the tegmen and the sipho. The tegmen has a basal lobe from which extends a median protuberance, the sipho (often mistakenly called the penis), and a pair of lateral parameres. The sipho is long, thin and tubular, and rather atypical for a beetle. At its tip lies a rather variable and complex structure. In many species, this structure may be inflated into a bowl during copulation. Here sperm is packaged into a spermatophore before being released into the female. The tegmen comprises a basal plate, a basal lobe, paired parameres and a trabes (Fig. 2.12). The parameres

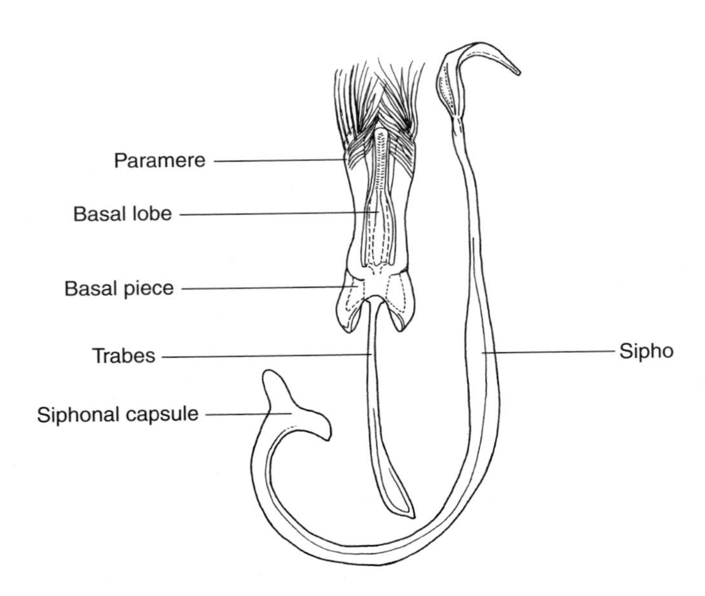

Fig. 2.12. Male reproductive system.

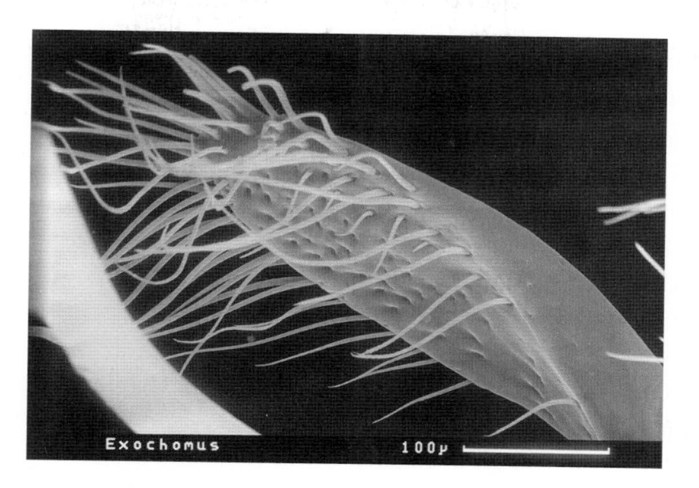

Fig. 2.13. Long, movable setae on the paramere.

usually carry long, movable setae (Fig. 2.13). The precise structure of the tegmen and sipho are relatively constant within a species, but very variable between species and, in consequence, have been widely used in taxonomy to provide reliable, species-specific, diagnostic features. The hard parts (i.e. those that are not dissolved by potassium hydroxide) of the female reproductive organs are also used for identification. These consist of the sperm storage organ, or spermatheca, the sperm duct and the infundibulum. Because the examination of these structures requires the death of their bearer, I have always encouraged the use of features visible on live beetles for identification whenever possible.

When the abdomen is viewed from above, with the elytra and hind wings moved to one side, the spiracles can be seen. These appear as small holes, one pair situated close to the lateral edges of each of the first five abdominal segments. The spiracles are the external ends of the internal respiratory tubes, known as tracheae. The tracheae transport oxygen directly to all the tissues of the insect, via a complex network of tubes, lined with cuticle that is continuous with that of the body wall. The spiracles may be opened or closed, allowing regulation of air supply and water loss, and preventing water or foreign particles from entering the tracheal system.

Most of a ladybird's digestive tract lies within the abdomen. The alimentary canal is longer than the body length, and thus has folds along it. These are mainly in the hindgut, although there is also some folding of the midgut, particularly in plant and fungal feeders. Generally, the length of the gut of aphid feeders and pollen feeders is shorter than that of herbivorous and mycophagous species, or of ladybirds that specialise on scale insects (coccids) or other groups of Hemiptera, such as adelgids and psyllids.

Reproductive System

A detailed discussion of the biochemistry, physiology and internal anatomy of coccinellids is outside the scope of this book. Many authoritative works dealing with these aspects of insect biology exist, and may be referred to. I will, however, consider one internal system, the reproductive system, in some detail, as it bears on other sections of this book. The internal reproductive organs of ladybirds (as distinct from the genitalia) are different in males and females. Essentially, in both, they comprise a pair of gonads, a system of ducts, and various storage structures.

In females, the system consists of a pair of ovaries, two lateral oviducts that join to form the median oviduct, the vagina, the bursa copulatrix, a sperm duct, a spermatheca, and an accessory gland (Fig. 2.14). The number of ovarioles, or egg tubes, in an ovary varies both within and among species. Usually, within a species, larger females have more ovarioles, allowing them to lay more eggs per clutch (Dixon and Guo, 1993). Examining 54 species from 28 genera, Robertson (1961) found a minimum number of 15 in *Stethorus pusillus*, and a maximum number of 51 in *Coccinella 7-punctata*. An individual ovariole consists of a terminal filament, a germarium and a vitellarium. The terminal filaments of the ovarioles are bunched together and suspend and anchor the ovary in the abdominal cavity. At the base of these lies the germarium, a cylindrical structure containing the primordial germ cells, some of which become eggs, or oocytes, while others become nutritive cells, or trophocytes. The vitellarium of a mature female contains a row of developing cells in various stages of development, the smallest and least developed lying closest to the germarium.

As the eggs develop, they distend the ovariole into a series of follicles, or egg chambers, each egg becoming enclosed in a sheath of follicular epithelium,

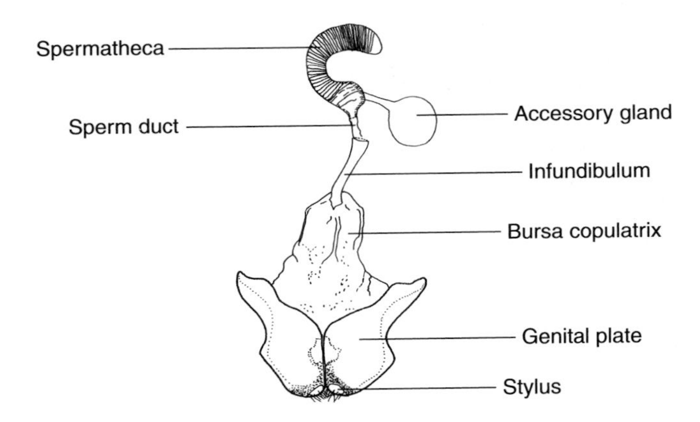

Fig. 2.14. Female reproductive system.

which lays down the chorion (outer shell) around each egg. Eggs are released from the ovarioles into the egg-calyx, from where they pass, via the lateral oviducts and median oviduct, into the vagina. They are laid from the external orifice of the vagina. Below the main cavity of the vagina lies a large bulbous structure, the bursa copulatrix, which is an expanded muscular portion of the vagina. During copulation, sperm are deposited in the bursa copulatrix, are passed through a sclerotised, funnel-like structure, the infundibulum, into the sperm duct and thence to the spermatheca, where they are stored.

The male reproductive system comprises a pair of testes, two lateral ducts, the vasa deferentia, which join to form a median duct, and the ejaculatory duct. Each testis contains many follicles, the number varying within and among species. Sperm are formed and stored in these follicles, which may be round or oval. They are released into the curved tubular vasa deferentia, and thence, via the ejaculatory duct, to the sipho (Fig. 2.15). In most species, sperm are released into an oval receptacle, the spermatophore, which is formed within the open swollen bowl end of the sipho (Fig. 2.16). Thus sperm are usually parcelled up before being transferred to the female. The spermatophore is formed while the end of the sipho is lodged in the female's bursa copulatrix, and here the spermatophore is deposited. Both males and females have accessory glands, attached to the bursa copulatrix in females and to the ejaculatory duct in males. Materials produced by these accessory glands are involved in spermatophore production and sperm transport. At some time during or after copulation, contraction of the muscular walls of the bursa copulatrix squeezes the spermatophore, thereby ejecting the sperm through the infundibulum

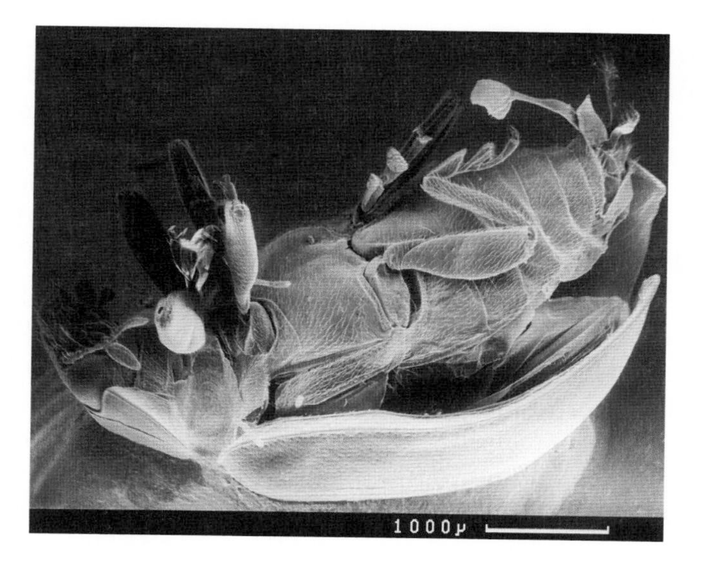

Fig. 2.15. Male reproductive system: the sipho (top right).

Fig. 2.16. Male reproductive system: spermatophore formed within the open swollen bowl end of the sipho.

and sperm duct into the spermatheca. In a few species, such as *Cheilomenes 6-maculatus*, sperm are passed freely into the female's reproductive tract.

In the coccinellids, the external morphology of the sexes is more or less similar. Females are generally larger than males, but the size ranges overlap to a considerable extent; hence, size is not a reliable criterion for sex determination. Obviously, the sex of a specimen can be determined by dissection. However, for many studies, the ability to determine the sex of living insects efficiently and accurately is crucial. One method is to keep ladybirds under conditions conducive to mating. Then, usually, the one on top can be designated as a male. However, even this method is not completely reliable, and it is inappropriate when virgin ladybirds are required. In a few species, underside markings are sex-specific. Thus, for example, male *Coccinula sinensis*, *Harmonia testudinaria* and *Propylea japonica* have small cream markings at the midline of some underside thoracic segments that are absent in females. However, few species have such convenient sex-specific traits. Consequently, Randall et al. (1992) described the underside abdominal segments of 22 species of coccinellid, highlighting sex-specific differences that could be seen with a x10 hand magnifying lens. In most species, the distinguishing traits involved the shape of the ventrally visible, more posterior abdominal segments and the relative sizes of the flexure bands. The flexure bands are areas of thin cuticle between the ventral segmental plates of the abdomen, which give flexibility to the abdomen. Generally, the bands are larger and more obvious in males than in females. This is presumably because males require greater abdominal flexibility than females during copulation (Fig. 1.3), when the male has to curve the rear end of his abdomen under and around to gain access to the female's copulatory opening.

Mating can be seen as the start of the ladybird life cycle. Following mating, female ladybirds will lay eggs.

Eggs

Typically, ladybird eggs are elongate, oval and smooth (Fig. 2.17), although those of the Chilocorinae and some species (of *Stethorus*) are stumpier and almost round. Eggs of some Scyminae have a short thread-like protuberance. A few species, such as *Chilocorus rubidus*, lay eggs singly, directly under scale insects on which the resulting larvae feed (Pantyukhov, 1968a). However, most species lay eggs in batches of between two and 100. The eggs are laid upright and close together. Eggs range in colour from white in some Scyminae, through a range of yellows to dark orange, and become grey before hatching as the developing embryo becomes visible through the shell. In a few, the eggs have a distinctly greenish tinge, and in some, such as *Calvia 14-guttata*, the outer surface of the eggs is covered in a lattice pattern of a reddish waxy substance.

The outer shell, or chorion, of ladybird eggs tends to be stronger than that of most beetles. This may reflect the fact that many coccinellid species lay their eggs in exposed positions, whereas most other beetles give their eggs some protection by laying them in holes or crevices in plant stems, wood, fungi, animal dung, or in the soil. However, the chorion is comparatively weak compared to that of groups such as the butterflies and moths and the true bugs. The chorion consists of several layers, usually including a thin wax layer and a series of layers containing protein. In some Chilocorinae, the waxy outer layer is absent. Directly inside the chorion is the thin vitelline membrane, a product of the egg itself, which may be added to during embryonic development.

Fig. 2.17. Typically, ladybird eggs are elongate, oval and smooth.

At fertilisation, the egg consists largely of yolk, a mixture of fatty acids and proteins. The egg nucleus lies in a thin layer of cytoplasm surrounding the yolk. In most but not all ladybirds, the egg is completely covered by a thin transparent membrane. In some species, the surface of this membrane is granular; in others it is smooth. Ricci and Stella (1988) suggest that this membrane prevents eggs from drying out. The chorion contains specialised pores or canals known as micropyles. These micropyles appear to serve two purposes: allowing entry of the spermatozoa for fertilisation, and allowing oxygen to diffuse into the egg. In *Adalia 2-punctata*, the micropyles are arranged in two rings of between 40 and 50, ending in very short tubular processes at the end of the egg. Other species have a single ring, and in *Platynaspis luteorubra*, the micropyles occur in clusters at both ends of the egg. Members of the Chilocorinae have trumpet-shaped micropylar structures in addition to the tube-like micropyles (Ricci and Stella, 1988).

The development of ladybird embryos follows the general pattern of other beetles. After fertilisation, the embryo, or zygote, divides repeatedly, the resulting cells producing a continuous layer surrounding the yolk, called the blastoderm. Some of the cells move to the posterior end of the egg. These segregated cells are destined to give rise to the primary sex organs and germ cells. As the blastoderm develops, the cells along the mid-ventral line thicken, giving rise to the germ-band. This develops and differentiates, producing all the tissues of the embryo, with the exception of the primary sex organs. The rest of the blastoderm does not develop further, but remains as a thin layer surrounding the embryo and yolk. Continuing its expansion, the germ-band becomes enclosed in amniotic folds that develop from its periphery. These folds grow towards one another, eventually fusing so that the germ-line lies within an amniotic cavity, the amniotic and remaining blastoderm layers cushioning the developing embryo.

Early in the development of the embryo, a series of transverse crevices become visible, those at the head end becoming evident first. These furrows are the beginning of segmental differentiation. Twenty segments are formed, of which the front 6 will form the head, followed by the 3 thoracic segments and 11 abdominal segments. Each segment, apart from the first, develops lateral appendages. However, some of these only develop to a small extent, and are reabsorbed before the embryo hatches. The appendages of the second head segment develop into antennae, while those of the fourth, fifth and sixth head segments become the mandibles, maxillae and labium, respectively. The three pairs of appendages on the thoracic segments become legs, while those on the final abdominal segment form the cerci (or urogomphi). There are 10 abdominal segments, one of the initial embryonic segments disappearing. The cells that segregated off during blastoderm development migrate forwards in two groups and become established as primordial gonads in the body cavity.

Development continues, the nervous system, respiratory or tracheal system, alimentary canal and the integument all becoming differentiated. The yolk is gradually used up, and the embryonic membranes are finally ruptured and reabsorbed. Now the embryo is ready to hatch from the egg.

The rate of embryonic development of most aphidophagous (aphid-eating) coccinellids is exceptionally rapid when compared to that of other beetles of a similar size. Comparison of the duration of the egg stage of aphidophagous and coccidophagous ladybirds shows the latter to be, on average, twice that of the former (Dixon et al., 1997). As aphid- and coccid-feeding species tend to belong to different tribes, this may simply reflect different phylogenetic constraints. However, analysis of five aphidophagous and seven coccidophagous scymnines showed that within this single genus, the aphid feeders spent less time in their eggs than did the coccid feeders (Dixon, 2000).

If phylogenetic constraint does not explain the rapidity of embryonic development in aphidophagous coccinellids, then what does? Two hypotheses seem tenable. Dixon (2000) argued that the nature of prey is an important determinant of development rate. Species that feed on prey with a rapid rate of reproduction, such as aphids, can develop more rapidly than those that feed on more slowly reproducing prey, such as coccids. Indeed, the rates of development of all the immature stages of aphid-feeding species tend to be faster than those of coccid feeders (Dixon, 2000). Because embryonic and pupal developments, as well as larval development, are faster in aphidophagous than coccidophagous species, the difference in development rates cannot be attributed merely to differences in the nutrient value of the two types of prey. Dixon (2000) therefore argued that speed of development is adaptive.

The second hypothesis is also adaptive, and is the first of a number of traits of aphidophagous coccinellids that may have evolved as a consequence of their cannibalistic tendencies. Neonate ladybird larvae of most species eat any unhatched eggs in their clutch before dispersing from their egg batch. This behaviour is much more pronounced in aphidophagous than coccidophagous or phytophagous ladybirds, partly because aphids are harder for neonate larvae to catch and subdue than are coccids, and partly because of the way that eggs are laid. Coccidophagous ladybirds tend to lay their eggs in small groups, of one to five, under the scales produced by their prey. Phytophagous species usually lay their eggs in groups, but the batches are looser, adjacent eggs often not touching. Only the aphidophagous species routinely lay eggs both in large batches and touching, so that a newly hatched larva can reach other eggs from its position on its own egg. Neonate larvae consume not only infertile or inviable eggs in their clutch, but they also attack, kill and eat any slow-developing embryos that have yet to hatch. This habit will apply a strong selection pressure for embryos to develop as fast as possible and hatch as soon as they have a reasonable chance of survival (Majerus, 1994a). It is probable that both of these factors have played a part in the evolution of the rapid embryonic development seen in aphid-feeding species.

Larvae

The first task for the fully developed embryo is to force its way out through the eggshell (Fig. 2.18). This is achieved by the use of egg-bursters, which are specialised cuticular structures on the head or thorax of the embryo. The egg-bursters make the initial break in the egg chorion, and then the larva uses blood pressure to enlarge the break. Egg-bursters seem to have no other function than aiding the young larva's escape from its eggshell, for they are lost at the first larval skin-shed, or ecdysis.

The main function of the larval stage is to eat and grow. Ladybird larvae are atypical beetle larvae, in that they eat the same food as their adult counterparts. However, there the similarity between larvae and adult ladybirds ends. The larvae of the Coccinellini have elongate bodies (Fig. 2.19). This is also true for other tribes, such as the Sticholotidini, although to a lesser extent. The larvae of most Chilocorinae and Epilachninae would perhaps best be described as 'dumpy elongate' (Fig. 2.20). In the Scyminae and Ortaliinae, a dumpy elongate body shape is disguised by a series of fleshy protuberances (Fig. 2.21). The larvae of species of Platynaspidini and those within the subfamily Scyminae of the tribe Aspidimerini are most unusual, being hemispherical, presumably an adaptation to life with ants. Useful keys to the larvae of

Fig. 2.18. Emergence of first instar *Harmonia axyridis* larvae from eggs. (© Gilles San Martin.)

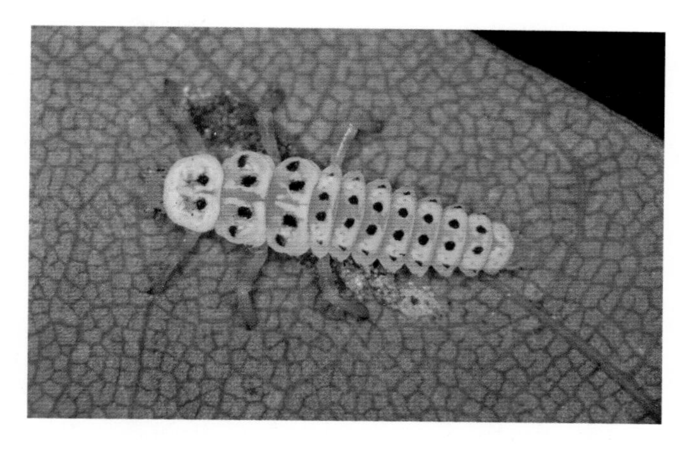

Fig. 2.19. Elongate body of *Halyzia 16-guttata* larva.

Fig. 2.20. Dumpy elongate larva of *Henosepilachna argus*.

ladybirds are given in Hodek (1973), Lesage (1991), Rees et al. (1994), Gordon and Vandenberg (1991, 1993, 1995) and Roy et al. (2013).

Typically, the body of a ladybird larva is arched above and flat underneath. It is rather matt in appearance. The dorsal and lateral surfaces are covered in distinct bumps, giving a rough or spiky appearance. In the Scyminae and Ortaliinae, the body is covered by a wax-like exudation. The head is approximately as long as it is broad, making it squarish with rounded corners. For these species of *Hyperaspis* and *Platynaspis*, it is much wider than it is long. The head capsule is usually hard, dark, and partially or completely sclerotised, although in members of the Scyminae (tribe Scymnini), sclerotisation is slight. There are usually well-developed 'fault lines' on the head, although these are lacking in third and fourth instar larvae of the tribes Scymnini, Platynaspidini and Hyperaspidini. The antennae have up to three segments, the norm

Fig. 2.21. *Scymnus interruptus* larva. (© Mat Kitchener.)

being a short first segment, a longer second segment with a spine, and a short third segment. The mouthparts comprise dark mandibles, maxillary palps, the galea and labial palps. In carnivorous species, the triangular mandibles are armed with one or two teeth at the tip, and have a smooth molar region. Phytophagous species have three, four or five teeth in this position. In non-carnivorous species of the largely carnivorous subfamily Coccinellinae, adaptations to alternative foods are clearly derived from the basic predatory design. Thus, the mycophagous *Tytthaspis 16-punctata* and the phytophagous *Bulaea lichatschovi* each have a row of teeth along the molar region of the mandible that help to grind fungal spores and plant material. The maxillary palps have three segments, except in the tribe Noviini, which have two segments. The galea may be adorned densely with short, thin setae (e.g. *T. 16-punctata*), or short, thick setae (e.g. *Psyllobora 22-punctata*), or sparsely with long setae (e.g. *C. 7-punctata*). The labial palps may have one or two segments.

The thorax has three segments, each bearing a pair of four segmented legs. The ratio of leg length to body length is variable and adaptive, leg length being correlated with food type and habitat. For example, most aphidophagous species have longer legs than those that hunt and feed on more sedentary food such as coccids, or mycophagous or herbivorous species (compare Fig. 2.22 with Fig. 2.23). The abdomen has ten segments, the last segment bearing a sucking organ, the anal cremaster, which can attach to the substrate. Most of the abdominal segments bear a row of six fleshy bumps that end in hairs, or setae. The first eight abdominal segments, together with the mesothoracic segment, each bears a pair of spiracles. Many larvae have hairs and protuberances of various types scattered on the thorax and abdomen.

Fig. 2.22. *Hippodamia 13-punctata*. (© Gilles San Martin.)

Fig. 2.23. *Exochomus 4-pustulatus*. (© Gilles San Martin.)

The basic colour of larvae is fairly constant within a species. Many of the larger species have grey, blue-grey or grey-brown larvae, more or less marked with white, yellow, orange or red spots on the abdomen. Older larvae are more brightly patterned than young larvae. In most of the genus *Psyllobora*, the ground colour is pale grey, marked with broad yellow longitudinal stripes, giving them an overall yellow appearance, spotted by black tubercles. Larvae of the tribes Chilocorini and Coccidulini tend to be some shade of brown, often with characteristic markings that make field identification possible. For example, larvae of *Chilocorus 2-pustulatus* have a series of six pale dots on the first abdominal segment, looking like a whitish belt. The waxy covering of species within the tribes Scymnini, Hyperaspidini and Ortaliini larvae gives them a whitish hue (Fig. 2.21).

The larvae have four stages, or instars, although five have been recorded in some species within the tribe Chilocorini (Yinon, 1969). At the end of each instar the larva sheds its skin, or ecdyses. In preparation for ecdysis, the larva stops feeding and attaches itself to the substrate by means of the anal cremaster. The splitting of the larva's old skin, or cuticle, begins at the head, and continues along the back. The skin is shed backwards, and left attached to the substrate, the cuticular linings of the foregut, hindgut and tracheal system being shed still attached to the external cuticle.

Perhaps the most crucial part of this process is the separation of the epidermal cells (those lying directly beneath the cuticle) from the cuticle itself. Division of the epidermal cells and the laying down of the new cuticle immediately follow this separation. The secretion of moulting fluid facilitates the separation of the epidermis from the old cuticle by the epidermis. This fluid contains the enzymes protease and chitinase, which dissolve the inner layer of the old cuticle, the products being reabsorbed through the epidermis. The actual shedding of the now-detached old skin involves sequential contractions of the abdominal muscles and, once the legs are free, active pulling out of the old skin (Fig. 2.24). The process of ecdysis, from the first splitting of the old skin to final detachment, takes up to an hour, or sometimes longer. The new skin is initially pale, soft and flexible (Fig. 2.25). Following ecdysis, a larva will take in air to increase its volume as the new cuticle hardens, allowing more room for subsequent growth before the next ecdysis.

The final larval moult, turning the larva into a pupa, is preceded by a prolonged period of quiescence, which is rarely less than 12 hours, even in the tropics, and may last several days. The larva stops feeding and attaches itself by its anal cremaster to a stem, leaf, or some other substrate. It decreases in length and assumes a hunched position. This is the pre-pupal stage (Fig. 2.26).

Fig. 2.24. *Coccinella 7-punctata* in the process of ecdysis (shedding of larval skin). (© Gilles San Martin.)

Fig. 2.25. Pale, soft and flexible skin of newly emerged *Anatis ocellata* larva.

Fig. 2.26. Pre-pupal stage of *Harmonia axyridis*.

Pupae

The pupae of coccinellids are somewhat variable in form, depending on the extent to which the final larval skin is shed. In the Coccinellinae, the pupa is uncovered (Fig. 2.27), the larval skin being shed right back to the point of attachment to the substrate. In the Chilocorinae and some Coccidulinae, such as species in the tribe Noviini, the larval skin splits along the dorsal surface, and gapes open, leaving the pupa partially covered, but visible (Fig. 2.28). However, in others, such as *Coccidula scutellata*, the skin is shed back. Some species of the tribes Scymnini and Hyperaspidini pupate within the final larval skin, which, although detached from the epidermis of the pupa, is not fully ruptured until the adult beetle emerges (Fig. 2.29).

Fig. 2.27. *Harmonia axyridis* pupa.

Fig. 2.28. Partially covered pupa (Chilocorinae).

In the Coccinellinae, the skin of the larva splits from the front of the thorax, backwards along the dorsal midline. The skin is worked backwards by muscular contractions of the abdomen and occasional upward jerks of the pre-pupa. The newly formed pupa is typically pale yellow all over, but darkens and becomes patterned as the pupal case hardens. A key to pupae of North American coccinellid subfamilies and tribes is given in Phuoc and Stehr (1974).

Pupae vary in colour and markings, both among and within species. Some are quite striking, such as those of *Halyzia 16-guttata*, which are jet black with

Fig. 2.29. Scymninae pupation within the final larval skin, which is not fully ruptured until the adult beetle emerges.

Fig. 2.30. *Halyzia 16-guttata* pupa. (© Gilles San Martin.)

a series of bright yellow markings (Fig. 2.30). Others are various shades of white, orange, grey, brown and black, with most of those that shed the final larval skin being considerably patterned. Some of the variation in the colour and markings seen within these species is inherited (Majerus, 1994a), but in many species temperature and humidity are also influential. For example, Hodek (1958) showed that under hot, dry conditions (35°C, 55% humidity), *C. 7-punctata* larvae produced pale orange pupae, while at lower temperatures and higher humidity (15°C, 95% humidity) dark brown pupae were produced.

Pupae of many species are formed on the part of the plants that the larvae were feeding on. Some species, such as *C. 6-maculatus*, tend to pupate in exposed positions on the tops of leaves; others prefer to pupate under leaves. Larvae of a few species, such as *C. 7-punctata*, often select their pupation site away from aphid colonies, possibly to reduce the risk of cannibalism by larvae at this vulnerable stage. They typically select a site in an exposed position in the sun. Some coccinellids pupate singly, but many pupate in groups. The reason for this behaviour has not been investigated.

Although the pupal stage extrinsically appears to be a quiescent phase, the pupae of most coccinellids are not completely immobile. When disturbed, the pupae of many tribes have the ability to flick their anterior end up and down rapidly a number of times. This behaviour is apparently an automatic response to certain kinds of tactile stimuli and is thought to be an anti-parasitoid adaptation.

Internally, the pupal stage is far from quiescent. Within the pupa considerable structural transformation takes place. Many of the larval tissues and organs are broken down. The products of this breakdown are used in the development of adult organs, such as antennae, mouthparts, elytra, flight wings, wing muscles and reproductive organs, from imaginal discs that have remained rudimentary since embryogenesis. This process is temperature dependent, usually taking four to ten days in temperate climates, but considerably longer in cold weather, or at high altitudes or latitudes. Conversely, in the subtropics and tropics, the pupal stage of some aphidophagous species can be as short as 38 hours.

Once the transformations within the pupa are complete, the adult beetle is ready to emerge.

The Taxonomy of Ladybirds

The systematic classification of the coccinellids began in the eighteenth century with Carl Linnaeus, the instigator of modern biological classification and nomenclature. Linnaeus, in 1758, described and named many of the larger and more obvious European species. Fabricius, Degeer, Thunberg, Herbst, Rossi, Kugelann, Goeze, Pontoppidan and Scriba added further species in the latter part of the eighteenth century. In 1846, Mulsant reorganised the classification, erecting a number of new genera that are still valid. These include *Adalia, Harmonia, Halyzia, Vibidia, Myrrha, Propylea* and *Anatis*. Mulsant's 1846 reorganisation was the first significant treatment of the Coccinellidae, a group that he named Sécuripalpes. Mulsant's 1850 monograph of the world Coccinellidae, with its 1853 and 1866 supplements, still stands as the basis for modern coccinellid classification. However, this classification was highly artificial, the major division of the group being based purely on the presence or absence of hairs on the dorsal surface.

George Crotch, who also worked extensively on this family of beetles, significantly revised Mulsant's system (Crotch, 1874). He erected six subfamilies, including the Chilocorides and the Coccinellidae. The genus *Platynaspis* was here first combined with *Chilocorus* and several other genera. The Coccinellidae were divided into tribes, the plant-eating Epilachnides, and the carnivorous Coccinellides. Crotch's collection of ladybirds from around the world is now housed in the Zoology Museum, University of Cambridge.

At about the same time, but independently of Crotch, Chapuis (1876) produced a classification based on Mulsant's system, but dividing the family into two subfamilies, the carnivorous Coccinellides aphidiphages and the herbivorous Coccinellides phytophages. In 1899, Ganglbauer split the Coccinellides aphidiphages into two subfamilies, the Coccinellinae, containing the vast majority of the carnivorous genera, and the Lithophilinae, containing just one genus, *Lithophilus*. He also named the phytophagous subfamily the Epilachninae.

Among other coccinellid taxonomists of particular note is Weise, who wrote extensively on the family from 1878 until his death in 1930, before the Coccinellidae section of the Junk Catalogue was published. However, Korschefsky, one of his protégés, and a coleopterist of considerable ability, completed the work, which was published in 1931 and 1932. Korschefsky retained Weise's three subfamilies and divided the Coccinellinae into 20 tribes. Mader (1926–1937) produced a detailed catalogue of all the then known colour pattern variations of European and Asian ladybirds, most of them being afforded Latin names. More recently, Sasaji, in *The Phylogeny of the Family Coccinellidae (Coleoptera)* (1968), conducted a full revision of the group, paying more attention than any previous author to the evolutionary relationships of the different groups in the family. In this phylogenetic classification, Sasaji erected seven subfamilies. These included the Lithophilinae and the Epilachninae of previous classifications, while the Coccinellinae were divided into five subfamilies. The seven subfamilies were further divided into tribes. This was subsequently further revised (Fig. 2.31).

Phylogenetic analyses are useful in bestowing a long-term evolutionary perspective on the relationships among different taxa in a group of organisms, and among different groups. At its best, a modern phylogenetic classification should be based on a variety of different types of comparative evidence, including appropriate morphological and anatomical details of all stages of the life cycles, chromosomal compositions, analysis of enzymes and other proteins by means of gel electrophoresis, and molecular genetic analyses. Sasaji's (1968) treatment was based on the comparative morphology of coccinellid larvae and adults. Care must be taken during such comparative analyses that the traits considered are appropriate. For instance, mouthparts are unlikely to provide particularly suitable material, for the characteristics of the

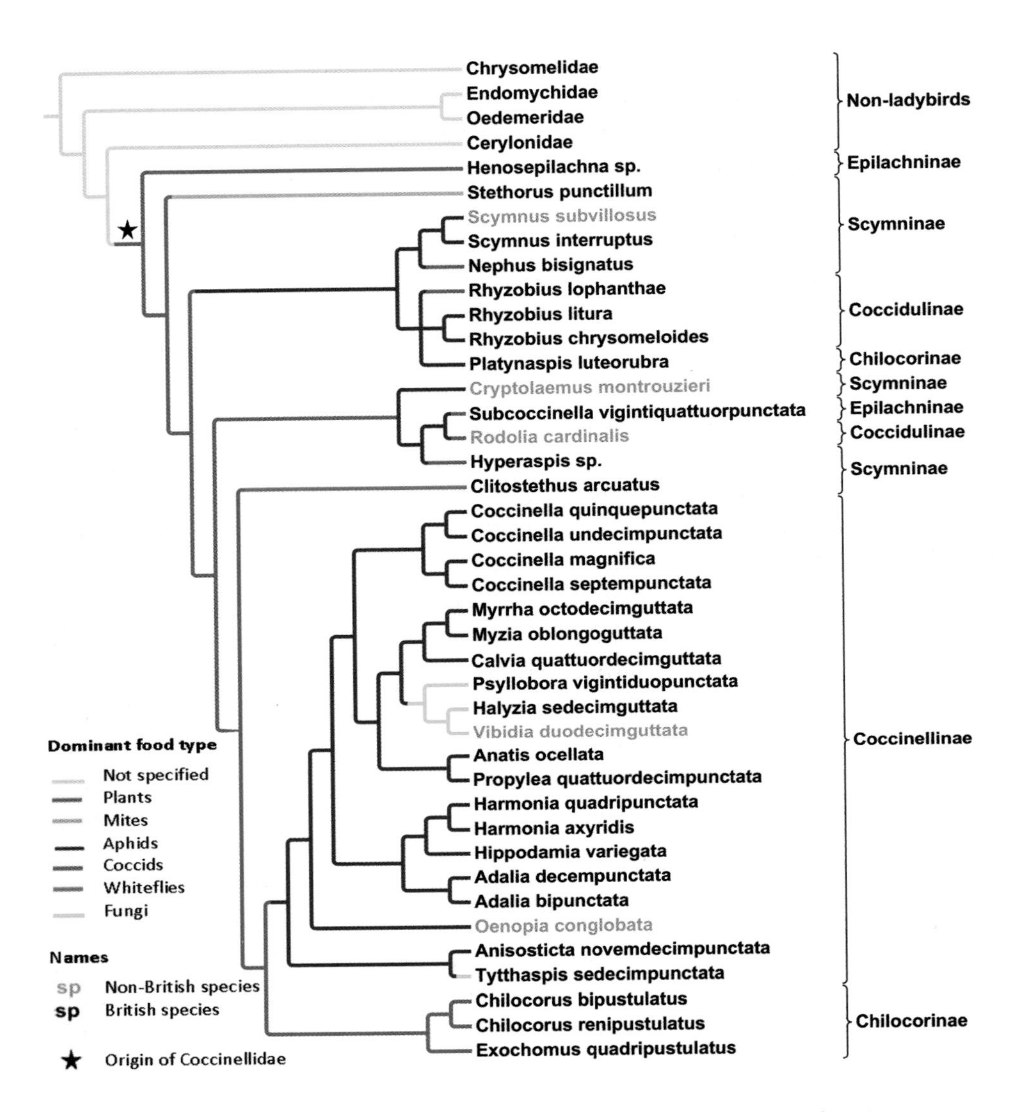

Fig. 2.31. Phylogeny of the Coccinellidae found in Britain based on comparisons of DNA sequences (revised from Magro et al., 2010; © Roy et al./Pelagic.) (For more information on phylogenetic relationships, see Nedvěd and Kovár, 2012.)

mandibles of coccinellids appear to reflect diet rather than evolutionary affinity. Highly variable traits, such as colours and patterns, are also inappropriate for phylogenetic analyses, as they are simply too dynamic in evolutionary terms. Of more use are those characters that are relatively constant within species, but are variable between species, and show distinct evolutionary trends. The genitalia of coccinellids fit these requirements well, and have been used more than any other features both in species diagnosis and in the construction of phylogenetic classifications. The male penis has proved to be valuable in this respect. In primitive members of the tribe Coccinellini, the penis tube is a simple, short, thick structure, ending in a point. In more advanced members of the tribe the penis is longer, bulb-ended, and adorned with various embellishments in the form of grooves, flagella and other protuberances.

Sasaji's 1968 classification has subsequently been revised a number of times (Gordon, 1971, 1985; Chazeau et al., 1989; Duverger, 1989; Booth et al., 1990; Fürsch, 1990; Lawrence and Newton, 1995; Kovár, 1996), with some subfamilies being combined or additional subfamilies being proposed. Furthermore, the phylogenetic positions of a number of tribes and genera have been moved between the higher taxonomic levels. However, these revisions have also relied largely or exclusively on morphological features, usually of adults, but also of larvae (Savoiskaya and Klausnitzer, 1973) and pupae (Phuoc and Stehr, 1974). DNA is central to a consideration of evolutionary relationships, as it is the principal molecule that is passed from one generation to the next through evolutionary time. Differences in the sequence of nucleotide bases, particularly in non-coding or selectively neutral regions of the DNA molecule, can be used to investigate the closeness of the evolutionary relationships among species. Closely related species will have DNA with very similar nucleotide sequences; distantly related species will show more sequence differences.

The advances in DNA technologies provide new and powerful tools to those studying the taxonomy of the ladybirds and the evolutionary relationships among the varied features shown by this diverse family of beetles. These tools are now being used to investigate the evolutionary history of species in the family (von der Schulenburg et al., 2001a; Nedvěd and Cihakova, 2004; Magro et al., 2010). Robertson et al. (2008) confirmed that the Coccinellidae is a monophyletic group, that is, ladybirds are more closely related to each other than they are to other beetle families. Magro et al. (2010) analysed five genes from each of 61 species in 37 genera to re-evaluate the evolutionary history within the ladybird family. They found some discrepancies at the subfamily level: while the subfamily Coccinellinae was found to be monophyletic, subfamilies Coccidulinae, Epilachninae, Scymninae and Chilocorinae are paraphyletic. As an example, *P. luteorubra* (a European species that has a close association with ants), which currently falls within the Chilocorinae, was found to be more closely related to *Rhyzobius* species, in the Coccidulinae, than it is to other members of the Chilocorinae (Magro et al., 2010). As well as enabling a revision of the evolutionary history of ladybirds, these technological advances will allow scientists to answer questions such as,

'Why are female *A. 2-punctata* so promiscuous?' 'How do habitat specialisations evolve?' and 'Why do some ladybirds only produce daughters?' Increases in our knowledge of ladybird taxonomy, ecology, behaviour and evolution, through the use of these tools and more traditional methods, will provide more information to help answer the question of 'what is a ladybird?' over the next decade.

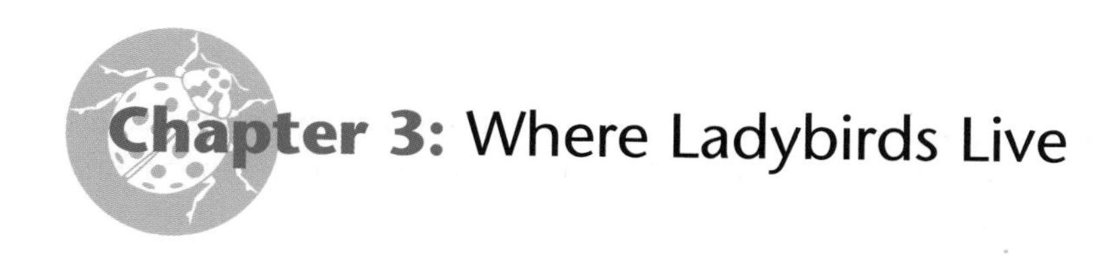

Chapter 3: Where Ladybirds Live

Geography, Habitat and Microhabitat

If I am asked the question 'where can I find such-and-such a ladybird?', my response is based on my knowledge of the geographic distribution of the species, the types of habitats it prefers, the trees and plants within these habitats upon which it is usually found, and where on the trees and plants it usually resides. In addition, my answer may depend on the time of year, as the habitat and microhabitat preferences of many species change with the seasons.

The distribution of a species can be considered at the different levels of geography, habitat and microhabitat. Here I consider the geographic distributions, habitat and microhabitat preferences of ladybirds during the months when ladybirds are active and breed. Chapter 6 deals with the behaviour of ladybirds during their dormant periods.

Habitat and microhabitat preferences usually determine current geographic distributions, not vice,versa. For example, in Europe, the range of *Aphidecta obliterata* increased with changes in forestry practice that resulted in the growing of large plantations of needled conifers for timber. These provided plentiful food and favourable habitats for these adelgid feeders.

Geographic distributions are affected by changes in the availability of suitable habitats; the habitats that ladybirds use are not usually affected by geographic distributions. Conversely, in the past, geography, or more specifically topography and changes in abiotic environmental conditions, have played a part in the evolution of many ladybird habitat preferences and their adaptations to particular habitats. It is because of this that I will begin by considering the factors that have influenced the geographic distributions of coccinellids, before discussing habitat and host plant preferences.

Geographic Distributions

Ladybirds may be found on all the continents of the world with the exception of Antarctica. In considering the geographic distributions of ladybirds, we can take two perspectives: the first descriptive, the second causal. The first of these approaches would lead to a series of maps of the known distribution of each species, and give some indication of which ladybirds live where, at least on a broad scale. The second may allow some understanding of why ladybirds live where they live and not elsewhere. I will briefly consider both approaches.

Ladybird Distribution Maps

Over the last century, there have been numerous attempts to define the geographic distributions of a vast array of organisms. Ladybirds have not been ignored, and many surveys have been conducted on various geographic scales. In Britain, for example, Hawkins led a team that surveyed a single county, Surrey, over a 20-year period (Hawkins, 2000). In France, Le Monnier and Livory (2003) mapped the distributions of ladybirds of La Manche, the north-western region of Normandy, producing a superbly illustrated book of the results. On a larger scale, the Cambridge Ladybird Survey attempted to produce current maps of ladybird distributions in England, Scotland and Wales (Majerus, 1995). These maps can be contrasted with maps produced by the Biological Records Centre, which include all records, through time, of British ladybirds (Roy et al., 2011). Similar contrasts can be made between surveys for parts of North America, such as that for Alberta compiled by Acorn (Acorn, 2007), that for the whole of Canada from a survey in the 1990s run by the Canadian Nature Federation and involving some 3000 participants, and the maps produced by Gordon for all the coccinellids of North America north of Mexico, largely from museum specimens (Gordon, 1985). The Lost Ladybug Project (www.lostladybug.org) is a web-based project that maps ladybird sightings from across North America, relying on members of the public to submit records, which are verified from photographs by experts.

There are good and bad features of all these mapping schemes. The value of relatively short-term recording schemes is that they give, in the first instance, a contemporary picture of distributions, and they are sufficiently time limited to allow comparative surveys to be undertaken at a later date so that changes can be monitored and interpreted. An additional benefit is that these surveys generally include members of the public and engage their interest and enthusiasm. This is particularly valuable when children become involved, for it not only fires their natural curiosity, often leading to a life-long interest in natural history, or even a career in biology, but also because children frequently discover things that experienced entomologists may miss. To give just a single example, the host plant preference of *Halyzia 16-guttata* for sycamore, *Acer pseudoplatanus*, in Britain was discovered by a 12-year old girl, Zoë Williams, who did not appreciate that this introduced tree was previously considered of little entomological interest (Majerus and Williams, 1989).

The negative side of these surveys is that coverage is usually uneven and incomplete, due to variations in human population densities and recording activity. Furthermore, errors may occur when less experienced recorders are used, and as such, surveys are largely based on field observations and possible mistakes cannot be checked by later examination of collected specimens. This is not true of specimen-based maps or those that rely only on data gathered by experienced entomologists. However, these also have their flaws. Firstly, these maps rarely give an indication of changes in distributions over time. Secondly, the coverage is often sparse, with records widely spaced. Here the assumption is often made that, at least for the more secretive species, if a few dispersed records exist, a species will occur throughout the region between them. For example, Gordon (1985) gives a map for *Anisosticta borealis* that shows it to have a continuous distribution from the coast of Alaska to the south-east corner of the Hudson Bay, and including the north of Alberta. However, John Carr, who has collected extensively in northern Alberta, never found it there. He asked Gordon about his map. Gordon answered that the map was prepared by a technician who merely drew a line between two other records outside Alberta (J.L. Carr, pers. comm. to John Acorn). Dot maps, where each dot represents a record, are better, but unless different symbols are used for different periods of time, they do not indicate temporal changes in distributions.

The internet and increases in digital photography may help to broaden the recording base, without jeopardising stringency. In 2005, the first online recording survey was introduced in Britain, with records only being verified if accompanied by a specimen or a photograph of sufficient quality to allow the survey organisers to make a definite identification. New statistical modelling techniques are enabling more robust analysis to account for differences between years and places than was previously the case (Isaac et al., 2014).

The Causes of Ladybird Geographic Distributions

Trying to explain why different species have the geographic distributions that they have is problematic because these are largely historical issues. Consequently, we can only speculate, on the basis of circumstantial evidence relating to climatic changes, vegetative changes, geographic topography and plate tectonics, how ladybirds come to be distributed as they are. However, some of this evidence, particularly when drawn from comparative analyses of a variety of taxonomic groups, is compelling.

It is pertinent to make a point here. When conditions change, mobile organisms, such as ladybirds, tend to move rather than adapt. Adaptation to specific local habitats is most likely to occur when individuals of a species are unable to move further because dispersal is restricted by geographic barriers, such as seas or mountains, or by absence of habitats that offer necessities for life, such as food. Thus, we would expect to see different responses to environmental change by coccinellid populations on the great continents compared to those on isolated oceanic islands or archipelagos. On the continents, unless some barrier, such as a mountain range or desert, bars migration, we would expect populations to move rather than adapt. On isolated islands, ladybird populations faced with environmental change would have no option other than to adapt to the new conditions.

This means that on large land masses, dispersal is the fundamental biological influence on current coccinellid faunas; Darwinian evolution is not. Examination of fossil material from the Palearctic and Arctic regions has shown that species in Quaternary deposits (up to 2.4 million years old) are morphologically identical to living species. In earlier deposits, from the Late Miocene (5.7 million years ago), some species fall outside the range of morphological variation of living species, but are so similar to them that they may be ancestral to current species.

Appreciably, not all ladybirds will react to environmental change in the same way, because not all ladybirds start from the same position. Some species can cope with environmental change better than others. In general, an organism may solve the problem of where to live in one of two ways. Either it may become adapted to a wide range of environmental conditions, and so be able to exist over a vast geographic range (eurytopic), or it may become adapted to one or a small number of habitats (stenotopic). If stenotopic, the limits of a species' total geographic range may or may not be affected by the habitat specialisation. However, on a local scale, the specialist will have a more patchy distribution than the habitat generalist. Two closely related species, *Coccinella 7-punctata* and *Coccinella magnifica* illustrate this well. *Coccinella 7-punctata* is a generalist, while *C. magnifica*, while living in a range of habitats, always lives in association with ants. In Europe and Asia, the longitudinal and latitudinal limits of the distributions of these species are not too dissimilar. However, within this range, *C. 7-punctata* has an almost continuous distribution and is abundant over much of this range. Conversely, *C. magnifica* has a much disrupted distribution and is only locally common where appropriate species of ants occur.

Specialist species will also find successful movement to new suitable habitats more difficult than will generalists. Once adapted to the precise conditions offered by a particular habitat, selection will favour the evolution of behaviours that restrict a species to the habitat to which it is adapted. Thus, ladybirds that have become closely adapted to one or a small number of specific habitats must find it more difficult to change habitats than those with less specialised adaptations.

Considering these matters, we might try to divide ladybirds into groups on the basis of the causes of their geographic distributions. However, the categories will be different if we consider matters on a global scale, or a more local scale, such as a country. Thus, on a global scale, the distribution of a species today might be a consequence of movements of the Earth's tectonic plates, or climate changes during and between ice ages. On a local scale, factors such as food availability, the distribution of particular host plants and micro-climate may have the greatest influence. Here, on a global scale, we are concerned with historical biogeography, while on a local scale, ecological biogeography has the greatest effect.

In considering the present geographic distributions of ladybirds, we are largely concerned with the relatively recent geological past: the Quaternary Period that began only about 2.4 million years ago and continues to the present day. This span of Earth history has been characterised by numerous and intense climatic oscillations. At latitudes outside the tropics, these oscillations saw swings from Mediterranean to subarctic, or temperate to high arctic conditions. Across North America, Europe and Asia, regions that now, during the current warm period (interstadial), host temperate continental or continental maritime climates were, at the height of the glacial periods, polar deserts. In North America, the ice-sheets extended down into the United States. In Europe, they extended to the Alps and beyond. Even in Britain, with its mild maritime climate, more than two-thirds of the land was covered by an ice-sheet. To the south of these glacial wastes, tundra-like landscape with an arctic flora and fauna dominated.

At lower latitudes, in the tropics, the oscillations in temperature were less severe. However, these regions also saw climatic change, with wet periods interspersed by drier ages, during which the tropical rain forests that, until man's recent activities dominated immense areas of the tropics, were reduced greatly in extent to rain forest refuges at higher altitudes or near rivers and other wetlands.

The last of the great glacial phases reached its maximum about 18 000 years ago and, at that time, the coccinellid fauna in the northern hemisphere above a latitude of 50°N was extremely impoverished and concentrated within very few northern refugia, such as Beringia, or the ice-free area around Mountain Park in the Rocky Mountains (Belicek, 1976). With the retreat of the ice from 14 000 years ago, the insect faunas gradually moved north and with them came many arctic species, such as *Hippodamia arctica*, which was among the leaders in the northward recolonisations.

The cycles of climatic oscillation during the Quaternary period have resulted in many beetles changing their geographic ranges on an enormous scale, particularly with respect to latitude. Species have tracked climatic change, moving in such ways that the environments in which they live remain more or less constant. They have maintained environmental constancy, not geographic constancy. It is important to realise that these movements were not the result of active intent. Individuals did not move south because it was getting cooler, or north because it was getting warmer, as do migratory birds, for example. The temperature changes were too small, too unpredictable, and too slow, compared to the generation times of the species concerned, to be detectable. Rather, the geographic range in which a species could exist changed. Those individuals in areas where the climate was deteriorating, from that species' point of view, simply died if the climate became too adverse. At the other extremity of the species' climatic range, the geographic range would be extended, if the species spread into regions where the climate, previously unfavourable for the species, was ameliorating. Certainly, should geographic or habitat barriers bar the way of such advance, colonisation of new regions would not be possible, and local extinctions would result, unless specialised adaptations to local conditions evolved. This pattern has been likened to 'walking the plank'. A species moves along the plank as the climate changes. If climatic changes reverse before the end of the plank is reached, the species retraces its steps. If climate changes continue in the same way, the end of the plank will eventually be reached, when no suitable habitats are available for colonisation. In theory, the species may then teeter on the end of the plank, as selection, acting upon the variation in the gene pool, adapts the species to the changing local conditions. In practice, however, the rate of climatic change may be too rapid, or the amount of appropriate genetic variation available too small, for the species' balance on the end of the plank to be maintained indefinitely. Biologically, falling off the plank is extinction!

On land masses with wide, more or less continuous, latitudinal dimensions, such as Eurasia with its connection to Africa, or the Americas, the plank is very long, and extinctions will have been scarce. Here speciation will also be relatively rare. The frequency of large-scale climatic fluctuations during the Quaternary period, coupled with the mobility of insects, must have continuously broken down geographic barriers between populations, thus preventing divergence of gene pools (Coope, 1970). Significant divergence would occur only in circumstances where populations were fully isolated by habitat or geographic barriers, as in caves, on islands, or in very specific and disjunctly distributed habitats from which movement to other favourable locations was all but impossible. In these isolated habitats, environmental change must have been endured on the spot, with adaptive evolution to local conditions being the only alternative to extinction. This type of 'habitat islands' scenario was rare in North America, or continental Eurasia, but it did occur. For example, the satyrid butterfly genera *Erebia* and *Oeneis* are two of the most speciose in higher latitudes (<40°N) in the northern hemisphere. Most of the species are alpine or arctic in their distributions. Some have very restricted geographic distributions, and are highly specialised with respect to the altitude range within which they occur. The pattern of distribution and altitude ranges of these species suggests that, in the past, populations evolved adaptations to very specific altitudes when they gradually ascended mountains as the climate warmed during interstadials, there to be stuck at the top of one or a small number of mountains, isolated from other populations. It is possible that some arctic-alpine coccinellids evolved their specialist adaptations to cold climes in the same way.

During the northward recolonisation, some coccinellids did evolve to some extent, and in some cases, there is evidence that northward migration was associated with genetic changes sufficient to promote speciation, as in the genus *Chilocorus* in North America (Smith, 1959, 1966).

The recent history of the coccinellid fauna of Britain has been most closely studied, and illustrates the factors that affect species turnover. In the context of zoogeography of relatively dispersive species, Britain is part of continental Eurasia, and should not be thought of as an island. There was, after all, a land bridge to the continent as recently as 7000 years ago.

Most of the current British coccinellid species reached England during a relatively short period after the last glaciation, around 10 000 years ago, but before the land bridge to the continent was breached, 7000 years ago. The presence of the English Channel for the last seven millennia has been important in restricting colonisation by a number of species that occur on the continent in habitats similar to those present in Britain today. Certainly, species such as *Calvia 10-guttata*, *Oenopia conglobata* and *Cynegetis impunctata* could probably establish in Britain, if they could breach the barrier of the English Channel. Indeed, *Harmonia 4-punctata*, *Henosepilachna argus* and *Harmonia axyridis*, are examples of species that have arrived and established in the last 100 years. *Harmonia 4-punctata* is a needled conifer specialist that arrived and established in East Anglia in the 1930s, has subsequently spread to most of mainland Britain, and is still extending its range (Majerus, 1994a; Roy et al., 2011). *Henosepilachna argus* is a more recent arrival and its initial colonisation in south London was probably unintentionally man-aided. It is a plant-eater, feeding on cucurbits, such as *Bryonia dioica*. The species is now firmly established and is spreading slowly (Hawkins, 2000; Roy et al., 2011). It will be instructive to follow the spread of this host plant specialist vegetarian as a comparison with the aphidophagous, host plant specialist *H. 4-punctata*. The most recent arrival, *H. axyridis*, which was first recorded in 2004, probably reached Britain both by flying across the English Channel, and aided by humans, on flowers, vegetables and in packing cases from Europe, North America and possibly Asia (Majerus et al., 2006). This species and its impacts on other ladybirds in Britain and elsewhere will be discussed in more detail in Chapter 10.

For many species in the current British coccinellid fauna, colonisation after the last glaciation was not their first occurrence in Britain (Majerus, 1994a). For example, Coope and Angus (1975) recorded five current species, *C. 7-punctata, Coccinella 11-punctata, Hippodamia 13-punctata, Anisosticta 19-punctata* and *Scymnus frontalis*, from southern English organic silt deposits, radio carbon dated to a temperate interlude in the middle of the last glaciation, about 43 140 years before present. Consideration of a total of 248 identified fossil beetles at this site alone provides strong evidence of an intense climatic oscillation in the middle of this glaciation; for a short period conditions were as warm as, or perhaps warmer than, those of southern England now (Coope and Sands, 1966). Analysis of the ecology and niche ranges of these beetles provided the first good evidence of this short sharp warm interlude (Coope and Angus, 1975). More recent oscillations of this type are known. For example, between 13 000 and 14 000 years ago, the climate fluctuated from conditions as warm as now, down through several lesser oscillations to an episode of arctic conditions, before a sudden temperature amelioration returned the climate to conditions as warm as the present day, about 10 000 years ago.

The importance of movement in response to climatic change as a major influence on the coccinellid fauna of Britain is beautifully illustrated by two species that do not occur now in Britain. One, *H. arctica*, has been found to be common in fossiliferous organic lenses in the Midlands, dated to approximately 40 000 years ago, in the Upton Warren Interstadial (Fig. 3.1). According to Coope (pers. comm.), the species was widespread in Britain from this time until about 25 000 years before present, when it disappeared as temperature declined for the duration of the Late Weichselian Glaciation, and conditions were too cold for it. Just over 13 000 years ago, the species reappeared for a few hundred years, apparently invading from the south, moving northwards across Britain, and disappearing to the north as the climate warmed. *Hippodamia arctica* again invaded and became common between 11 000 and 10 000 years ago during the cooler, Younger Dryas period (Loch Lomond Stadial) (Osborne, 1971). It then disappeared again and has not returned since. In Europe, *H. arctica* is now restricted to very northerly latitudes in Scandinavia.

The second species, *Anisosticta strigata*, was also common during the Upton Warren Interstadial. However, there is no evidence that this species has reappeared since the Late Weichselian Glaciation. In Europe, it is currently confined to cooler regions of Scandinavia. Other fossil coccinellids from during the Upton Warren Interstadial reinforce the picture of a cohort of subarctic species being present in Britain during this period. Perhaps most unexpected of all, remnants of the ladybird, *Ceratomegilla ulkei* (Fig. 3.2), were obtained from a lens of organic silt within a terrace of the River Thames, again dating to about 40 000 years ago (Briggs et al., 1985). This ladybird is now

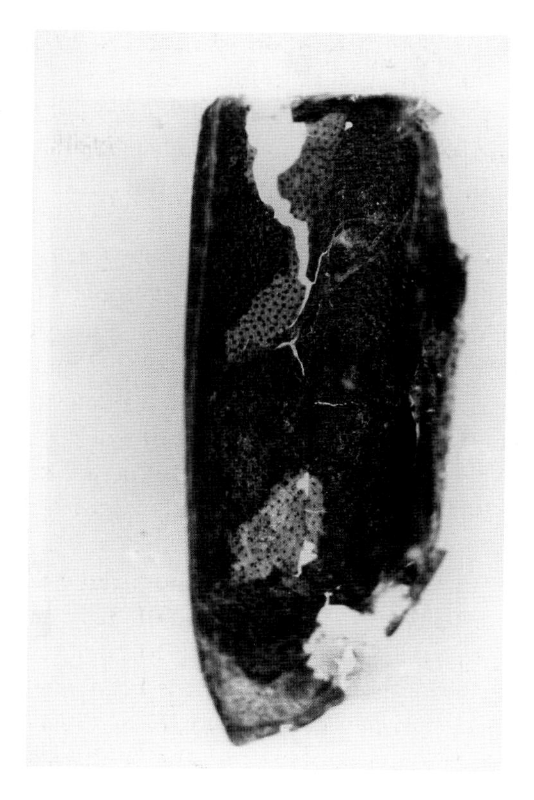

Fig. 3.1. *Hippodamia arctica* elytron found in fossiliferous organic lenses, dated to approximately 40 000 years ago.

confined to north-west Canada (British Columbia, Ontario, Yukon Territory, North West Territory) and eastern Asia (North Kazakhstan, Altai, Transbaikal, Mongolia, West China), generally at high altitude.

Interestingly, *C. 11-punctata* was present during the Upton Warren Interstadial, from 40 000 to 25 000 years ago, but all remnants found during this period have been assigned to the form *confluens*, which is now only common towards the northern latitudinal limits of this species' range.

The information that has been gained by the study of ladybirds and other beetles in silt deposits has many applications. I well recall my first visit to Russell Coope's laboratory in 1992. There, Russell and Peter Osborne invited me to look down a microscope at a sample of insect fragments washed out of a clay deposit. There lay fragments of elytra, pronota, thoraxes, heads, legs, mandibles and abdomens, and Peter and Russell began reeling off names of what was in the sample, while I just looked at the plethora of bits and pieces in amazement.

The lessons to be learnt from studies of the past have great biological importance. For example, the finding that mobility, rather than adaptability

Fig. 3.2. *Ceratomegilla ulkei.*

to local conditions, is paramount in the long-term survival of most species, has great implications for conservation of all groups. It shows the importance of long-term and continuous latitudinal wildlife corridors, which facilitate north–south dispersal. In the long term, nature reserves are islands of conserved habitat. If surrounded by unfavourable habitats – monoculture agricultural land or urban sprawls – they will only be of long-term value to those species that have a sufficient dispersal range to travel from one conserved island to the next, jumping unfavourable habitats. Inevitably, the favourable habitats that nature reserves provide for the species that currently live in them will become unfavourable in the future, as a result of natural or man-induced environmental changes, such as climatic oscillations, pollution or global warming.

Immigration

The lists of coccinellids recorded from a particular country usually include a few species that have only been reported on one or a small number of occasions. Such coccinellids may simply be species that are rarely recorded because they are secretive, or that they are naturally rare in that country, or they may be rare vagrants or accidental importations. Inclusion of this latter category of species on country lists is controversial. Thus, for example, *Vibidia 12-guttata* has been included on some British lists on the basis of four nineteenth century records, and a single observation from Dorset in the 1920s, which probably came from imported goods (Majerus, 1994a). This mildew-feeder is orange, with 12 white spots. It occurs commonly in Europe, favouring deciduous woodland. Most authorities suggest that the few British records represent natural vagrants.

Conversely, the first British records of *Cryptolaemus montrouzieri* in the wild, recorded in north-east England in 1991, were almost certainly escapes (Constantine and Majerus, 1994). This Australian species has been transported to many other parts of the world to control scale insect pests. It has been commercially available in Britain since about 1980 and has been used in butterfly houses and by organic greenhouse gardeners. This species has, in my view correctly, not yet been added to the British list. However, a winter 2007 record of the species supplied to the UK Ladybird Survey suggests that this species may now be able to overwinter in Britain outdoors.

As we transport more and more produce rapidly around the world, we run the risk of transporting ladybirds, and many other small creatures with them. Records of the central and southern American *Eriopis connexa*, of the Jamaican *Procula douie*, and of the South African *Cheilomenes lunata*, all in fruit brought from English supermarkets in the last 20 years, show that accidental importations do occur. However, such species should not be added to a region's faunal list unless they become established (evidence of successful reproduction).

That said, when a species is found in a locality for the first time, it should certainly be recorded, with the decision of whether it is added to the region's faunal list being left until it is known whether the species becomes established. The current interest in assessing the effect of climate change on species' distributions, and particularly the movement from low to high latitudes as a result of global warming, makes what may or may not be spurious records of immigrant species highly noteworthy. All such records should be published without delay, so that the range expansion of a species that does become established in a new region, and the consequences the expansion has on other species, can be fully monitored from first colonisation.

Population Explosions

Ascertaining the exact geographic distribution of a species is of value in understanding the habitat specialisation of that species, and its environmental constraints, which, in turn, may be important considerations for both its potential use in pest control, and its own conservation. However, distribution maps are only of limited use in the long run, because the geographic distributions are themselves dynamic. Ladybirds can fly and do disperse. One feature of ladybird biology is particularly confounding for distribution mapmakers. However, near real-time mapping is becoming possible with online recording systems such as iRecord (www.brc.ac.uk/iRecord). Intermittently, the numbers of some aphidophagous ladybirds explode. Then, huge numbers of ladybirds take to the air, moving hither and thither, often in great swarms. Generalists are most prone to these population explosions. The precise causes of such explosions have rarely been critically analysed. However, in temperate and Mediterranean climates, they usually occur during hot summers after mild winters and warm wet springs. Considering European explosions, Majerus and Majerus (1996) suggested that the causes were a combination of low winter mortality, an abundance of food during the late spring/early summer reproductive period and then hot weather allowing rapid larval development.

The magnitude of some ladybird population explosions is difficult to conceive unless one has personal experience. The last great ladybird year in Britain was 1976. Majerus (1994a) cited a number of anecdotal reports of the immense swarms of ladybirds recorded in the July of that year. Two examples will give some flavour. Mr A. Paine wrote: 'Regarding the events of 24th July 1976, the figures from our own front garden totalled approximately 17 500 insects. There are 50 or so gardens along Spriteshall Lane alone, and there appeared to be no diminution in ladybird numbers anywhere. Therefore the estimated total for these gardens was over half a million, all being seven spots as far as could be seen. When you consider that this is for just one lane in one town, it gives some idea of the phenomenal numbers that must have been involved.'

A flying enthusiast described an unusual flight: 'On 11 July 1976, I was piloting a light aircraft about 20 miles north of Manchester, at 1500 feet, when I flew into a large swarm of ladybirds. It was like flying into birdshot. I put down safely at East Midlands Airport to clean the canopy and check the air intakes. There were hundreds of the little beasts, alive, walking all over the plane.' The use of new technology for detecting insects, such as Vertical Looking Radar, has revealed that it is not unusual to find ladybirds at altitudes above 1000m (Jeffries et al., 2013).

Perhaps most staggering is the statistic that, during late July and early August 1976, at least 400 miles of tide-line on the south and east coasts of England consisted of little but dead or dying ladybirds. If all were *C. 7-punctata*, i.e. the largest species involved, a conservative estimate gives a figure of 23 654 400 000 beetles in the tide-line at any one time. To give some idea of the magnitude of this figure, it means the number of ladybirds in the tide-line was well over three times the number of people who now live on Earth – and this figure is for a single day, and takes no account of the ladybirds that stayed on dry land, or those that drowned at sea and were never washed up. In Britain, similar but more local explosions of ladybirds have occurred since 1976, notably in 2009.

The causes of the 1976 explosion in Britain have been well researched (Majerus, 1994a; Majerus and Majerus, 1996). Despite reports in the national press at the time, there is no evidence to support the view that the ladybirds were immigrants. The increase in numbers was the culmination of a series of weather conditions, beginning in the summer of 1975. That summer was unusually long, warm and sunny. The winter that followed was mild, and led into a warm wet spring that, in May, gave way to one of the hottest summers then on record. These weather conditions were close to optimal for ladybirds. Their numbers increased substantially during 1975, and the majority of an abnormally large overwintering population survived the mild winter that followed. In the spring, as they emerged from their winter retreats, they found plenty of food, as aphids had also survived the winter well and reproduced rapidly on the lush spring vegetation. The peak in aphid populations coincided with the initial hot weather, when the ladybirds began to reproduce. This led to high ladybird fecundity and low larval mortality, and by the end of June, enormous numbers of new adults were hatching from pupae. However, in the first half of July, conditions began to deteriorate. Plants began to dry off and wither. Aphid populations started to decline rapidly. Moreover, huge numbers of their predators and parasites, not just ladybirds, but hoverfly larvae, lacewings, predatory beetles and bugs, and parasitoid wasps also decimated these prey populations. By mid-July, the aphids were all but gone, and billions of ladybirds were hungry. Seeking food, they took to the air in huge swarms, which flew across the countryside, partly on the prevailing winds, until they reached the sea. Here they stopped and, now starving hungry, tried to eat anything that could give them a little nourishment.

It was, of course, these starving ladybirds that bit people. In their desperate attempts to find food, the ladybirds nipped people to discover whether they were edible. Ladybirds do not sting in the normal sense of the word, but in biting, they inject a tiny droplet of a digestive enzyme, to help soften up food, before imbibing it. Our own immune system reacts against this foreign chemical, producing a stinging sensation, and a small bump, which lasts for an hour or so. Naturally, people are not good eating for ladybirds, so they moved on in their desperate search. Sadly, a great many never found sufficient food, and huge numbers drowned at sea, starved to death in the late summer, or died the following winter, having been unable to build up sufficient fat reserves. Many of those that survived did so largely through cannibalism.

Ladybird population explosions are often spectacular and so have been reported with irregular frequency since biological records began. The species involved in large explosions are almost invariably aphid-eating generalists.

This is a consequence of the much greater variations in the availability of aphids than other types of ladybird food, such as coccids, adelgids, spider mites or fungi. Most reports are from temperate and Mediterranean regions, reflecting the predominance of aphidophagous species in these climes. Within these regions, massive increases in population densities, and vast swarms have been reported from many parts of the world, including North America, many parts of Europe, the north African coast, South Africa, central and western mainland Asia, Japan, Australasia, and southerly parts of Argentina and Chile.

The pertinence of these population explosions to a consideration of ladybird distributions lies in the dispersal behaviour shown by these huge populations when they run short of food. Taking to the air, and rising to high altitudes on thermals, these ladybirds can travel large distances before coming back to Earth. Should food be available where they do return to the ground, these ladybirds may survive and eventually breed. They may therefore, by this behaviour, extend their range. The irregularity of population explosions and the unpredictability of the dispersal of ladybird swarms mean that it is difficult to sensibly add the potential effects of these expansions into models of distribution or population demography. However, they undoubtedly do have an effect, and parts of some of the current distributions, particularly in the case of disjunct distributions, are undoubtedly a consequence of such events.

Habitat Favourability

Dispersing ladybirds have to be able to survive and reproduce where they end up if they are to contribute to future generations. Their ability to do so will depend on the favourability of the habitat that they land in. If they land in an unfavourable habitat, they may seek to move again. However, should they land in a habitat that provides food and appropriate environmental conditions, they may stay and reproduce. I therefore turn now to a consideration of habitat favourability and habitat specialisation.

Scale and perspective are persistent problems when considering habitat specialisation. Think of a small oak woodland. This may be just a part of the habitat in which a pair of owls live and forage. The individual tree in which this pair nests, however, may effectively be the whole habitat for many insects that are born and die on that tree. Furthermore, some of these insects may pass through their entire lives on just a very small part of the tree, say the underside of an individual leaf. Consequently, an appreciation of a scale appropriate to the organism under consideration is important in developing any useful classification of habitats.

In light of this difficulty, Southwood (1977) put forward a simple habitat classification based on the favourableness of a habitat to a particular organism in terms of both time and space. He defined a favourable habitat as one that allowed a species to both survive and breed. With respect to time, four classes were recognised, these being constant (always favourable), seasonal (periods of favourable and unfavourable conditions alternate in a regular temporal pattern), unpredictable (favourable periods of variable duration are interspersed by periods of unfavourable conditions of variable length), and ephemeral (a favourable period is followed by an indeterminately long unfavourable period). In terms of space, Southwood erected just three classes: continuous (the area of favourable habitat is more extensive than the dispersal range of the organism), patchy (favourable patches are close enough to each other for organisms to move from one to another), and isolated (a limited favourable area that is too far from other favourable areas for the organism to migrate to). The time and space classes can be combined to give just twelve habitat classes. A population cannot survive over long periods in either a patchy ephemeral or an isolated ephemeral habitat, leaving just ten categories that can sustain life.

This generalised type of habitat classification is very different from a descriptive habitat classification based on the pervading environmental factors (e.g. hot desert) or the dominant species of organism present (e.g. maize field or papyrus swamp), for it is based on the organism under consideration. To *Exochomus 4-pustulatus*, a scale insect-infested pine tree would be a patchy seasonal habitat. It is patchy because other suitable pine trees would be within its dispersal range, and seasonal because conditions become unfavourable for reproduction during the winter months. Under this classification, an ash tree in a woodland dominated by this species in northern France, and an olive tree in an olive grove in southern France, would also be classified as patchy seasonal habitats, although the periods of unfavourability would vary with latitude.

Here then, Southwood's classification works well. *Exochomus 4-pustulatus* is adapted to patchy seasonal habitats. This gives us some understanding of the geographic variation in the habitats in which this ladybird is found. However, for those that wish to find this ladybird in a particular geographic area, some indication of the host plants on which it is most likely to occur is generally of more use than simply saying that it should be sought in patchy seasonal habitats. This leads us to a consideration of host plant preferences.

Host Plant Specialisation

Most ladybirds fly well, and many species feed upon ephemeral prey. They may thus periodically take to the air, particularly when a food supply has been exhausted, and in seeking food elsewhere may land on almost any plant. Indeed, ladybirds of one species or another may be found in virtually every type of common terrestrial habitat. Some species, such as *H. axyridis* and *Adalia 2-punctata*, show relatively little in the way of habitat or host plant preferences. Others have some limitations, but are still found in a wide range of conditions. Thus, for example, *Coleomegilla maculata* and *Hippodamia variegata* are largely confined to low vegetation, whereas *Calvia 14-guttata* and *Eocaria muiri* live mainly in the trees. Still others are highly specialised. A number of species, such as *Macronaemia episcopalis* (Fig. 3.3), *Anisosticta bitriangularis* (Fig. 3.4) and *A. 19-punctata* are confined to wetland habitats and are adapted to living on reeds. Thus, some species may be considered habitat generalists while others are specialists.

Food specialisation is primary among the factors that influence the habitat or host plant choices of a species. Habitat generalists tend to have a wide range

Fig. 3.3. *Macronaemia episcopalis.*

Fig. 3.4. *Anisosticta bitriangularis.*

Fig. 3.5. *Myzia oblongoguttata*. (© Gilles San Martin.)

of foods on which they can feed and reproduce, while many habitat specialists are particular about what they eat. We may consider *H. axyridis* as a generalist, as it has been recorded preying upon over 60 species of aphid, as well as scale insects, psyllids, adelgids and lepidopteran eggs and larvae, and will reproduce successfully in captivity on a range of artificial foods. Conversely, *Myzia oblongoguttata* (Fig. 3.5), a conifer specialist, requires aphids of the genera *Schizolachnus* or *Cinara* for reproduction (Majerus, 1993). Similarly, Eastop and Pope (1966, 1969) studied the small coccinellid *Scymnus auritus* over a six-year period. Fully 99% of the specimens they recorded were found on *Quercus* infested with the aphid *Phylloxera glabra*. Furthermore, they found a close correlation between the abundance of the ladybird and its prey. In these cases, it is probable that food specialisation has evolved secondarily to habitat specialisation. Certainly, *M. oblongoguttata* has a number of specific adaptations, not least in its colour pattern, to living on mature needled conifers. Once it had become habitat specific, its prey specialisation may have evolved secondarily with selection favouring increased capture and digestive abilities in respect of the large *Cinara* aphids that commonly frequent these conifers. Alternatively, the prey specialisation may have evolved as an optimal habitat indicator.

Although food availability is undoubtedly a major determinant of habitat choice, other factors, such as climate, vegetation stratification, altitude, soil or substrate type, and the presence of other organisms, may also be influential in some species. To take just one of these factors as an example, a number of species of coccinellid appear to be confined to high altitude. Thus, *Bothrocalvia lewisi* and *Bothrocalvia albolineata* are confined to mountainous parts of India, while the same is true of *Coccinella alta* (Fig. 3.6) in Canada and the United

States, and *Spiladelpha barovskii kiritschenkoi* and *Coccinella reitteri* in Kazakhstan. In some cases, adaptations to the conditions afforded by these specialised habitats are obvious. For example, many of these high-altitude species have poorly developed wings and rarely fly as a result of the jeopardy that lies in that activity in habitats dominated by high winds.

When considering habitat and host plant preferences, it is implied that these are active behavioural preferences. *Anatis ocellata* (Fig. 3.7) and *Anatis labiculata* (Fig. 2.10A–C) breed on pine trees in the spring and summer because they actively search out these types of tree, and when they find them, if suitable food is available, they choose to stay there. An alternative hypothesis is that these ladybirds are well adapted to living on pine trees and cannot

Fig. 3.6. *Coccinella alta.* (© John Acorn.)

Fig. 3.7. *Anatis ocellata.* (© Gilles San Martin.)

survive elsewhere in the summer. They might, therefore, be passively restricted to pines; those that land on and stay on the pines after their dispersal from overwintering refuges survive; those that land on other plants die. Circumstantial evidence suggests that this latter scenario is false. Firstly, neither of these *Anatis* species is strongly restricted by food. In the laboratory, they will mate and breed readily on a wide variety of aphid species. Secondly, in the spring, when they first become active, they can commonly be found on a wide range of plants and trees, but copulation and oviposition rarely occur before they have found pines. Thirdly, in late summer and autumn, these ladybirds may often be found feeding on aphids on many species of deciduous trees. As aphid numbers decline on pines late in the summer, the newly emerged ladybirds disperse to find food to increase their fat stores for the winter. They do so apparently without detriment.

Through the mechanism of natural selection, a species' habitat plays an influential role in the evolution of its adaptations. Several species of coccinellid with strong preferences for reeds show adaptations to living on these plants. *Anisosticta 19-punctata*, *A. bitriangularis* and *M. episcopalis* have the body rather more flattened than most species of ladybird, allowing them to find shelter in the winter by squeezing between the flat leaves of dead reeds. In such situations, the buff ground colours and black spots or stripes of these species make them highly cryptic (Fig. 3.8), and they can even swim, after a fashion. These ladybirds often overwinter in reeds growing in water, and if the water rises, so that they are threatened with flooding, they move to a higher position. If the stem becomes totally submerged, they simply float off the top with legs splayed wide and moving slightly, until these movements, or the water current, bring them into contact with a piece of emergent vegetation.

Fig. 3.8. *Anisosticta 19-punctata* winter camouflage colouration. (© Gilles San Martin.)

They then climb up, dry off, and seek another sheltered position. Tests with *A. 19-puncata* and three species that usually live away from water, *C. 7-punctata*, *Psyllobora 22-punctata* and *A. obliterata*, in which ladybirds were dropped into water, showed that only *A. 19-punctata* was capable of directed swimming. Typically, the response of this species to landing in water was to use motions of its legs to rotate through one or two complete circles, before starting to swim off in the direction of the nearest vegetation. The swimming stroke was alternated both within and between sides. Thus, as legs one and three on the right side and leg two on the left side moved backwards, the other legs were drawn forwards close to the beetle's body. The other three species all floated well, at least for a time, but their movement was undirected and they did not appear to assess their position relative to the nearest vegetation.

Changes in Habitat Preferences

The conclusion that some ladybirds have particular adaptations for living on a specific type of plant need not imply that ladybirds cannot come to use other host plants, or evolve novel host plant preferences. For example, *H. 16-guttata* is a common and widespread species that, in Britain, has a strong preference for *A. pseudoplatanus* trees. Both larvae and adults feed on the hyphae of powdery mildews (Fig. 3.9). Yet, prior to 1987, when this preference was first discovered (Majerus and Williams, 1989), the species was considered to be scarce in Britain, and an indicator of relict deciduous woodland. It is, of course, possible that the association between *H. 16-guttata* and *A. pseudoplatanus* has existed for a long time and has previously been overlooked because this tree has received little attention from coleopterists. Alternatively, *H. 16-guttata* may

Fig. 3.9. *Halyzia 16-guttata* feeding on mildew.

have only recently begun to use mildews on *A. pseudoplatanus* as food, perhaps just in the last three or four decades. This question is difficult to resolve, although Professor Tony Dixon (pers. comm.) believes that a novel preference has evolved, for he does not recall finding *H. 16-guttata* on *A. pseudoplatanus* in the 1950s and 1960s, when he was conducting research on aphids on this tree. The novel preference hypothesis gained further support in 2004, when considerable numbers of *H. 16-guttata* were found breeding successfully on mildew on *Fraxinus excelsior* (Mabbott, pers. comm.; Majerus, pers. obs.). From data obtained during the Cambridge Ladybird Survey (1984–1994), it is clear that the common use of mildews on this tree by *H. 16-guttata* has arisen since 1994.

If *H. 16-guttata* has widened its preferred habitat in Britain over the last few decades, this represents a relatively rare example of a species, previously reported to have strong habitat preferences, relaxing them to make use of foods that would have been present for some time. This is distinct from the more or less regular changes in habitats used by many aphidophagous generalists throughout the year. The cycles of changes, which differ for many of these species from region to region (reviewed in Majerus, 1994a), are a normal part of the local adaptation of these species. The habitat changes are a reflection of regular increases and decreases in habitat favourability, due largely to the seasonal timing of changes in aphid densities on different types of vegetation.

The species that face regular alterations in habitat favourability in a particular region can be contrasted with species that have been moved from one part of the world to another for use in pest control. Such species have had to face changes. Many have failed, but some have established with remarkable success. These include species such as *Rodolia cardinalis* imported from Australia into California in the nineteenth century to control scale insects on citrus trees, and both *C. 7-punctata* and *H. axyridis*, introduced into the United States to control a variety of aphid pests. Of these three, the success of the latter two in adapting to American conditions is perhaps not surprising as both have a wide dietary range. Conversely, *R. cardinalis* has a rather narrow food range, thus its success in California may seem a little unexpected. However, as it was introduced intentionally to control one of its normal prey species, an Australian scale insect, the cottony cushion scale, *Icerya purchasi*, which had been introduced into the United States accidentally, it would have been well adapted to the diet available to it.

Habitat Categorisation

The examples of species changing habitats highlight another problem encountered when trying to answer the question of which ladybirds live in which habitats. The habitat preferences of ladybirds are likely to be dynamic, often depending on the availability of primary or alternative food. Furthermore, the habitats used can vary unpredictably from one season of the year to another. Finally, the habitat preferences of species with wide geographic ranges vary over distance as dominant vegetation types and major plant species change.

Previously, as a result of survey work, I classified the larger species of British coccinellid into seven categories in respect of the habitats or host plants that they regularly breed in, or on (Majerus, 1991a). These habitat categories (Table 3.1) reflect, in part, the predominant vegetation types in Britain. On a more global scale, some of these habitat divisions are inappropriate, and certainly there are many types of habitat that are not included; Britain, for example, is not known for its hot deserts.

Having said that these categories are not fully appropriate to a global consideration of ladybird habitat preferences, they need be changed relatively little to make them appropriate, with the main changes being to reduce the emphasis on plant/tree type. An adapted set of habitat classes is given in Table 3.2.

Assigning most species of coccinellid to one of these categories presents some difficulties as the habitat preferences of most species are only known for

Table 3.1 Habitat Categories for British Ladybirds (Majerus, 1991a)

1. **Generalists**: breeding in a wide range of herbaceous and arboreal habitats (e.g. *Adalia 2-punctata*)
2. **Generalists with environmental constraints**: breeding in a wide range of herbaceous and arboreal habitats, but limited by climate, humidity, soil type, presence of other non-food organisms, or some other environmental factor (e.g. *Coccinella magnifica*)
3. **Herbaceous generalists**: breeding on a wide range of low-growing vegetation, but rarely, if ever, on trees (e.g. *Coccinella 11-punctata*)
4. **Arboreal generalists**: breeding on a wide range of trees, including both broad-leaved deciduous and conifer species, but rarely, if ever, on low herbage (e.g. *Adalia 10-punctata*)
5. **Broad-leaved deciduous woodland specialists**: breeding on a wide range of broad-leaved deciduous trees, but rarely, if ever, on conifers or low-growing herbaceous plants (e.g. *Calvia 14-guttata*)
6. **Conifer specialists**: breeding on a range of coniferous trees, but rarely, if ever, on broad-leaved deciduous trees, or low-growing herbaceous plants (e.g. *Aphidecta obliterata*)
7. **Host plant restricted specialists**: breeding on only a small number of plant species (e.g. *Myzia oblongoguttata*)

Table 3.2 Habitat Preference Categories of Coccinellids, from a Global Perspective

Category	Example species
1. **Generalists**: breeding in a wide range of herbaceous and arboreal non-extreme habitats. Such species tend to have a range of essential foods that occur in a range of habitats	*Hippodamia convergens, Coleomegilla maculata, Harmonia axyridis, Propylea 14-punctata, Propylea japonica, Adonia variegata, Adalia 2-punctata*
2. **Generalists with abiotic environmental constraints**: breeding in a wide range of non-extreme herbaceous and/or arboreal habitats, but limited by climate, humidity, soil type, or some other abiotic environmental factor	*Coccinella 5-punctata, Hippodamia 13-punctata, Hippodamia episcopalia, Coccinella 11-punctata*
3. **Generalists with biotic environmental constraints**: breeding in a wide range of non-extreme herbaceous and/or arboreal habitats, but limited by presence or absence of some other, non-food organism	*Platynaspis luteorubra, Coccinella magnifica, Cleidostethus meliponae*
4. **Herbaceous generalists**: breeding on a wide range of low growing vegetation, but rarely, if ever, on trees	*Coccinella 7-punctata, Coccinella transversoguttata, Coccinula sinensis, Coccinula crotchi, Micraspis frenata, Tytthaspis 16-punctata*
5. **Arboreal generalists**: breeding on a wide range of trees or shrubs, but rarely, if ever, on low herbage	*Calvia 14-guttata, Eocaria muiri, Adalia 10-punctata, Harmonia testudinaria, Halyzia 16-guttata*
6. **Host plant restricted specialists**: breeding on only a small number of plant species	*Scymnus suturalis, Myrrha 18-guttata, Coccinella hieroglyphica, Myzia pullata, Macronaemia episcopalia, Anisosticta bitriangularis*
7. **Extremists**: confined to breeding on a range of plant species in extreme habitats, such as high altitude, high latitude or hot desert	*Coccinella alta, Coccinella reitteri, Hippodamia arctica, Ceratomegilla ulkei*

part of their geographic range. To facilitate such assignation, examples of the type of species that would be placed in each category may be helpful. These are given in Table 3.2.

Most of these examples are species that occur in temperate or Mediterranean regions, reflecting a greater amount of survey work in these areas. Yet, even here the assignation of a species to a habitat class is based more on data showing that a species has been observed in certain habitats or on particular types of plant than on an understanding of the factors that lead a species to have the habitat/host plant preferences that they have. This is not surprising for two reasons. Firstly, the causes of habitat/host plant preferences are frequently complex. Only in a few of the most specialist species, such as the myrmecophile *C. magnifica* and the host plant specialist *Coccinella hieroglyphica*, which feeds on beetle larvae, are the causes of habitat choices understood. Secondly, as already intimated, evolution does not stop once habitat/host plant preferences arise, with the result that it is frequently difficult to ascertain whether a particular correlation between some trait of a species and its habitats or host plants is causal or consequential on its preferences.

Although the factors that affect host plant and habitat use by ladybirds have only been shown for a few species, we may consider the factors most likely to play a role.

Food Availability

The main factors that have determined the evolution of and now maintain habitat/host plant preferences of most ladybirds are few. Of greatest importance is the food available to a ladybird, which may have an influence in a variety of ways. Dietary requirements obviously influence where a species lives and breeds. Species with wide food ranges tend to have a greater habitat/host plant range than species that specialise on just one or a few food types. For aphid-feeding ladybirds, the size, density and age composition of prey colonies may also be important. Thus, for example, *C. 7-punctata* and *A. 2-punctata* will only lay eggs at high aphid densities, while others, such as *C. 5-punctata* and *Propylea 14-punctata*, can breed at very much lower densities (Honek, 1979, 1982; Hemptinne and Dixon, 1991). Some species of coccinellids, such as *Coccinella californica* and *A. 2-punctata*, will only oviposit on colonies of aphids with a high proportion of young nymphs, unless aphids are very abundant.

As habitat structure influences the diversity, density and population dynamics of ladybird prey species, habitat preferences are intimately entwined with the habitat/host plant preferences of ladybirds. This may be realised if agricultural and more natural habitats are compared. Crop plants that are grown over large areas at high density dominate many agricultural habitats.

Table 3.3 Comparison of the Number of Species of Ladybird Found Regularly in Agricultural and Semi-natural Habitats in Cambridgeshire and Suffolk, UK (1997–2002) (Majerus, unpubl. data)

Agricultural habitats		Semi-natural habitats	
Habitat type	Number of coccinellid species	Habitat type	Number of coccinellid species
Sugar beet	5	*Typha* reed-bed	8
Vicia faba	5	*Calluna* heathland	8
Oil seed rape	3	Chalk grassland	7
Wheat	3	*Urtica* (nettle) bed	13
Barley	3	*Fraxinus* woodland	8
Malus orchard	6	*Quercus*/*Corylus* woodland	12
Pinus sylvestris plantation	5	Open, self-seeded *Pinus sylvestris* woodland	19

Such habitats have little floral diversity. However, they frequently play host to very high-density populations of one or two species of plant pests, such as aphids, coccids or spider mites. These may in turn attract large numbers of ladybirds, particularly mobile generalist species, such as *Hippodamia convergens*, *Coccinella 3-fasciata* and *C. californica*. Yet, such habitats, which tend to be short lived due to harvesting and crop rotation, generally have lower coccinellid species diversity than more stable and florally diverse habitats (Table 3.3). The ways in which ladybirds find their food, and thus are attracted to certain habitats, are discussed further in Chapter 4.

Microclimate

Both temperature and humidity affect the behaviour of ladybirds. In general, microclimatic factors influence the distribution of particular species within habitats. However, if this is the case, it is also likely that these factors influence the favourability of habitats. In crop fields, temperature varies with the density of plants; high-density stands generally show less fluctuation in temperature through 24 hours and are relatively cooler in the warmest part of the day. Smith (1971), working in maize crops, showed that the difference in temperature could be as much as 8.6°C. He also showed that *Coccinella 9-notata* and *Coccinella transversoguttata* preferred the higher temperatures in low-density stands, while *C. maculata* and *H. 13-punctata* sought shelter from the hottest

Fig. 3.10. *Hippodamia sinuata* basking on a flower.

temperatures in the denser stands. Similar results were obtained in Europe by Honek (1979, 1982), where, in cereals, *C. 7-punctata* and *C. 5-punctata* preferred warm conditions in sparse stands, while *P. 14-punctata* was more tolerant of lower temperatures in dense crops.

Microclimatic preferences are reflected in the behaviour of some ladybirds at certain times of the day or seasons of the year. For example, some species, such as *Hippodamia sinuata* (Fig. 3.10) and *C. 7-punctata*, climb to exposed positions high on plants early on sunny mornings, following cool nights, to bask in the sun and warm up. *Exochomus 4-pustulatus* also shows this behaviour from February to April in northern parts of its range, but not later in the year or in southern Europe. In *C. maculata*, Benton and Crump (1981) observed a regular pattern of climbing upwards in the vegetation in the morning, and returning to the lower strata in the afternoon.

The effect of humidity was demonstrated by Ewert and Chiang (1966). The vertical distribution of three species of ladybird, *H. convergens*, *H. 13-punctata* and *C. maculata*, in maize and barley fields, was partly a result of variation in humidity through the crop, with higher humidity lower down in the crop, showing that the preferences of the three species reflected their tolerance to desiccation. Humidity may also affect ladybird sex ratios (Nedvěd and Kalushkov, 2012).

Plant Species and Characteristics

Both the species of plant and some features of its morphology, growth pattern and chemistry may affect ladybird presence. The effect of plant species has been shown in studies of the relative abundance of ladybirds on different plants that host the same aphid species. For example, Tamaki et al. (1981) showed that when the aphid *Myzus persicae* was present on three species of crop plant – sugar beet, broccoli and radish – *C. transversoguttata* occurred at highest density on sugar beet, while *Scymnus marginicollis* was most abundant on radish. The reasons for these preferences are not known, and the structure of the different plants, or the microclimatic conditions that they offer, may be involved. However, a third possibility is that *M. persicae* may sequester different chemicals from the different plants, and so their palatabilities to these two coccinellids may vary. This is certainly suggested from observations of feeding *Aphis fabae*, collected from a variety of wild plants, to laboratory cultures of various adult ladybirds. For some species of ladybird, the host plant from which aphids were collected affected feeding rate, reproductive rate and longevity.

The morphology of plants also has an influence on their suitability as hosts to ladybirds. This is because structures such as hooked hairs may cause injury to ladybird larvae. Thus, Putman (1955) found that the hooked trichomes on leaves of *Phaseolus coccineus* wounded larvae of *Stethorus pusillus*, with death frequently resulting. Indeed, on this plant, adult longevity was also reduced.

In other cases, physical damage is not caused directly by the plant, but features, such as hairs, may reduce the rate at which adults can forage. For example, Banks (1957) observed that *P. 14-punctata* moved more slowly over the hairy leaves of potatoes than the smooth leaves of broad bean. The foraging rate of *Delphastus pusillus* is also affected by the hairiness of leaves; on *Poinsettia*, the rate of whitefly consumption was greater on plants with few hairs on the leaves than on those with many hairs (Heinz and Parrella, 1994). Similar results were obtained for *H. convergens* on tobacco (Belcher and Thurston, 1982).

A plant's structure may also influence its suitability to a particular ladybird by affecting the likelihood of the coccinellid staying on the plant. Both leaf hairiness and leaf smoothness may have an effect. For example, larvae of *C. 7-punctata* captured and consumed *Acyrthosiphon pisum* at a faster rate on hairy leaves of *Vicia faba* than on the smooth leaves of *Pisum sativa*, because the larvae fell from the latter more often (Carter et al., 1984). Other studies have shown similar correlates between plant morphology and the likelihood of falling from the plant (Kareiva and Sahakian, 1990; Grevstad and Klepetka, 1992; Frazer and McGregor, 1994). However, the reactions of ladybirds to plant structures are not always the same. While larvae of *C. maculata* have difficulty with the hairs on cucumber plants, and frequently fall off, larvae of *Cycloneda sanguinea* and *A. 2-punctata* do not (Gurney and Hussey, 1970).

In coniferous woodland, some species of ladybird appear to have preferences dependent on the age of the trees. Perhaps most striking of these is *Myrrha 18-guttata*, which shows a strong preference for the crowns of mature pines

Fig. 3.11. *Mulsantina picta.*

(Bielawski, 1961; Klausnitzer, 1968; Majerus, 1988), with immature stages rarely being found on young trees or the lower boughs of old trees. Other conifer species, such as *Scymnus suturalis*, *Mulsantina picta* (Fig. 3.11), *Myzia pullata*, *M. oblongoguttata*, *A. ocellata* and *Anatis mali*, show similar preferences, but to a lesser degree. Conversely, *Scymnus lacustris*, *Scymnus nigrinus* and *C. transversoguttata* prefer young pines, while some species (e.g. *H. 4-punctata*, *E. 4-pustulatus*) do not seem to have age-related host tree preferences.

The health of plants and their chemical composition may affect ladybirds through their prey. Honek and Martinkova (1991) showed that aphid abundance on maize was strongly influenced by both competition with the weed grass *Echinochloa crus-galli* and soil quality, with aphid densities being very much higher in the absence of competition and on good soils. This was reflected in higher densities of *C. 7-punctata*, *C. 5-punctata* and *P. 14-punctata*.

The Evolution of Habitat Preferences

To consider the way in which habitat preferences evolve, we have to return to the question of changes in habitat preferences. Although the causes of habitat shifts in insects generally have received considerable attention, those in the ladybirds have been, until recently, largely unexplored. In other insects, the main factors that influence habitat changes have been identified as resource availability, linked to the level of competition for resources, and the presence or absence of enemies in the form of predators, parasites and parasitoids. Sloggett and Majerus (2000a) considered the roles of these factors in the evolution of habitat preferences in predatory ladybirds.

Dietary Diversity

In the coccinellids, the presence of essential prey certainly has an influence on habitat usage. However, the link is not always very strong. For example, in Britain, 24 of 26 predatory ladybirds tested will feed, breed and complete development on the aphid *A. pisum*, yet only eight of these have been recorded feeding on this aphid in the wild. In the case of *M. 18-guttata*, which has a very pronounced preference for living in the crowns of mature pines, it is unlikely that *A. pisum* would ever be encountered, but the beetle thrives on this species in the laboratory. The ability of many ladybirds to eat, in captivity, prey that they do not use in the wild may be coincidental, or it may be an evolved response to the need to be able to survive on atypical food when more usual prey is scarce. A third possibility is that it results from evolutionary lag, whereby habitat specialists have retained an ability to feed on a more catholic diet from a time when their ancestors used a greater variety of habitats. Whichever is the case, the diversity of potential appropriate prey is wider than that which is actually used in the field. This gives many ladybirds some flexibility, and opens the possibility for the occurrence of habitat shifts that do not depend on the simultaneous evolution of an ability to use a novel food type.

Factors Affecting Habitat Shifts

Many ladybirds do not remain in a single habitat throughout the year. Aphid feeders, as noted, are highly mobile, no doubt a reflection of the ephemerality of their prey, the populations of which often crash dramatically. Late in the summer, preferred aphid prey species often become scarce due to dispersal to winter host plants, predation or parasitism. Although ladybirds are rarely breeding at this time, they do feed to accumulate sufficient food reserves to allow survival through winter dormancy, with some individuals seeking prey well into the winter months if their reserves are insufficient (Barron and Wilson, 1998). The ability to feed on a wide range of non-optimal prey is advantageous at this time. Indeed, the same is true for some species when they emerge from winter dormancy, for in the spring many aphid species take considerable time before they have reached densities exploitable for ladybirds (Hemptinne and Desprets, 1986). At these times, therefore, many ladybirds feed on high-risk or suboptimal-quality diets (Forbes, 1883; Clausen, 1940; Hagen, 1962), often in unusual habitats (Iperti, 1965; Sloggett and Majerus, 2000b).

It seems fanciful to suggest that this behaviour, in which ladybirds move to less preferred habitats periodically, when preferred habitats become less favourable, is of recent origin. Thus, in considering the evolution of habitat preferences, we may consider that ladybirds, at the population level, will encounter a range of habitats. Should the conditions offered by a non-preferred habitat prove more favourable than the preferred habitat, a habitat shift, with subsequent evolution of traits appropriate to the new habitat, may occur. Favourableness may involve many factors, but in ladybirds, those considered most influential are availability of food, competition for food and the distribution of enemies.

The Influence of Prey Availability on Habitat Shifts

Just by considering the prey and habitat preferences seen in different species of ladybird, it is obvious that the adoption of new prey types and new habitats has occurred many times (Sloggett and Majerus, 2000a). However, it is difficult to assess the role of prey availability on habitat shifts in individual cases because much of the information needed to assess past changes is simply not available. Not only are the lists of essential and acceptable prey of most species incomplete, but to assess evolutionary history, detailed phylogenetic information is also needed. In short, one has to know what evolved from what and when. The easiest cases to consider are those involving shifts to highly unusual habitats, which are clearly different from those of closely related taxa. In this category could be included the evolution of myrmecophily, that being the habit of living in close association with ants, in species such as *C. magnifica* or *Platynaspis luteorubra*, or the change to non-homopteran diets in species such as *C. maculata*, *Aiolocaria hexaspilota* and *C. hieroglyphica*. Harder to interpret are more general changes, such as increased habitat specialisation as seen among some members of the genus *Adalia*.

Coccinella magnifica, which lives in close association with ants (Fig. 3.12), has been closely studied. In an elegant series of field observations and experiments, Sloggett (1998) considered the evolution of myrmecophily in this species. He showed that several species of ladybird, including the closely related *C. 7-punctata*, feed on aphids tended by the wood ant *Formica rufa* (Wellenstein, 1952), on *Pinus sylvestris*, during periods of aphid scarcity. Of six ladybirds found in pine woodlands during periods of aphid scarcity, two species, *M. 18-guttata* and *A. ocellata*, showed virtually no tolerance of ant presence, while a third species, *H. 4-punctata*, exhibited little tolerance. *Coccinella 7-punctata*, although being found on ant-tended aphid colonies, was much more common where ants were absent. Interestingly, *M. oblongoguttata*, which was much more common away from ants during periods of abundance, was found as commonly in the presence of ants as in areas from which ants were absent when aphids were scarce. The sixth species, *C. magnifica*, was found almost exclusively with ants, irrespective of aphid abundance.

Fig. 3.12. *Coccinella magnifica.*

Three points may be deduced from these results. Firstly, when aphids were scarce, four of the species were less abundant in the presence of ants than in their absence, despite the fact that aphids were more common on trees with ants. Secondly, *M. oblongoguttata*, which is most finicky in its diet, only breeding in the presence of particular aphids (Majerus, 1993), showed considerable tolerance to ants, perhaps out of necessity given its extreme dietary specialisation. As alternative food sources are unavailable, it has evolved to show some defence against and tolerance of ant attacks. Thirdly, in Britain, *C. magnifica* is entirely restricted to living in the forage range of *Formica* ants, as many observers have noted (Sloggett and Majerus, 2000b).

Perhaps most pertinent to a consideration of the myrmecophilous habit of *C. magnifica* is that its close relative *C. 7-punctata* has some tolerance of *F. rufa* during periods of aphid scarcity. Donisthorpe (1919–1920) also wrote of *C. 7-punctata* 'experimenting in a myrmecophilous existence'. Furthermore, Bhatkar (1982) observed large groups of this ladybird in the vicinity of *Formica polyctena*, while various workers have reported other *Coccinella* species (*C. 11-punctata*, *C. transversoguttata*, *C. 3-fasciata*) with ants, particularly in late summer (Bradley and Hinks, 1968; Bhatkar, 1982). It thus seems feasible that the non-myrmecophilous ancestors of *C. magnifica* may have occasionally had to prey upon ant-tended aphids, and thus selection was imposed on these ancestors to evolve some degree of tolerance to ants. Additional selective advantages to myrmecophily may have enhanced the behaviour over time. These may have included more efficient use of particular prey species that are frequently ant-tended, reduced energetic costs associated with prey switching (Hattingh and Samways, 1992), reduced requirement for hazardous migrations, reduced competition with other aphid predators and reduced densities of ladybird predators and possibly parasitoids (Sloggett and Majerus, 2000b).

This case can be contrasted with that of *C. maculata. Coleomegilla maculata* has a broad dietary range (Hodek and Evans, 2012), possibly as a consequence of ancestral feeding behaviour during periods of aphid scarcity. It eats not only aphids, but also pollen (Conrad, 1959; Smith, 1960; Benton and Crump, 1981), non-homopteran insects (Putman, 1957; Groden et al., 1990), nectar and honeydew (Majerus, pers. obs.). The habitat choices of this ladybird have been linked to its diversity of diets (Andow and Risch, 1985; Groden et al., 1990). It has been argued that pollen is a more recent dietary item because ladybirds fed exclusively on pollen have slower larval development, take longer to mature their ovaries, and lay fewer eggs than those that feed on aphids (Smith, 1960; Hodek et al., 1978; Hazzard and Ferro, 1991).

From Generalist to Specialist or Vice Versa?

In the genus *Adalia, A. 2-punctata* is a generalist feeder and is found on a diverse array of trees, shrubs and herbaceous plants, while its close relative *Adalia 10-punctata* is largely confined to trees and shrubs (Honek, 1985; Majerus, 1994a). One can address the question of whether the ancestor of these two species was a generalist or specialist by considering a third species in the genus. *Adalia 4-spilota* is less closely related to either of the other two species than they are to each other. It is found on both trees and low-growing vegetation, thus sharing the habitat generalism of *A. 2-punctata*. The simplest explanation of the habitat preferences of these three species is that they all evolved from a common ancestor that was a habitat generalist, and that a narrowing of habitat preference occurred in the *A. 10-punctata* lineage, after the split from *A. 2-punctata*. Although this pattern of evolution cannot be proven, it is the most parsimonious explanation of the information available.

We may then ask what *A. 10-punctata* could gain through such specialisation. Habitat generalists certainly gain some advantages through the wider range of suitable prey that they are likely to encounter. However, there are costs for generalists. Firstly, dispersal between host plants requires energy and imposes risks. Secondly, prey switching may involve costs as a consequence of the differences in the chemistry of prey. Certainly, in generalist species of the coccid-feeding genus *Chilocorus*, Hattingh and Samways (1992) found that when the prey species of females changed, this was often followed by a temporary dip in the rate of egg production. Thirdly, specialists that have a narrower range of food may evolve a suite of digestive enzymes that are geared to and process this narrower range of prey more efficiently than can generalists. This is because the digestive systems of generalists have to cope with diverse prey types. This increased efficiency may allow specialists to use lower aphid densities than generalists can. Indeed Honek (1985) reports that *A. 10-punctata* is often found breeding in habitats with lower aphid densities than *A. 2-punctata*. It is likely, therefore, that the evolution of greater habitat specialisation in *A. 10-punctata*, compared with *A. 2-punctata* and *A. 4-spilota*, involved a trade-off between the cost of having access to prey in a narrower range of habitat types and the benefits from less frequent host plant and prey switching, plus more efficient prey processing, allowing successful use of low-density prey.

Enemy-free Space

Most coccinellids have chemical defences and, as a result, they tend to be attacked by specialist parasites and parasitoids rather than by generalist predators (Ceryngier et al., 2012). Some specialist parasites and parasitoids can infect high proportions of some populations (Iperti, 1964; Disney et al., 1994; Geoghegan et al., 1997). Thus, perhaps these enemies of ladybirds have influenced the evolution of the habitat preferences of the ladybirds.

Little work has been conducted on the role of enemy-free space in the evolution of habitat or host plant preferences in the coccinellids. This is partly because knowledge of the range of enemies of ladybirds is incomplete. However, some strands of evidence do suggest a role for enemy-free space, although it is unlikely that this role is as widely important as that of prey availability (Sloggett and Majerus, 2000a). Evidence in support of a role for enemy avoidance comes from studies of two myrmecophilous ladybirds. In *P. luteorubra*, Völkl (1995) found evidence that by living with ants, the levels of infection by the host-specific parasitoid wasp *Homalotylus platynaspidis* were reduced. Similarly, Majerus (1994a) reported that infestation levels of *C. magnifica* by the wasp *Dinocampus coccinellae*, in conifer woodland in the presence of *Formica* ants, were less than a fifth of those reported from *C. 7-punctata* in ant-free conifer habitats close by. However, in this case, Sloggett (1998) has argued that the low prevalence of *D. coccinellae* in *C. magnifica* is a consequence of the extremely repellent chemistry of this ladybird rather than a deterrent effect of the ants this ladybird lives with.

Another strand of evidence comes from work on spiders, which are a major cause of ladybird mortality. In a simple investigation into spider population density around *F. rufa* nests, it was shown that spiders were between 5 and 11 times as common over 100 metres away from *F. rufa* nests as they were 5 to 10 metres from the nests (Majerus, 1989). The relative dearth of spiders near the nests may be a direct consequence of ant aggression, or could be the result of ants driving away the potential prey of the spiders. Either way, a low density of spiders would undoubtedly be beneficial to *C. magnifica*, which at these sites was only found within 100 metres of a *F. rufa* nest. The same situation may exist in relation to other predators of ladybirds close to ants' nests.

Habitat Preferences in Ladybirds: Future Research

In research on habitat or host plant preferences, two questions should be addressed. Firstly, and obviously, one needs to ask why a species lives in certain habitats or on certain plants. The second question, which is often forgotten, is why a species does not live in other habitats or on other plants. In other words, we need to ask what makes a particular habitat favourable while others are unfavourable. Species with highly specialised ecologies are easiest to interpret.

In the case of *C. magnifica*, we have some idea of why it lives with ants and not elsewhere. They do so to utilise a food source protected from other predators by an aggressive guardian, against which they are themselves well defended. They may also be protected from their own potential enemies by these aggressive predators. But why does *C. magnifica* not live anywhere else? Here we must speculate. Possibly *C. magnifica* is a poor competitor or lacks efficient defences against predators and parasites. However, if this is so, more fundamental questions must then be asked: why are they bad competitors? Why are their defences inadequate?

The reason that *C. magnifica* is either a bad competitor, or has inefficient defences, may follow from an evolved specialisation to life with ants. Scent mimicry, one of the systems by which the ladybirds deter ant attacks, may have an energetic cost. If so, resources will not be available for other functions, such as production of toxins, or fighting ability. This is a direct cost of immunity. A more indirect cost may be that when living in the proximity of *Formica* nests, the selection pressures that might promote the evolution and maintenance of strong defences against a range of predators and parasites are reduced, because the ants keep these enemies at bay. If any cost is incurred in having these ancestral defence systems, the systems are likely to be lost. If encounters with potential predators or parasites become rare events, selective disadvantages, incurred by their costs, will outweigh the selective advantages from their maintenance.

Studies investigating the efficiency of *Lasius niger* in defending *A. fabae* colonies on *Cirsium arvense*, against *A. 2-punctata*, *C. 7-punctata* and *C. magnifica* larvae and adults, give some credibility to this idea. Unexpectedly, the ants attacked *C. magnifica* ferociously, biting at them and squirting them with formic acid. However, remarkably, neither the adult beetles nor the larvae seemed to take any notice of the ants. Possibly *C. magnifica* has lost its fleeing response to ant attacks, because the *Formica* ants, with which it normally lives, do not attack it (Majerus, 1994a).

In the early 1990s, a colleague, Dr John Barrett, devised an interesting analogy. If *C. magnifica* are the populace of the United States, then the ants could be seen as a Reaganesque 'Star Wars' system, spreading a powerful defensive umbrella over an area surrounding their territory. Presence of this defensive system negates the need for more old fashioned, conventional defences, and the costs of these can be saved. One is safe as long as one stays under the umbrella, but not if one strays.

For some of the other more specialist species, we can deduce with reasonable certainty that their shunning of many habitats is a consequence of adaptations that tie them closely to specific habitats. Thus, species such as *M. oblongoguttata* and *C. hieroglyphica*, with their precise dietary requirements, will not breed in the absence of their preferred prey, which is only found on a narrow range of plants or trees. Similarly, some of the species that prefer to live in wetlands have precise adaptations to life in the reeds.

Thus, we have some understanding of why some of the most specialised ladybirds live where they live and not elsewhere. However, for more habitat generalist species, we know little about why such species are absent from certain habitats or plants that appear to present favourable conditions in respect of food and other resources.

To gain a greater understanding of the reasons for habitat usage of these species, hypothesis testing, by careful observation and experimental study, will be needed. Systems that are either temporally or spatially dynamic with respect to habitat usage offer the greatest chance of understanding. Two systems, one involving spatial, the other temporal, change in habitat usage seem to offer scope for work in this area.

In continental Europe and Asia, *C. 5-punctata* is a generalist, being found in a wide range of habitats, including diverse urban and suburban habitats, in rural grasslands, cereal and asparagus crops, hedgerows and both coniferous and deciduous woodland (Klausnitzer, 1969; Majerus, 1994a). However, in Britain, it is highly habitat specific, living and breeding exclusively on, or close to, unstable river shingles (Majerus and Fowles, 1989). Perhaps this high degree of habitat specificity is the result of the species being on the edge of its range in Britain. The rationale is that the conditions that *C. 5-punctata* faces on the edge of its range are only just within those allowing its survival and reproduction. The increased habitat specialisation in Britain allows the species to survive by becoming precisely adapted to the local conditions offered by a particular habitat.

Edge of range habitat specialisation is well known. The swallowtail butterfly, *Papilio machaon*, is widely distributed in many habitats in Europe and Asia, its larvae feeding on many species of umbelliferous plants. In Britain, it is now confined to one lowland area of East Anglia. Here eggs are laid only on *Peucedanum palustre*, although larvae from this locality can be reared in captivity on many umbellifers. By evolving a host plant-specific oviposition strategy, this population restricts itself to an optimum habitat at the edge of its range where it is just able to survive. *Peucedanum palustre* is not a more nutritional food than alternatives; it simply serves as an indicator of optimal habitat type.

This edge of range effect is a tenable hypothesis for *C. 5-punctata*, although it is unclear why unstable river shingle banks might provide optimal habitat. However, various hypotheses might be tested. For example, unstable river shingles are regularly flooded, particularly during the winter. *Coccinella 5-punctata* survives these inundations by overwintering in air pockets under larger stones on the shingle banks, or by moving a metre or two off the shingle, to overwinter on *Ilex europeus* growing on the stable banks of the river cuts. The annual flooding of the shingle banks may be beneficial to *C. 5-punctata* in one or more of four ways. Firstly, the flooding is likely to remove from the area other species of aphid predator that would compete with *C. 5-punctata* for food. Secondly, in the same way, the flooding may reduce the densities of predators and parasitoids; hence, the shingle banks represent relatively enemy-free space. Thirdly, those plants, mainly water-tolerant shrubs, such as *Salix* spp., or annual plants that grow on the shingle banks in the summer, are stressed when flooded and are unable to contribute so many resources to their own defence against herbivores. The result is rather stunted plants on the shingle banks that are frequently heavily infested with aphids that are then available as prey for *C. 5-punctata*. Finally, the shingles alongside rivers may have been a habitat in which aphids were present in times of general aphid scarcity. Owing to the proximity of water, vegetation would have stayed lush for longer in dry periods, allowing aphids and their predators to continue to breed when vegetation away from rivers and lakes had dried off. Certainly, some habitat generalists, such as *C. 7-punctata* and *H. variegata*, move to damper habitats during the hottest, driest periods of the year (Hodek et al., 1966; Majerus, pers. obs.). It is not difficult to conceive a monitoring programme and series of experiments to test these alternative explanations.

Since the establishment in North America of, firstly, *C. 7-punctata*, and more recently *H. axyridis*, many aphidophagous ladybirds in the United States and

Canada have faced severe competition for food. A number of reports show that the distribution and abundance of many endemic species have been in decline, the primary cause being competition for food from these introduced generalists. This situation provides an opportunity for coleopterists to monitor and experiment upon the effects of interspecific competition and niche usage in the wild. The introduction of these two extremely successful aphid predators into North America has added a strong new selection pressure that is likely to cause the evolution of greater habitat specialisation, or habitat/ dietary shifts in some of the native species of ladybird, assuming that they survive, and probably in many other aphid predators and parasitoids. Here is a prime opportunity to observe evolving habitat preferences.

Chapter 4: What Ladybirds Eat

The Diversity of Ladybird Diets

A sycamore tree stands by a road in Luxembourg. On the trunk, *Exochomus 4-pustulatus* and *Chilocorus renipustulatus* larvae are feeding on large white scale insects. On leaves above, larvae and adults of *Adalia 2-punctata, Adalia 10-punctata* and *Calvia 14-guttata* are feeding on sycamore aphids, while next to them adult and neonate larvae of *Halyzia 16-guttata* graze hyphae and sporangia of powdery white mildews. The ladybird fauna on this single tree gives some indication of the diversity of ladybird diets.

The relationships between ladybirds and their food have attracted considerable attention, because of the economic importance of some of their prey. We regard ladybirds as beneficial because most eat plant pests, such as green fly and black fly (aphids), scale insects (coccids), psyllids and so-called woolly aphids (adelgids). Many eat other foods as well. For example, for some predatory species, beetle or moth larvae are the main diet. Over a third of ladybird species are not carnivorous, feeding instead on leaves, pollen, nectar or fungi. Differences in the adaptations of organisms that have to hunt, catch and subdue prey, and those that eat inanimate food, make it convenient to discuss predatory ladybirds separately from vegetarian and mycophagous species.

Predatory Ladybirds

The dietary specificity of many ladybirds varies during the year. When eggs are being matured and laid, females of many species feed on a more restricted set of foods than at other times, to maximise nutritional contributions to their offspring. However, when feeding up for the winter, replenishing reserves following dormancy, or when preferred prey becomes scarce, many carnivorous ladybirds will feed on a diverse range of foods. They will catch and eat a variety of live insects and other arthropods, scavenge on dead organisms, eat honeydew, feed from extra-floral nectaries, drink sweat, consume a variety of plant materials including pollen (Figs. 4.1–4.3), nectar (Fig. 4.4), sap and resin (Fig. 4.5), and they may even bite humans.

Fig. 4.1. *Tytthaspis 16-punctata* consuming pollen. (© Gilles San Martin.)

Fig. 4.2. *Mulsantina hudsonica* feeding on pollen.

Ladybirds exist within a complex and taxonomically diverse guild (that is the group of species interacting in one way or another around shared prey) comprising parasitoids, pathogens and other predators. Intraguild predation (IGP) is a term used to describe the trophic interaction between two species that share a host or prey (Rosenheim et al., 1995), such as is the case when one aphidophagous ladybird eats another aphidophagous ladybird. IGP, alongside competition and cannibalism, is considered to have a major influence on the structure of aphidophagous guilds (Lucas, 2005). The eggs, larvae and pre-pupae of ladybirds are particularly vulnerable to predation (Majerus, 1994a) but are protected to some extent by chemical and physical defences. Some ladybirds are more effective intraguild predators than others. *Harmonia*

Fig. 4.3. *Coccinella 7-punctata* consuming pollen.

Fig. 4.4. *Tytthaspis 16-punctata* larva consuming nectar.

axyridis is widely considered to be a strong intraguild (possibly top) predator, while *A. 2-punctata* is a common intraguild prey (Ware et al., 2009). This is thought to partly explain the observed decline in *A. 2-punctata* in a number of European countries following the arrival of the invasive *H. axyridis* (Roy et al., 2012). There have been many empirical studies (Roy et al., 2008a; Alhmedi et al., 2010; Hautier et al., 2011; Katsanis et al., 2013; Brown et al., 2015) and reviews (Roy and Pell, 2000; Lucas, 2005; Pell et al., 2008) on IGP, particularly within the field of biological control, and it is likely to be the focus of continued research interest.

Fig. 4.5. *Coccinella 7-punctata* and *Anatis ocellata* consuming resin.

Prey Categorisation

Hodek (1973) attempted to formalise the amazing diversity in coccinellid foods. He created four categories of prey: essential prey, accepted but inadequate prey, rejected prey and toxic prey.

Essential prey were defined as those that promote egg maturation and oviposition and allow complete development. Accepted but inadequate prey prolonged survival, but did not allow eggs to mature or larvae to develop fully. Rejected prey were those which, when attacked by ladybirds, were immediately rejected and thereafter left alone. Toxic prey were those, which, if eaten, killed ladybirds or significantly reduced their fitness. While Hodek's categories were a useful tool, the fuzzy transitions between them, and the limitations of putting variable biological interactions into a system of precise classes, limited their value (Majerus, 1994a).

Hodek (1996a) thereafter revised his classification. He merged his first two classes into accepted prey, discussing under this heading variation in suitability of accepted prey, levels of food specificity and so-called 'mixed feeding'. Additionally, he introduced a new class called 'alternative (substitutive) prey'. This seems a cogent approach and is the one that I shall follow here.

Accepted Prey

Many lists of the prey species of different ladybirds have been published (see Hodek and Evans, 2012, for review). Most of these have flaws because they are either incomplete or include prey accepted in captivity that would not normally be encountered in the wild.

While Hodek and Evans (2012) provide additional detail on this subject, a summary list of the essential foods of coccinellids is presented in Table 4.1. This shows the wide variety of prey of coccinellids. For example, most species of *Nephus* and *Diomus* specialise on Pseudococcinae and Coccinae scale insects. Most members of the tribes Chilocorini and Coccidulini feed on coccids, although they will also eat aphids when necessary. Species within the genus *Psyllobora* are specialists on mildews, while the Epilachnini eat plant material. The Stethorini feed mainly on plant-eating mites of the family Tetranychidae; *Stethorus pusillus*, for example, specialising on *Phyllacotes* mites. One species, *Cleidostethus meliponae*, in East Africa, has been found only in nests of the bee *Melipona alinderi* and, being both wingless and blind, appears to have become adapted to life in nests of this bee (Salt, 1920).

Other groups of coccinellids show a greater diversity of diets. Thus, in the tribe Coccinellini, although most species are aphidophagous, examples of phytophagous, and mycophagous species are also present, as are species that specialise on other prey, such as other sap-sucking insects and the eggs and larvae of beetles (Table 4.2).

Many predatory coccinellids have relatively wide ranges of essential food, although, in practice, these ranges are often limited by the habitat and host plant preferences of the ladybirds. An exception is *Coccinella hieroglyphica*. Attempts to breed and rear English stocks of this species on *Acyrthosiphon pisum* or *Aphis fabae*, in captivity, proved fruitless. In Europe, *C. hieroglyphica* mainly eats aphids (Kanervo, 1940; Klausnitzer and Klausnitzer, 1972). However, Hippa et al. (1978) report that in southern Lapland it feeds on eggs and larvae of several species of chrysomelid beetles. They showed that although aphids were consumed and ladybird development could be completed on some aphids, such as *Myzus persicae*, larval mortality was lower and pupal weight higher if larvae were reared on *Galerucella sagittariae* (Chrysomelidae) larvae.

Hodek (1973) considered the relative suitability of different prey types to specific coccinellids, the suitability being judged in terms of fecundity, rate of larval mortality and rate of larval development. Of particular interest are studies where *A. 2-punctata* and *Coccinella 7-punctata* were fed a variety of prey, including *A. fabae*. Blackman (1965, 1967) provided data on larval development time, larval mortality rates and adult weight at emergence that showed that *A. fabae* and *Aphis sambuci* are relatively poor food for *A. 2-punctata*, as manifest by relatively long larval development time, high larval mortality, and the small adults produced, despite both being eaten by *A. 2-punctata* in the wild. The cabbage aphid, *Brevicoryne brassicae*, was even less suitable. Furthermore, *A. 2-punctata* females laid nearly three times as many eggs when fed *M. persicae*, as when fed *A. fabae*, and the fertility was close to 90% when fed the former, but only 55.9% on the latter. To *C. 7-punctata*, *A. sambuci* was

Table 4.1 Main Essential Foods of the Coccinellidae (After Hodek, 1996a: compiled from Balduf, 1935; Sasaji, 1971; Savoiskaya, 1983; Klausnitzer and Klausnitzer, 1986; with additional information from Dixon, 2000 and Majerus, unpubl. data). (Note tribes refer to classification prior to Magro et al., 2010.)

Taxonomic level	Group	Dietary preference
Subfamily	**Sticholotidinae**	
Tribe	Sticholotidini	coccids, Diaspidinae
Subfamily	**Microweiseinae**	
Tribe	Sukunahikonini	coccids, Diaspidinae
Tribe	Microweiseini	mainly coccids of the subfamily Diaspidinae (*Aspidiotus*, *Chinaspis*)
Tribe	Serangiini	aleyrodids
Subfamily	**Scymninae**	
Tribe	Stethorini	phytophagous mites, Tetranychidae
Tribe	Scymnillini	Aleyrodidae
Tribe	Scymnini	coccids (62%), aphids (24%), extensive specialisation in some genera
Genera	*Clitostethus*, *Lioscymnus*	aleyrodids, aphids
Genera	*Diomus*, *Nephus*	Pseudococcinae, Coccinae
Genera	*Sidis*, *Parasidis*	*Pseudococcus*
Genus	*Cryptolaemus*	Pseudococcinae
Genus	*Pseudoscymnus*	Diaspidinae
Genus	*Platyorus*	aphids
Subgenus	*Pullus* (*Scymnus*)	mainly aphids on shrubs and trees, but sometimes coccids and adelgids
Subgenus	*Scymnus* (*Scymnus*)	mainly aphids on low herbage, more rarely aphids on trees
Tribe	Platynaspidini	aphids
Tribe	Aspidimerini	aphids
Tribe	Hyperaspidini	mainly coccids (Coccinae, Ortheziinae, *Pseudococcus*, *Phenacoccus*, *Ripersia*)
Tribe	Ortaliini	psyllids, Flatidae
Subfamily	**Chilocorinae**	
Tribe	Telsimiini	coccids, Diaspidinae
Tribe	Chilocorini	coccids (75%), otherwise aphids and adelgids

Table 4.1 *(cont.)*

Taxonomic level	Group	Dietary preference
Subfamily	**Coccidulinae**	
Tribe	Coccidulini	Almost exclusively coccids, rarely aphids
Tribe	Rhyzobiini	Diaspidinae (51%), Coccinae (35%), Lecaniinae (14%)
Tribe	Exoplectrini	*Icerya* and closely related coccids
Tribe	Noviini	*Icerya* and closely related coccids
Tribe	Azyini	Dispidinae
Subfamily	**Coccinellinae**	
Tribe	Coccinellini	aphids (85%), also psyllids, coleopteran larvae, pollen, nectar, fungi, larvae of mycophagous coccinellids
Tribe	Synonychini	aphids
Genus	*Neda*	coccids
Genus	*Archaioneda*	coccids
Tribe	Psylloborini	mildews (Erisyphaceae)
Subfamily	**Epilachninae**	
Tribe	Epilachnini	Plant material

the least suitable, with larvae fed on *B. brassicae* also showing slightly increased development time and mortality, and lower adult weight. However, contrary to the response of *A. 2-punctata*, *A. fabae* was the most suitable prey for *C. 7-punctata*. Hariri (1966a, b) also showed *A. fabae* to be rather unsuitable food for *A. 2-punctata*; larvae fed on this aphid producing smaller adults with lower fat and glycogen content, and only 50% the fecundity of those fed on *A. pisum*. Given that *A. fabae* is poor food for *A. 2-punctata*, it is surprising that this ladybird commonly oviposits on broad bean and sugar beet infested with this aphid (Banks, 1955, 1956; Barczak, 1991). In both situations, *A. fabae* is used as an accepted food. Cabral et al. (2005) showed that *Coccinella 11-punctata*, when fed on *A. fabae* or *M. persicae*, have similar development times, longevity and fecundity, but that those fed on *Aphis proletella* were slow to develop, had high pre-imaginal mortality, shorter adult longevity and did not lay eggs. Similar results, showing variation in the suitability of different aphids to *A. 2-punctata*, have been shown by Kalushkov (1998).

Table 4.2 Dietary Diversity in the Coccinellini

Species	Diet	References
Coccinella 7-punctata	Over 100 species of aphid recorded	Dixon, 2000; Majerus, pers. obs.
Coccinella hieroglyphica	Larvae of chrysomelid beetles, e.g. *Lochmaea suturalis*	Hippa et al., 1984; Majerus, 1995
Calvia 15-guttata	Larvae of chrysomelid beetles	Kanervo, 1940
Calvia 14-guttata	In Canada, mainly psyllids; in UK, mainly aphids	Sem'Yanov, 1980; Majerus, pers. obs.
Harmonia conformis	Mainly psyllids, some aphids and coccids	Majerus, pers. obs.
Aiolocaria spp.	Larvae of beetles	Iwata, 1932; Savoiskaya, 1983; Kuznetsov, 1997
Neocalvia spp.	Larvae of ladybirds of the tribe Psylloborini	Camargo, 1937
Tytthaspis 16-punctata	Mainly hyphae and spores of Erysiphaceae mildews, but also pollen, nectar and aphids	Dauguet, 1949; Turian, 1969; Ricci, 1982; Majerus, 1994a
Bulaea lichatschovii	Plant material	Capra, 1947; Dyadechko, 1954; Savoiskaya, 1966
Olla v-nigrum	Psyllids	Majerus, pers. obs.

In general, aphidophagous ladybirds are less prey specific than are coccidophagous ladybirds (Jalali and Singh, 1989; Kairo and Murphy, 1995; Strand and Obrycki, 1996). There are probably two reasons for this. Firstly, aphids tend to be more prone to population crashes than coccids. This is a consequence of the alternation of host plants used by many aphids and the dramatic declines in aphid populations that result from the build-up of aphid predators and parasitoids through the season. The ephemerality of aphids means that aphid-feeding adult ladybirds frequently move from one host plant to another when the prey densities, or the age of aphid colonies, become suboptimal. By comparison, coccid populations have more constant size and age characteristics, and do not migrate en masse to alternative host plants. Therefore, coccidophagous ladybirds have less need for a range of prey species than do aphid-feeding ladybirds. Secondly, the defences of coccids, which tend to be relatively sedentary, usually involve physical traits, such as the tough dorsal covering or scale, or chemical traits in the form of waxy secretions or toxins. Conversely, aphids are fairly mobile and use this mobility to avoid predators (Dixon, 1958). Thus we would expect coccid feeders to evolve foraging strategies, feeding behaviours or detoxification mechanisms that breech the defences of the coccids. Habitat/host plant specialists evolve traits appropriate to their specialisation that increase the degree of specialisation and restriction. Likewise, having specialist foraging traits allows a ladybird to use a well-defended prey that is attacked by few other predators, and increases the specialisation of that ladybird to that prey species.

Considering another comparison between prey types, *C. 14-guttata* feeds in the wild on both aphids and psyllids. It can reproduce and complete development when either type of prey occurs in the absence of the other. However, when both are present, psyllids are preferred. Indeed, Sem'Yanov (1980) showed that psyllids enabled faster larval development and produced greater pupal weight and higher fecundity than aphids. Therefore, the prey preference shown by *C. 14-guttata* is adaptive, leading to higher fitness.

Mixed Feeding

While *C. 14-guttata* will feed on both psyllids and aphids, but shows a preference for the former if available, some ladybirds feed on a mixed diet routinely. Mixed diets usually include animal, plant and/or fungal material. Examples include *Tytthaspis 16-punctata* (Fig. 4.6), which feeds on grass pollen, flower nectar, the spores and conidia of fungi, mites and thrips (Ricci, 1982; Ricci et al., 1983); *Rhyzobius litura* (Fig. 4.7) feeding on aphids, pollen and fungal spores and conidia (Ricci, 1986); *Illeis galbula* (Fig. 4.8), which feeds on fungal conidia and hyphae and pollen (Anderson, 1982); and *Coleomegilla maculata*, a highly polyphagous species that feeds on aphids, mites, (Putman, 1957), moth eggs (Ables et al., 1978; Risch et al., 1982), beetle eggs (Hazzard and Ferro, 1991) and pollen (Hodek et al., 1978). In these cases, a mixed diet may be optimal. By feeding on a mixture of food types, these ladybirds may be obtaining a balanced diet (Hodek, 1996a). Such choice has been shown for moose (*Alces alces*) on the shores of Lake Superior that feed on both the leaves of deciduous forest trees and on plants growing under water in shallow lakes. The deciduous leaves are energy rich, but poor in sodium; the reverse is true of the aquatic plants. Moose have a minimum sodium requirement and their stomachs can contain a limited amount of food. Belovsky (1978) showed that moose ate a mix of deciduous leaves and aquatic vegetation that both maximised energy intake and ensured that their sodium intake never dropped

Fig. 4.6. *Tytthaspis 16-punctata.*

Fig. 4.7. *Rhyzobius litura.* (© Gilles San Martin.)

below the required minimum. Unfortunately, research has yet to be conducted comparing the effect of mixed diets and single diets on ladybird fitness, thus this balanced diet hypothesis awaits tests in these beetles.

Alternative (Substitutive) Foods

Alternative foods are defined here as those that ladybirds will feed on in the wild, but that do not permit maturation of germ cells or full development of immature stages. They are not essential foods for predatory species. However, the demarcation between accepted and alternative foods is indistinct, as some

Fig. 4.8. *Illeis galbula.*

foods, commonly eaten before or after dormancy but rarely used in the reproductive or developmental periods in the wild, at least in captivity, will allow egg maturation and larval development.

When essential prey is scarce for carnivorous ladybird adults or larvae, or when adult ladybirds are building up or replenishing fat stores, glycogen, lipids and fluids before and following dormancy, ladybirds will consume a wide variety of foods. They will attack and eat many other types of insect, including the eggs and larvae of lacewings, hoverflies, chrysomelid beetles and moths, nymph and adult frog-hoppers, shield bugs, small beetles, adult crane-flies, grasshoppers and immature roaches. They will also eat the eggs of a wide variety of insects; field records include instances of Lepidoptera, Coleoptera, Diptera and Hemiptera eggs being consumed. In captivity, *H. axyridis* has been cultured for many generations on eggs of the flour moth *Ephestia kuehniella* (Ettifouri and Ferran, 1993), while Olszak (1986) showed that larvae of *Propylea 14-punctata* could complete development on eggs of the moth *Sitotroga cerealella*. I also have field records of ladybirds killing and eating spiders, mites and worms. Furthermore, they scavenge, feeding on the corpses of any reasonably sized arthropods they encounter. Indeed, some ladybirds have been reported feeding on fly corpses caught in spiders' webs. And in desperation, when starving, they bite into human flesh to test whether we are edible. In my experience, ladybirds find humans unpalatable, therefore we should perhaps be included in the rejected prey category!

Ladybirds also regularly feed on honeydew produced by aphids, and some essentially predatory species graze the spores, hyphae and conidia of mildews growing on leaves or on honeydew. Many carnivorous coccinellids feed on plant pollen, nectar from flowers or extra-floral nectaries, and plant sap or resin from wounds in tree trunks.

Undoubtedly, these alternative foods are often of great importance to the reproductive potential of coccinellids. Although such foods do not promote germ-cell development or reproductive behaviour, they allow fat reserves to be built up prior to unfavourable periods, such as winter or a dry season. The additional resource reserves thereby obtained increase the likelihood of survival through such periods. A ladybird that dies during the winter cannot reproduce in the spring!

Rejected Prey

Many ladybirds eat aphids, yet they do not eat all aphids indiscriminately. A preference for particular aphids may depend on variation in the nutritive quality or defence capabilities of the aphids, or on the habitat/host plant preferences of the ladybirds, which will affect the likelihood of encountering different aphid species. Moreover, some species of aphid are rejected as food by most, or in some cases, all aphidophagous coccinellids.

Hodek (1973), who first raised rejected prey as a class of food, does not define it precisely, and an ambiguity thereby arises. Rejected prey could include those prey species that will always be refused by coccinellids, even to the point that the coccinellids starve to death, if no other food is available. For example, Gagne and Martin (1968) demonstrated that *Hyperaspis* species and *Anisosticta bitriangularis* refused the aphid *Schizolachnus piniradiatae* and starved in the absence of other food. Alternatively, rejected prey could be taken to be those prey species that are avoided when other more preferred species are available. For example, Sloggett (1991) showed that *C. 7-punctata* rejects the aphid *Dactynotus aeneus* in the presence of either *A. pisum* or *Microlophium carnosum*, but accepts *D. aeneus* in the absence of alternative foods. On the other hand, *A. 2-punctata* offered this but no other aphid, starved to death. In this latter case, it appears that the defences of *D. aeneus*, in particular its smearing of attackers with 'wax', are sufficient to deter *A. 2-punctata* completely, but only repel *C. 7-punctata* when other food is available. The 'wax' produced by *D. aeneus* and some other aphids consists of an emulsion of lipid droplets in water (Edwards, 1966). Once smeared onto an attacker, the lipid droplets coalesce and crystallise to form a hard waxy plaque. If *A. 2-punctata* are smeared with this 'wax', they break off their attack and spend considerable time cleaning themselves. Thereafter, they are reluctant to attack *D. aeneus* again.

Hodek (1973) included, in his section on rejected prey, both examples of prey always rejected by coccinellids, and those that are rejected only when other prey are available. To my mind, the second definition, in which rejected prey are those refused when other prey are available, is more useful than the first.

There are several reasons why certain coccinellids reject some prey species. Some aphids and coccids are distasteful to ladybirds. Others are more or less toxic. Telenga and Bogunova (1936) reported that *H. axyridis* refuses the cabbage aphid in the field and, while George (1957) also noted that this aphid is avoided by coccinellids, it is occasionally eaten by *C. 7-punctata*. Some potential prey have strong physical defences, being covered by hard plates or waxy secretions. Finally, some palatable prey are rejected if they have previously been parasitised or are suffering from advanced fungal infections (Quezada and de Bach, 1973; Hoelmer et al., 1994). In this context it is notable that there is variation between species, for Roy et al. (2008a) found that *H. axyridis* accepted cadavers of *A. pisum* infected with *Pandora neoaphidis* much more readily than did *C. 7-punctata brucki* (Fig. 4.9).

The interactions between three common ladybirds and the aphid *Hyalopterus pruni* are interesting. Larvae of *A. 10-punctata*, which pierce the exoskeleton of this aphid with their mandibles, immediately back away from it, and thereafter reject it as soon as they touch it with their palps (Dixon, 1958). Hawkes (1920) noted that larvae of *A. 2-punctata* would eat this aphid, but that the aphid produced 'a grey-green mealy exudation which fills the stomata of the larvae and so kills them'. In complete contrast, Hodek (1959) found *C. 7-punctata* able to eat this aphid, despite its waxy covering, and included it as an essential food of *C. 7-punctata*. There are many other examples of prey that are rejected by some coccinellids, but accepted by others. *Chilocorus rubidus* rejects *Chionaspis salicis* and *Lepidosaphes ulmi*, yet both are accepted prey for other species of *Chilocorus* (Pantyukhov, 1968a). *Aphis nerii* is eaten by *Scymnus frontalis*, *Hippodamia variegata* and *Chilocorus nigritus*, but is rejected by *A. 2-punctata*, *C. 7-punctata*, *Hippodamia 11-notata*, *Leis dimidiata*,

Fig. 4.9. *Harmonia axyridis* larva feeding on cadaver of *Acyrthosiphon pisum* infected with the fungal pathogen *Pandara neoaphidis*.

Coccinella repanda and *P. 14-punctata* (Iperti, 1966b; Tao and Chiu, 1971). This aphid attacks milkweeds (Asclepiadaceae) and oleanders (Apocyaneae). Oleanders contain high concentrations of cardiac glycosides, such as oleandrin and neriin, which are sequestered by the aphids. Malcolm (1990) extracted and identified 25 cardenolides in *Nerium oleander*, and found 17 of these in aphids and 20 in honeydew.

The host plants of aphids and coccids undoubtedly affect their palatability to some coccinellids. For example, *C. 7-punctata* rejects *A. nerii* if the aphid fed on *N. oleander*, but accepts it if it fed on *Calotropis procera*. Furthermore, larvae of *H. variegata* develop normally if fed on *A. nerii* from either *Cynonchum acutum* or *N. oleander*, but produce adults with reduced wings or lacking wings when fed on *A. nerii* from *Cionura erecta* (Pasteels, 1978). Similarly, *Rodolia cardinalis* rejected its normal prey *Icerya purchasi* from *Spartium* or *Genista* species (Savastano, 1918), and this behaviour was observed even when the coccids were offered in isolation from the plants (Poutiers, 1930). These plants contain defensive chemicals, such as the alkaloid spartein and the yellow pigment genistein, and sequestration of these provides the coccids with an effective chemical defence.

Interestingly, *R. cardinalis* also avoids *I. purchasi* if the coccid has been parasitised by *Cryptochaetum icerya*. This avoidance is so complete that the ladybirds starve to death if only parasitised prey are available (Quezada and de Bach, 1973). A similar case is reported for the small coccinellid *Delphastus pusillus*, which prey on whiteflies, such as *Bemisia tabaci*. Prey containing third instar and pupal endoparasites, such as *Encarsia* species, were avoided (Hoelmer et al. 1994).

Toxic Prey

One of the defences of aphids against predation by coccinellids is to accumulate toxic substances from their food plants, or manufacture and store toxic chemicals. The slow developmental rates and high mortalities of larvae fed on certain aphids are probably a consequence of the coccinellid having to deal with toxic substances. If so, these substances are, in effect, slow poisons. Some authors have proposed the alternative explanation that some aphids are lacking in general nutritive value, or lack certain important nutrients. For example, Hariri (1966b) argued that, for *A. 2-punctata*, *A. fabae* is less nutritious than *A. pisum*.

Blackman (1967) supports this view, adding that *A. 2-punctata* may have difficulty in ingesting this aphid. However, the toxicity of this aphid to *A. 10-punctata*, and its inclusion in the list of accepted/essential prey of *C. 7-punctata*, argues that this aphid may contain noxious substances, with different ladybirds varying in their ability to cope with these toxins. Indeed,

Hodek (1956, 1957) working with *C. 7-punctata* and *A. sambuci*, produced evidence arguing against rate of food intake or nutritive value being the cause of the low suitability of this aphid.

In other cases, aphid defensive chemicals may have a very much more dramatic effect. Dixon (1958) found that larvae of *A. 10-punctata* will attack and eat both *A. fabae* and *Megoura viciae*, but after a minute or two they reject the prey and regurgitate their gut contents. Despite this, some of the larvae died from this 'tasting', even when fed on suitable aphid prey thereafter. *Megoura viciae* is also toxic to *A. 2-punctata*, and Blackman (1965, 1967) found that in choice experiments neither larvae nor adults were able to differentiate between this aphid and non-toxic *A. pisum*, but those that ate *M. viciae* died soon afterwards.

The evolution of chemical defences by aphids undoubtedly imposes selective pressure upon their ladybird predators to evolve tolerance to these chemicals or to feed selectively. This development of chemical defences, and evolved resistance to these, comprises an evolutionary arms race between aphids and ladybirds. While *M. viciae* is currently leading the race against *A. 2-punctata* and *A. 10-punctata*, it is losing out to *C. 7-punctata*, which commonly feeds on this aphid in the field (Majerus, 1994a).

The list of aphids shown to be toxic to one or more ladybirds has been growing steadily over the years. In particular, aphids that are pests on crop plants have been tested. The brown citrus aphid, *Aphis citricida*, was shown to be toxic to larvae of *Harmonia dimidiata, H. axyridis, C. 7-punctata, Coccinella 8-punctata, C. repanda, Cheilomenes 6-maculatus, Synonycha grandis* (Tao and Chiu, 1971) and *Cycloneda sanguinea* (Morales and Burandt, 1985). *Aphis craccivora* is usually lethal to *H. axyridis* (Okamoto, 1961), *S. 11-notata* (Hodek, 1960) and *C. 7-punctata* (Azam and Ali, 1970). *Harmonia axyridis* was only able to feed on this aphid and survive if the aphids were taken from *Vicia sativa* or *Vigna catiang*, out of a total of nine plant species. When feeding on *Robina pseudoacacia*, the toxicity of *A. craccivora* has been attributed to the amines canavanine and ethanolamine, which are sequestered from the food plant (Obatake and Suzuki, 1985). However, both larvae and adults of *Propylea japonica* appear to have some resistance to these toxins (Hukusima and Kamei, 1970).

Here, as in some cases of rejected prey, the toxicity of an aphid may vary as a result of its host plant. Thus, the level of toxicity of *Macrosiphum albifrons* to *C. 7-punctata* depends on the cocktail of quinolizidine alkaloids that the aphids can obtain from the various species of lupins that they feed on. Aphids from bitter lupins, such as *Lupinus albus, Lupinus angustifolius* and *Lupinus mutabilis*, which contain high concentrations of lupanine, caused all *C. 7-punctata* larvae to die. However, *C. 7-punctata* developed normally when fed on *M. albifrons* from the sweet lupin, *Lupinus luteus*, in which the total alkaloid content is similar to that in the bitter lupins, but in which lupanine is largely replaced by spartein.

Some aphids also use a form of chemomechanical defence, as already mentioned in the case of the waxing behaviour of *D. aeneus*. Palmer (1914) noticed that *Macrosiphum* aphids could effectively ward off *Hippodamia convergens* by smearing the attacker's mouthparts with a sticky substance. The same behaviour was observed in the interaction between *H. pruni* and *A. 2-punctata* (Hawkes, 1920). The aphid *Acyrthosiphon nipponicum*, which feeds entirely on *Paederia scandens*, secretes droplets from its siphunculi containing a glycoside, paederoside, which is a powerful detergent. Ladybirds that bite into one of these aphids show immediate stress. They drop the prey, salivate excessively and spend a considerable time cleaning their mouthparts thereafter (Nishida and Fukami, 1989).

Prey Recognition, Capture and Consumption

We now turn to the question of how predatory ladybirds find, recognise, subdue and consume their prey. Early researchers, such as Thompson (1951), believed that, as coccinellids had a full complement of sense organs, they could perceive prey at a distance. Hodek (1973), reviewing pertinent experimental work, took an opposing stance, arguing that sight and scent orientation were not involved in coccinellid foraging. He cited, for example, Banks (1957), who found that prey may be missed if only a few millimetres away, even when the coccinellids were downwind of the aphids. Hodek (1973) concluded: 'no discovery of prey is made by the coccinellid until actual contact occurs'.

More recent evidence leads to very different conclusions. To consider this evidence, we need to recognise that the finding of prey by adult ladybirds effectively involves three scales: location of an appropriate habitat, location of patches of prey within that habitat, and singling out of individual prey for attack. Little is known about how ladybirds find appropriate habitats. However, once in an appropriate habitat, some use sight and smell to increase their chances of locating prey patches. For example, Hattingh and Samways (1995) showed that *C. nigritus* were attracted to images of trees and leaf shapes. Khalil et al. (1985) demonstrated that *A. 2-punctata*, which often inhabits trees, is attracted to the taller of two objects, while *C. 7-punctata*, which rarely breeds on trees, is not. An interesting indirect piece of evidence comes from studies in which pink and green forms of *A. pisum* were offered to *C. 7-punctata* and *H. axyridis* on different backgrounds. Harmon et al. (1998) found that *C. 7-punctata* ate more of the form that contrasted most with the background, while *H. axyridis* ate more of the red morph, whatever the background.

Olfactory stimuli are important in prey patch location. The source of olfactory cues varies. *Coccinella 7-punctata* larvae spent longer searching ears of wheat with deposits of honeydew from the aphid *Sitobion avenae*, than clean ears (Carter and Dixon, 1984). Furthermore, female *C. 7-punctata* oviposit in the presence of chemical traces of aphids and honeydew, even in the absence of aphids (Evans and Dixon, 1986).

Stubbs (1980), using adult and final instar larvae of *C. 7-punctata*, showed that adults can detect prey by sight, and larvae can detect prey by smell. Nakamuta (1984) analysed video film of *C. 7-punctata* foraging behaviour and concluded that the ladybirds were attracted to prey at a range of 7 mm, but not at all in the dark, concluding that sight is important in prey detection. Vision has also been implicated in the attack behaviour of *Anatis ocellata*. This ladybird was imported into Michigan as a predator of larvae of the tortrix moth *Choristoneura pinus*. Allen et al. (1970) showed that at a distance of 1.3–1.9 cm from the prey, foraging adults stopped momentarily before moving forwards and quickly snatching it in their mandibles. Attack behaviour did not require previous physical contact and visual stimuli appeared important. Scent has been implicated in the searching behaviour of a number of ladybirds. In choice experiments, *S. pusillus* was more attracted towards vials containing prey, than to empty vials (Colburn and Asquith, 1970). Using more sophisticated olfactometric experiments, *C. nigritus*, *C. 7-punctata* and *Cryptolaemus montrouzieri* have been shown to be attracted by prey odours (Sengonca and Liu, 1994; Hattingh and Samways, 1995; Ponsonby and Copland, 1995). Furthermore, work has shown that some ladybirds are attracted to chemicals released by plants when they are 'wounded' by aphids (Al Abassi et al., 2000; Petterson et al., 2012). One wonders whether ladybirds have evolved a mechanism to respond to chemicals that plants coincidentally release when suffering herbivory, or whether plants have evolved a mechanism to 'call for help'. Recent research has shown that foraging ladybirds leave a trail of footprint chemicals that are detected by parasitic wasps, which also attack aphids.

Cooke (1987) made an intriguing observation. He noted *A. 2-punctata* adults standing at the exit holes of galls made by the aphid *Pemphigus spirothecae*, on *Populus italica*. It may sound far-fetched, but one wonders whether the ladybirds had learnt that food, in the form of aphids, may intermittently issue from the galls.

The consensus view now is that adults and larvae do not search for aphids in a haphazard or random fashion. Rather, their behaviour increases the likelihood of encountering prey. Initial hunting involves an extensive and nonsystematic search for prey. This search is not totally random as ladybirds are both attracted to light and walk against gravitational pull (Dixon, 1959; Kesten, 1969; Ponsonby and Copland, 1995); thus they have a tendency to walk upwards on plants. As many aphids show similar behaviour, the two are likely to meet. Hungry ladybirds are attracted to certain odours and are also attracted to some colours more than to others. Yellow and yellowish-green colours are most attractive to many. As aphids are often most prolific on young plant shoots, which are often yellowy-green, the attraction to this colour is adaptive. Furthermore, foraging ladybirds tend to follow the path of prominent leaf veins, and many species of aphid habitually form colonies close to such veins (Banks, 1957; Dixon, 1959; Vohland, 1996).

The basic foraging behaviour of hungry ladybird adults and larvae is strongly influenced by the structure of their host plants. Shah (1982) showed that larvae of *A. 2-punctata* could not search leaves covered with dense upright hook-shaped or glandular hairs, such as tomato and tobacco. Smooth leaves with a thick slippery waxy layer are also difficult for larvae, particularly young larvae, which are more or less confined to walking along protruding leaf veins. This is also true of smooth leaves with a non-slippery surface, such as sugar beet and broad bean, but to a lesser extent. Shah suggests that leaves with few scattered hairs, on which aphids are distributed evenly, are the most suitable for foraging larvae. Stadler (1991) noted that spatially complex plant structures such as buds, shoots and leaves, are searched more assiduously than simple parts, such as stems and petioles, and that it is these complex parts that aphids usually frequent.

There is one obvious difference between the foraging behaviours of hungry coccinellid adults and larvae. The adults can fly. It is probably this capacity that influences an adult's behaviour most when it alights on a new plant. It tends to search up and down the stem; briefly exploring side twigs and leaves as it goes. However, if it fails to encounter an aphid after a few passes, it will soon fly away to another stem.

Once larvae or adults encounter and consume an appropriate prey item, they change their behaviour, switching from non-systematic searching to intensive area-restricted searching. The rate of forward movement is slowed, turns to one side or the other become more frequent, and they begin 'casting' movements, whereby they move the front part of the body from side to side, increasing the area searched (Banks, 1957; Carter and Dixon, 1984; Podoler and Henen, 1986). This area-restricted searching following the finding of one prey is adaptive; most prey species are colonial and hence several are likely to be found in the same area. Furthermore, area-restricted searching is also elicited simply by the smell of prey (Heidari and Copland, 1992; Sengonca et al., 1995). The switch from random to area-restricted searching depends partly on the quality of prey encountered. Kalushkov (1998, 1999) found that *A. 2-punctata* switched after feeding on *Phorodon humuli*, a high-grade prey, but not after eating the low-grade *A. fabae*. Increased likelihood of location of a prey patch may also result from a simple response to a cue stimulus. Several ladybirds, such as *C. montrouzieri*, and species of both *Diomus* and *Exochomus*, slow their rate of forward motion, or even stop if they encounter the wax of mealy bugs (van den Meiracker et al., 1990; Merlin et al., 1996a, b).

Once a patch of prey has been encountered, a ladybird has to assess its quality. This assessment involves several considerations: the state of the ladybird itself, in the form of its degree of hunger, whether it is feeding up for or following dormancy, whether it is maturing gametes, and its sex; the suitability of the prey as food; the ease with which it can be captured and processed; the size and age composition of the patch, and the number of competitors using the patch. Of course, female ladybirds also need to assess prey patch quality for their offspring. In this regard, they have to make an assessment of the future dynamics of the patch. The assessment of a patch as a nursery for progeny will be discussed in the next chapter. Here I consider the assessment of prey patches as food for the adult or larva itself. Because adults may have multiple motives when assessing patch quality, most work on this question has involved larvae.

A larva's assessment of patch or individual prey quality will be in terms of the benefit that may be obtained. However, benefit may be a function of encounter rate, or average food intake, or net intake of energy. Furthermore, we may ask whether a larva makes an instant assessment of the benefit that may be gained, or assesses potential benefit over a longer period. At present, it appears that a major determinant of variations in larval foraging behaviour is simply the level of hunger. Carter and Dixon (1982) studied the duration of area-restricted searching behaviour in *A. 2-punctata* larvae, varying the degree of hunger of the larvae, the size of prey and the rate at which they encountered prey. They found that neither encounter rate nor average rate of food intake affected foraging behaviour, but that hunger did. Hungry larvae, after eating one prey item, maintained area-restricted searching behaviour for less time before reverting to random search, than did less hungry larvae.

Various authors have reported that neonate larvae are rather inefficient at finding and capturing aphid prey. Banks (1957) noted that first instar larvae of *P. 14-punctata* remain on their eggshells for 12–24 hours and then can search for food for up to 35 hours. If, however, suitable food is not found in this period, the larvae become inactive and subsequently die. The period of neonate larval survival in the absence of food varies between species and with temperature. For example, at 21°C, *A. 2-punctata* survive about 40 hours after dispersal from their eggshells, whereas, under the same conditions, larvae of *Coccinula sinensis* can live for more than 72 hours before dying (Elnagdy, pers. comm.). Dixon (1959) showed that the efficiency of first instar *A. 10-punctata* larvae in capturing first instar nymphs of *M. carnosum* was less than 20%, and with later aphid instars was lower still. Laboratory studies with a variety of other species, including *A. 2-punctata*, *C. 7-punctata* and *H. axyridis*, have led to the general belief that neonate ladybird larvae suffer high mortality from starvation as a result of failing to find or subdue prey (Dixon, 1959; Wratten, 1976; Majerus and Hurst, 1997).

Aphids may possess chemical and chemomechanical defences and also have the ability to either fight or flee. Dixon (1958, 1959) observed the interactions between *A. 10-punctata* larvae and adults and *M. carnosum*. He found that the aphids had a considerable array of defensive behaviours including kicking, pulling free, running away, dropping off the plant, or waxing. The precise response employed by an aphid depended on the type and size of the attacking predator. Fighting behaviours were mainly used against small larvae, while large larvae and adults elicited fleeing responses. Wratten (1976) obtained similar results with *A. 2-punctata*.

Despite the array of defences with which aphids are armed, neonate ladybird larvae are sometimes successful in subduing aphids very much larger than themselves. They do this by the simple expedient of climbing onto the backs of the aphids, and then 'riding' them with their mandibles embedded in the dorsal surface of their 'steed' (Fig. 4.10).

Fig. 4.10. Early instar larva feeding on aphid.

How a ladybird larva or adult consumes an aphid depends on the relative sizes of predator and prey. Both adults and larvae regurgitate fluid containing digestive enzymes from the gut and inject it into their prey. This partially digests the prey and the predigested food is then sucked back in. Small larvae suck out the contents of aphid prey, but larger larvae and adults may consume smaller aphids completely, and masticate large aphids to a considerable degree (Figs. 4.11, 4.12), discarding only legs and antennae. Some species, such as *Platynaspis luteorubra*, have specialised perforated mandibles, which are used to suck fluid out of the prey directly. The mite-eater, *S. pusillus*, sucks predigested fluid from its prey along grooves in the mandibles and galea. *Clitostethus arcuatus*, which feeds mainly on eggs, breaks the chorion with a single toothed mandible and sucks out the contents (Ricci and Stella, 1988).

The mandibles of coccinellids show interspecific variation, which seems to be adaptive, the foods of different species being reflected in the structure of these mouthparts (Capra, 1947; Ricci, 1982; Ricci and Stella, 1988). The mandibles of aphid-eating species have a double pointed incisor with the edge of the ventral tooth more or less serrated, or divided into many small tooth-like processes, and a strongly developed molar projection. This design

Fig. 4.11. *Harmonia axyridis* larva feeding on aphid.

Fig. 4.12. *Coccinella 7-punctata* feeding on aphid.

may be contrasted with the single-toothed incisor of most coccid-feeding species that have a single point to the incisor, which is used to penetrate the chitinous exoskeleton of scale insects (Samways and Wilson, 1988).

Cannibalism

One further source of food warrants detailed consideration. The survival of many predatory ladybirds, particularly aphidophagous species, is enhanced by their habit of eating other members of their own species. Aphid feeders are most prone to cannibalism because their prey is so ephemeral. The problem is exacerbated when ladybird numbers explode during particularly favourable conditions. Aphid populations then crash as a result of the pressure from ladybirds and other aphid predators. Vast numbers of ladybirds may then be left without their normal food.

In a biological sense, cannibalism is formally defined as the act of killing and eating an individual of one's own species. This definition is precise and excludes eating of conspecifics that are already dead, which is a form of scavenging. For example, humans surviving in extreme conditions by eating the corpses of others of their party, who have died through starvation or injury, have not engaged in true cannibalism.

Group and Kin Selection

Foraging ladybird larvae frequently attack ladybird eggs. As several species of predatory ladybird may occur together, the eggs that a larva encounters may be conspecific or a different species. Thus, ladybird eggs are prone to both cannibalism and predation by other species of ladybird. Surprisingly, cannibalism is more common than interspecific predation. Agarwala and Dixon (1992) and Petersen (1992) reported that both adults and larvae of *A. 2-punctata* find conspecific eggs more palatable than those of other species. While perhaps contrary to intuitive expectation – surely it would be better to eat eggs of another species than eggs of one's own species – the reason is not difficult to explain. Firstly, females will provision eggs with nutrients geared to the developmental requirements of their own species. As nutritional requirements differ from one species to another, conspecific eggs may be more nutritious than those of other species. Secondly, females protect their eggs by provisioning them with the defence chemicals they carry themselves. These defensive chemicals vary between species. For example, the main defence chemical is coccinelline in *C. 7-punctata*, adaline in *A. 2-punctata* and *A. 10-punctata*, hippodamine in *H. convergens* and *C. magnifica* and harmonine in *H. axyridis* (see Pettersson, 2012). Ladybirds can eat eggs provisioned with the defensive chemicals that they themselves carry more easily than those that contain chemicals they do not possess.

It is worth considering here why the intuitive expectation that it would be better to eat the eggs of another species, rather than one's own, is fundamentally wrong in evolutionary terms. Darwin's theory of evolution by natural selection is based on the struggle among individuals of a species to survive and reproduce. The emphasis on the individual is crucial. An inherited characteristic that provides its bearer with some advantage in the survival stakes will increase in the population through time. Its bearer will be more likely to survive and reproduce than will those lacking the trait. Offspring inheriting the trait will consequently have a higher likelihood of survival and reproduction. This will be true even when the characteristic is not good for the species, as perhaps in the case of ladybirds eating conspecific eggs. The point here is that selection acts on individuals, not on groups. While cannibalism may not be good for the population of ladybirds, it is good for the individual larva doing the cannibalising.

It is still common in many natural history commentaries to hear that certain behaviours have evolved for the 'good of the group', or the 'good of the species'. Examples, such as 'red deer stags competing for possession of a harem of females rarely do each other real damage, because if they did it would endanger the survival of the species'; or 'lemmings commit suicide by jumping off cliffs when their population density becomes too great for the food available, so that those remaining can survive', are still common-place. Yet the behaviour of the red deer stags can be explained by selection acting on the individual; it is better to accept defeat and live to fight another day. Moreover, lemmings do not jump off cliffs! (Those who have seen lemmings jump off cliffs on celluloid, as I did as a child, could not see the men off-camera herding and driving them to their deaths.)

The idea of animals behaving for the good of their group is seductive. One of the main proponents of the idea, Wynne-Edwards (1962), used the concept to explain observations of organisms that appeared to control their rate of reproduction and food consumption so that food did not run out. If resources were over-exploited, all members of the group would starve, therefore it was in the interests of all members of the group to keep population numbers correlated to food availability, by keeping reproductive output in check. One of the reasons that this idea has such a seductive hold on our thinking is that it is the way humans should, but currently, quite tragically, do not behave.

The major problem with the group selection idea is that any new mutation that caused its bearer to behave selfishly, rather than for the good of the group, would be successful and increase in the population at the expense of the altruists. Wynne-Edwards realised that altruistic groups might be invaded by selfish individuals, who would outcompete them. He circumvented this problem by proposing that groups of selfish individuals would be more prone to extinction than groups of altruists. While this may work in some highly specific circumstances, for example, the host–parasite interaction involving rabbits and the myxoma virus in Australia (Maynard Smith, 1989), in practice, selection acting on individuals is almost invariably a much more powerful force than is selection acting on groups. Group extinction is simply too uncommon. Add the possibility of migration of selfish individuals into altruistic groups, and the likelihood of group selection being an important evolutionary force becomes very slight (Williams, 1966; Maynard Smith, 1976). Indeed, from many experiments over the past 40 years, workers have concluded that selection acts at the level of the individual, not the group, unless members of a group are close relatives, and even this is contentious.

When group members are close relatives, altruistic behaviour may evolve because, although an altruist may put itself at risk by acting selflessly, if these actions increase the survival of its relatives, the genes controlling the altruistic behaviour will be selected for, as kin carry many of the same genes. In this kin selection, an individual can increase the likelihood of its genes being represented in future generations by helping relatives who carry identical copies of the genes by descent. Kin selection is common. Obvious cases concern parents putting themselves at risk to ensure the safety of their offspring, or honeybee daughter workers, who sacrifice their own chance of reproducing to aid their siblings.

In the case of cannibalism by ladybirds, the group selection idea has no credibility. Selection acts on the individual and favours cannibals because, in them, the risk of starvation is reduced.

Who Eats Whom?

Cannibalism in ladybirds can conveniently be divided into four categories: (i) of immature stages by unrelated larvae; (ii) of eggs by adults; (iii) of larvae, pre-pupae, pupae and adults by adults; and (iv) of eggs by neonate sibling larvae (Majerus and Majerus, 1997a). Each type occurs in different circumstances, and has different implications.

Cannibalism of Immature Stages by Unrelated Larvae

The rationale behind cannibalism of immature stages by unrelated larvae is easiest to comprehend. Such cannibalism occurs commonly whenever appropriate aphids become scarce, and occurs in some circumstances even when they are not scarce.

Eggs, ecdysing larvae and pre-pupae are the most vulnerable to attack by larvae. Ladybird eggs are heavily preyed upon by small conspecific larvae, even when aphids are present (Fig. 2.18). The reasons for this are three-fold. Firstly, as several females may oviposit in the vicinity of a suitable aphid colony, egg clutches can be in the forage range of larvae resulting from eggs laid just a day or two earlier. Secondly, conspecific eggs are a nutritious food source for young larvae, having been provisioned with just the nutrients essential for rapid early development. Thirdly, at a time when larvae may find it difficult to subdue prey because of their small size, conspecific eggs provide an easy meal that will not fight back or run away.

Egg cannibalism by older larvae also occurs (Fig. 4.13), but more rarely. This is because females tend not to oviposit close to prey colonies that are already being attacked by ladybird larvae. Indeed, it is likely that the strategy of laying eggs away from aphid colonies under attack from larvae has evolved primarily to reduce the risk of a female's eggs being cannibalised.

Of the other pre-imaginal stages, ecdysing larvae and pre-pupae (Fig. 4.14) are most vulnerable because, in these stages, the larvae are semi-static, being attached to the substrate at the posterior end of the abdomen, and because for a short time after moulting their cuticle is abnormally soft. They are, therefore, less mobile than active larvae, and their soft cuticle provides less protection than the hard cuticle of a fully formed pupa.

Larvae do attack other active larvae, and when this occurs, it is generally the larger larva that wins the encounter (Fig. 4.15) (Majerus, 1994a). Cannibalism of fully formed pupae is less common, and is generally confined to third or fourth instar larvae. This is partly because late instar larvae occur more commonly with pupae than do young larvae, and partly because the hard cuticle of

Fig. 4.13. Egg cannibalism by *Harmonia axyridis* larva.

Fig. 4.14. *Harmonia axyridis* predation of conspecific pre-pupa.

fully formed pupae provides a fairly effective defence against the mandibles of first and second instar larvae.

Agarwala and Dixon (1993) report that larvae of *A. 2-punctata* prefer to attack and eat non-sibling larvae rather than siblings. One has to wonder how larvae are able to assess their relatedness to other larvae that they encounter, particularly as, due to the extreme promiscuity of *A. 2-punctata* females, larvae will have few full siblings, most larvae produced by a female being half siblings (Majerus, 1994a; Haddrill, 2001). I will return to this question later.

Fig. 4.15. Two *Harmonia axyridis* larvae feeding on a *Coccinella 7-punctata* larva.

Cannibalism of Eggs by Adults

Observations of the cannibalism of eggs by adults fall into two groups: adult ladybirds eating eggs with no indication as to whether the adults were parent to the eggs consumed or not, and eggs being eaten by females that have just laid them.

The consumption of eggs by unrelated adults is relatively scarce in the wild. The few records that have been made are all from late in the breeding season when aphids have been scarce.

Why do adult ladybirds only indulge in egg cannibalism late in the breeding season? The reasons depend on both the population dynamics of aphids, and ladybird oviposition strategies. The dynamics of an aphid colony can fairly be described as 'boom and bust'. The extraordinary reproductive rate of aphids means that when a single, asexually reproducing female founds a new colony, numbers initially increase exponentially. This rate of increase eventually slows due to lack of space, deterioration in the host plant, or to high mortality imposed by predators, parasitoids and parasites. Frequently, these factors coincide, so that numbers in aphid colonies crash dramatically.

Female aphidophagous ladybirds select where and when they lay their eggs with care. They oviposit close to aphid colonies that are increasing in number, have a high proportion of early instar nymphs compared to late instar nymph or adult aphids, and are not yet heavily attended by ladybird larvae or other aphid predators (Hemptinne and Dixon, 1991). This caution is partly a consequence of the cannibalistic tendencies of ladybird larvae. However, the strategy may also be a consequence of two other factors: the small size of newly hatched ladybird larvae, and the demography of aphid colonies. Neonate ladybird larvae are very small when they first disperse from their egg clutches and do not contain much nutrient reserve. It is therefore critical that they find aphid prey sufficiently small for them to subdue rather quickly after dispersing from their egg clutch. If only large aphids are available, these usually have a sufficient armoury of defensive stratagems to avoid capture. Female ladybirds must therefore lay close to aphid colonies containing reproducing females, so that small aphid nymphs are available.

Rapid declines in aphid colonies towards the end of the season also shape female oviposition behaviour. Progeny from females that oviposit close to senescent colonies may find that their normal prey disappears before they complete development. As the timing of sudden declines in aphid colonies is unpredictable, the safest strategy for female ladybirds is to lay eggs close to colonies in their early stages.

Early in the breeding season, female ladybirds will only oviposit if they are able to feed sufficiently to mature their eggs. As adult and larval ladybirds feed on the same food, a female that can mature her eggs is likely to be in a situation appropriate for oviposition, that is, where there will be food for her offspring. The availability of food means that these females need not resort to egg cannibalism. Conversely, towards the end of the breeding season, the adult population will consist both of old reproducing adults, and young adults of the next generation feeding up for the winter. As the old females are close to death, but may still bear a significant egg load, they cannot afford the energy to seek out the diminishing optimal sites for oviposition. For these females, there is at least some possibility that eggs laid, even in unsuitable situations, may survive, while eggs that are never laid cannot contribute to the females' reproductive success. Consequently, old females will oviposit close to any aphid colony, irrespective of its age, composition or health. If eggs are laid close to newly eclosed adults, these young adults may resort to egg cannibalism if other food is scarce. The only real concern for the newly emerged ladybirds is to build up fat reserves for winter. Observations of eggs laid by old females being consumed by next generation adults exist for *H. axyridis* in Michigan, Tokyo and London, *Coleomegilla maculata* in New York State, and *A. 2-punctata* and *C. 7-punctata* in Cambridge, UK (Majerus, pers. obs.).

The cannibalism of eggs by their own parents is even more rare than cannibalism of eggs by unrelated adults, but it does occur in very precise and restricted conditions. These conditions occur early in the breeding season, when a period of favourable weather is followed by a period of severely unfavourable weather, usually cold. An initial increase in aphid numbers during the favourable period initiates the maturation of ladybird ovaries. If this is followed by a decline in aphid numbers when the weather deteriorates, the ensuing dearth of aphid prey may leave females with few energy reserves. One solution to this problem is to reconvert developed eggs into energy. Partly developed eggs can be reabsorbed internally. However, eggs ready to be laid cannot be reabsorbed in this way. Instead the female will lay one egg and then immediately turn around and eat it. She may repeat this behaviour many times.

Cannibalism of Larvae, Pupae or Adults by Adults

Adult ladybirds rarely eat immature stages apart from eggs. This is because they rarely occur together except when suitable aphid prey occur at high density. The reluctance of females to oviposit on plants already hosting conspecific larvae means that by the time larvae, pre-pupae and pupae are present at a site, reproducing females will have moved to other sites. Only when the first of the new generation begin to emerge from their pupae will adults be found at high density with immatures. Then, some cannibalism of immature stages by the newly hatched adults occurs if aphids are becoming scarce. Again, ecdysing larvae, pre-pupae and pupae are most at risk.

Cannibalism of adults by adults is very rare in most years. However, it becomes very common when ladybird numbers explode. Consequently, phenomenal increases in ladybird numbers result in almost total annihilation of aphids and alternative prey. In their searches for prey, ladybirds take to the air in immense numbers (Majerus and Majerus, 1996) and huge swarms form. These are most often observed on the edge of large water bodies because the combination of on- and offshore air currents here brings the ladybirds to ground. These 'plagues' of ladybirds consist mainly of young adults that have to feed up for the winter. They are thus desperate to find food and will try almost anything, to see if it is edible. They will certainly consume conspecific ladybirds. As swarms come to ground, those that have already landed and folded away their wings are relatively immune to attack. The pronotum, elytra and ventral abdominal plates provide considerable protection and prevent access to the softer parts of their bodies. However, these 'landed' ladybirds will attack those alighting before they can fold away their wings and close their elytra, and so gain access to the softer dorsal surface of the abdomen.

During major population explosions, such as those in Britain in 1959 and 1976, the result of adult–adult cannibalism may leave a scene of absolute carnage, with millions of corpses littering the ground. That these corpses are the result of cannibalism is evidenced by the fact that the soft parts have been consumed, the corpses consisting of just their hard external parts.

Sibling Egg Consumption and Cannibalism

The consumption of unhatched eggs in a batch by larvae from the same clutch that have already hatched is a common feature of the biology of

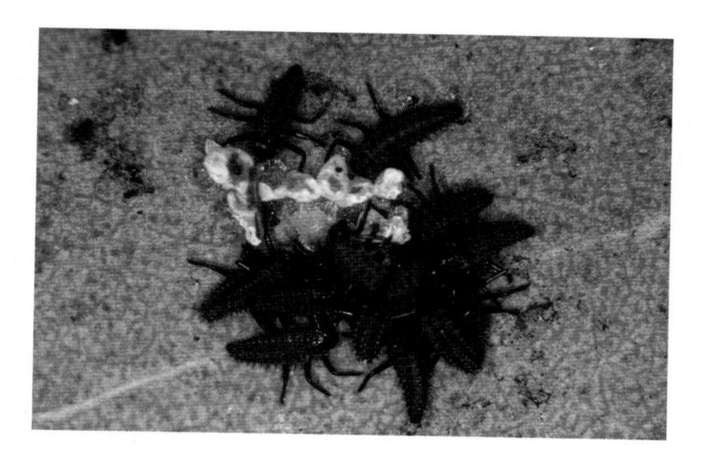

Fig. 4.16. Consumption of unhatched eggs by *Harmonia axyridis* larvae.

aphidophagous coccinellids (reviewed in Majerus, 1994a) (Fig. 4.16). This habit is fascinating because it has implications for many other aspects of ladybird biology. Two obvious questions that need to be addressed before these implications are considered are: which unhatched eggs do neonate larvae eat? and, why have neonate larvae evolved to eat unhatched eggs from their own clutch?

It is necessary to determine the nature of the unhatched eggs that are consumed, because the costs and benefits of egg eating will vary if just infertile or dead eggs, rather than viable eggs, are consumed. If neonates eat only infertile or already dead eggs, the consumption is not strictly cannibalism, because the neonate larvae have not killed conspecifics. Furthermore, such behaviour will impose no cost on the reproductive fitness of the parents of the cannibalisers. Rather, a fitness benefit results if larvae, by consuming infertile and dead eggs and thereby retrieving some of the otherwise wasted resources in these eggs, increase their own chances of survival. However, if neonate larvae eat eggs that contain live embryos that were simply developing more slowly, the lifetime reproductive success of their parents may be reduced. It is correct to say that the fitness of the parents may be reduced, rather than will be reduced by sibling egg cannibalism because a reduction is not a foregone conclusion. In some cases there are benefits to the parents of larvae that consume living sibling eggs when considered in terms of the number and quality of progeny that survive to adulthood. These benefits come in the form of a highly nutritious meal gained by the cannibalising offspring. If this cannibalism significantly increases the survival of those larvae that do hatch, the fitness increment may offset or exceed the fitness reduction that results from the loss of some eggs that otherwise would have hatched. Evidence of this fitness compensation has been demonstrated in various ladybirds that carry bacteria that cause some embryos to be killed.

Laboratory observations of several species, including *A. 2-punctata*, *C. 7-punctata*, *H. axyridis*, *A. 10-punctata*, *C. sinensis* and *P. japonica*, have shown that some of the eggs consumed contain developing embryos (Banks, 1956; Majerus, unpubl. data) and, in the first three of these, sibling egg cannibalism has been confirmed in the field (Majerus, pers. obs.). Infertile and dead eggs are also habitually consumed.

The Evolution of Sibling Egg Cannibalism

The reason why sibling egg cannibalism has evolved is simple. Eggs are nutrient rich and provide an easily reached and defenceless meal for small neonate larvae. Agarwala (1991) demonstrated the nutritional value of conspecific eggs, showing that larvae of *A. 2-punctata* survived longer on eggs than on an equal fresh weight of the aphid *A. pisum*. Furthermore, approximately three times the weight of aphids compared to eggs were needed to produce a fixed weight increase in fourth instar larvae. Similarly, the coccid-feeder *Exochomus flavipes* was shown to develop faster on conspecific eggs than on its usual prey, the coccid *Dactylopius opuntiae* (Geyer, 1947). Against this must be set reports that *C. 7-punctata* and *C. maculata* developed more slowly on conspecific eggs than on their usual diets (Warren and Tadic, 1967; Takahashi, 1987).

How then did sibling egg cannibalism evolve? Eating living eggs is probably simply an extension of the habit of newly hatched larvae of many insects eating part of their own eggshell. If extending this behaviour to any unhatched eggs increases one's chance of survival, the gene(s) inducing this behaviour will increase in frequency as long as the benefit to this gene is not outweighed by costs through the loss of some living eggs that contain the gene. The degree of cost will depend on the likelihood that other eggs in a clutch containing a cannibalising individual carry the cannibaliser gene. This likelihood depends in turn on the genetic dominance of the cannibalism gene, the sex of the carrier, and the mating system. Let us assume that a new cannibalism gene arises by mutation and that it is genetically dominant to the old non-cannibalism allele. In the early evolution of sibling egg cannibalism, when the cannibalism gene is rare, the population will consist mainly of individuals that carry two non-cannibalism alleles, with a few individuals carrying one cannibalism and one non-cannibalism allele. Most carriers of the cannibalism allele are thus likely to mate with non-carriers. If the carrier is female and heterozygous (carrying both a cannibalism and a non-cannibalism allele), then, on average, half the eggs in a clutch should receive the cannibalism allele. However, if the carrier is a heterozygous male, the proportion of eggs with the cannibalism allele may be lower. This is because female ladybirds are highly promiscuous, mating with many different males, and so the eggs in a clutch may not all have the same paternity. This is important for the dynamics of the initial spread of a cannibalism gene, because the lower the likelihood that the eggs being eaten by a cannibaliser are also carrying this gene, the higher the chance of the cannibalism gene spreading. Perhaps the easiest way to envisage this is to consider the ratio between cannibalism and non-cannibalism alleles in the eggs at risk in clutches producing cannibalisers. When rare, only a quarter of the alleles in a clutch with cannibalisers are likely to be cannibalism alleles. However, as the gene spreads through the population, the chance that cannibalised eggs are carrying the gene increases as it becomes more likely that some individuals are homozygous for the gene, or that several of the males that a female mates with are carriers of the gene. The costs to the cannibalism gene, resulting from its own expression, thus increase. This means that the fitness advantage of sibling egg cannibalism will be negatively correlated to the frequency of the cannibalism gene.

If the cannibalism allele is recessive to the non-cannibalism allele, the situation is different in the early stages of spread. As only individuals homozygous for the cannibalism allele now express the trait, when the allele is rare, the only egg clutches in which sibling egg cannibalism will be seen will be those resulting from matings between two carriers of the cannibalism allele. Such matings will give rise to egg clutches in which a quarter of the progeny are cannibalisers, while two-thirds of the remainder carry the allele. Hence, even in the early stages of spread, the proportion of the genes at risk that are cannibalism alleles will be high. Consequently, it is much more likely for a cannibalism-inducing allele to spread if it is genetically dominant than if it is recessive.

The Evolutionary Implications of Sibling Egg Cannibalism

Sibling egg cannibalism has a diverse array of evolutionary implications for ladybirds. It may be partly responsible for the high proportion of neonate larvae that die of starvation. The reason for this is simple. Slow developing embryos in a clutch of eggs may fail to hatch because larvae that have already hatched cannibalise them. Fast embryonic development and hatching from the egg as early as possible will be highly advantageous and will be strongly selected. This means that ladybird larvae will be as small as possible when they hatch, on condition that they retain a realistic chance of capturing prey. The result is a tenuous balance between being quick enough out of the egg not to be cannibalised by one's own siblings and being large enough to capture and subdue aphids.

The need for fast embryonic development may also be responsible for the high number of recessive lethal genes carried by some species of ladybird

(Lusis, 1947a). Such genes are responsible for the chronic levels of inbreeding depression in laboratory cultures of many aphidophagous ladybirds (Hodek, 1973; O'Donald and Majerus, 1985). They affect ladybirds in two main ways. Firstly, some of these genes only appear to be lethal. These are the genes that cause embryonic development to be slower than average. Embryos homozygous for one of these slow development recessives would hatch given time, but are killed and eaten by siblings that carry a fast development allele of the same gene and thus hatch earlier.

The second type of lethal gene is one that does the reverse of the first; its effect is to make the rate of embryonic development faster than average. Here eggs begin to develop normally, but fail to hatch, even when sibling egg cannibalism is prevented. It has been suggested that these lethal alleles are mutations that overload the developmental system. If selection imposed by sibling egg cannibalism has increased the rate of embryonic development to the maximum possible, cell division and differentiation systems will be under severe stress so that they cannot cope with any further increase (Majerus and Majerus, 1997a).

The presence of exceptional numbers of recessive lethal genes in some ladybirds poses an evolutionary conundrum, because selection should purge populations of disadvantageous genes. However, sibling egg cannibalism may be responsible for the maintenance of these deleterious genes in populations (Werren, 1987; Majerus, 1993). Assume that the genes are full lethals and are completely recessive, that is, heterozygotes develop at the same rate and have the same fitness as the dominant homozygotes. Homozygote recessives thus result only from matings between two heterozygotes. In a cross between two heterozygotes, one quarter of the progeny are homozygous recessive and therefore die. Neonate sibling larvae then eat these eggs. This gives these larvae a considerable survival advantage. Larvae from a mating between two heterozygous carriers of one of these lethal genes thereby gain because a quarter of their siblings die in the egg and may be eaten. As two-thirds of the hatching larvae will be heterozygous for the lethal gene, these carriers will maintain the lethal gene in the population whenever their fitness is at least double that of a normal larva from a normal clutch. The exact size of the advantage that is gained from sibling egg cannibalism is largely dependent on prey availability; thus the evolution and maintenance of these recessive lethals depends on both the cannibalism habit and on aphid density (Majerus, 1994a).

A further evolutionary consequence of sibling egg cannibalism in ladybirds is that it provides a special set of conditions that favour the evolution of male-killing.

The evolution of sibling egg cannibalism may be an example of an evolutionary feedback loop. The occurrence of a cannibalism gene through mutation and its spread due to selection favouring this selfish behaviour would in turn have imposed a pressure for embryos to develop and hatch quickly. Consequently, hatching larvae would have been progressively smaller and less capable of catching and subduing aphids. The benefit of an additional meal in the form of a cannibalised egg would have increased, leading to an increase in cannibalistic behaviour.

Non-carnivorous Ladybirds

Almost a third of coccinellid species are not normally predatory. These include the subfamily Epilachninae, in which both larvae and adults feed on plant material, and members of the genus *Halyzia*, which feed on fungi. In addition, one genus in the Coccinellinae, *Bulaea*, is vegetarian, feeding on pollen (Dyadechko, 1954; Bielawski, 1959; Savoiskaya, 1966), nectar and leaves (Savoiskaya, 1970). Finally, some species of Coccinellinae that sometimes feed on animal prey also feed on pollen and fungi. This latter group includes species such as *T. 16-punctata* (Ricci, 1982; Ricci et al., 1983), *I. galbula* (Anderson, 1982) and *R. litura* (Ricci, 1986).

The mandibular dentition of coccinellids that rarely or never feed on live prey differs considerably from that of predatory species. The mandibles of phytophagous Epilachninae are adapted to cutting through and grinding up plant tissue. The incisor region is armed with four or five large blunt teeth, each subdivided into smaller teeth of various sizes. The inner blade of the mandible is armed with a row of coarse teeth, and the molar projection is lacking.

Phytophagous species vary considerably in their food plant specialisation. *Subcoccinella 24-punctata* is quite eclectic in its choice of food, feeding on the leaves of many low-growing plants including clovers, vetches, plantains and grasses. In some parts of the world, it is a minor pest of alfalfa. Conversely, many epilachnid species feed on plants of just a single family, with the Curcubitaceae and Solanaceae being the most favoured. For example, *Henosepilachna 28-maculata* has a strong preference for potato, although it will feed on a variety of Solanaceae, including eggplants (Fig. 4.17). Both *H. 28-punctata* and *Chnootriba elaterii* feed on a variety of cucurbits and the latter, in particular, may reach pest densities.

A specialised form of phytophagy is represented by the pollinivory of species in the genus *Bulaea* (Capra, 1947). These species have specialised dentition for their pollinivorous habit, with combed mandibles that they use to collect the pollen grains. Many other species of coccinellid will feed on pollen, but for most, this is an alternative food, rather than an accepted food. Exceptions are *C. maculata* and *H. axyridis*, which can both complete

Fig. 4.17. *Henosepilachna 28-maculata* feeding on eggplant.

development and reproduce (at least to a limited extent), on an exclusively pollen diet, although they have a wide range of other accepted foods (Ewert and Chiang, 1966; Berkvens et al., 2008).

Hodek (1973) suggested that the many-toothed mandible of the Epilachninae is the ancestral condition, the mandibles of carnivorous species being derived from it. Conversely, the dentition of fungus feeders shows obvious affinities with that of predatory Coccinellinae. Members of the genera *Halyzia*, *Vibidia*, *Illeis* and *Psyllobora* feed on the hyphae, conidia and spores of lower fungi, or powdery mildews, particularly of the family Erysiphaceae. Individuals of these mildew-feeding species possess rows of small teeth for raking up fungal conidia and spores. In members of the genus *Halyzia*, the edge of the ventral tooth of the incisor subdivides into a row of smaller teeth, which diminish in size from the apex towards the base. In *T. 16-punctata*, a row of teeth lies along the inner edge of the mandible, between the incisor and molar projection (Minelli and Pasqual, 1977; Ricci, 1979; Samways et al., 1997). This species has a more catholic diet, feeding on spores and conidia of a variety of rusts and mildews (Turian, 1969), and also eating pollen, nectar and aphids, both in the larval and adult stages (Figs. 4.1 and 4.4).

The similarities between the dentition of mycophagous ladybirds and predatory members of the Coccinellinae lead to the deduction that mildew feeders have evolved a non-carnivorous habit and appropriate dentition secondarily. Mycophagy is shown in relatives of aphid-eating species, and may have evolved as a way of opting out of the risks inherent in feeding primarily on ephemeral prey. In times of aphid shortage, many aphid-feeding species

Fig. 4.18. *Psyllobora 22-punctata.*

consume other foods, including honeydew and mildews growing thereon. From this position, it is a small step to come to use mildews as the principal food, particularly as the occurrence of mildews on some plants is extremely regular and predictable. For example, the mildews that grow on *Heracleum sphondylium* and *Acer pseudoplatanus* are the main foods of *P. 22-punctata* (Fig. 4.18) and *H. 16-guttata*, respectively, in parts of Europe. The timing of reproduction in these species is precisely synchronised with the growth of these mildews (Majerus and Williams, 1989; Majerus, 1994a). This is another difference in coccinellids that is strongly influenced by food. While reproduction in these mycophagous ladybirds and many coccidophagous species is predictable, being synchronised with seasonal increases in the availability of preferred accepted foods, most aphidophagous species show much less predictability. Many simply reproduce when their ephemeral prey are abundant. It is to the reproductive strategies of ladybirds that I now turn.

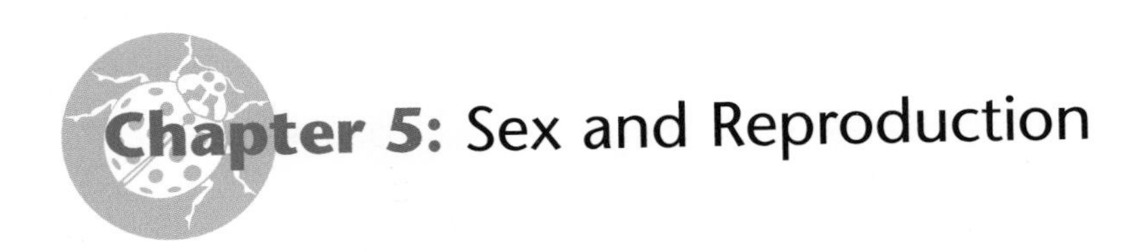

Chapter 5: Sex and Reproduction

The Aim of Life

The primary aim of life is to reproduce. In ladybirds, reproduction is exclusively sexual, a new zygote resulting from the fertilisation of an egg by a sperm. If a ladybird is to reproduce successfully, it must produce gametes, find a mate, persuade the mate to accede to copulation, copulate, and then, if female, lay eggs. These are the fundamentals of reproduction, with behavioural variations at each stage, both among and within species. The process may be further complicated by factors such as sperm competition, cryptic female choice, lack of genetic compatibility and the population sex ratio. This chapter considers the reproductive strategies of ladybirds, and the evolutionary forces that have shaped them.

Mating

The mating behaviour of ladybirds presents some interesting features, for in some species, both sexes have elements to their reproductive behaviour that at first sight appear at odds with what might be expected. To understand these features, it is necessary to appreciate the fundamental biological differences between the sexes. These concern the sex cells, or gametes.

The Difference between Males and Females

The gametes of most sexually reproducing organisms are of two types: either large and sessile or small and mobile. The former are eggs, the latter sperm. These are the ultimate definers of sex. Individuals that produce large sessile sex cells are female, while those that produce small mobile gametes are male.

The production of eggs and sperm involve an unusual type of cell division called meiosis. Here, the genetic material of a cell is exactly duplicated, and the cell divides twice to produce four daughter cells, each with half the genetic material of the parental cell. The products of meiosis are thus haploid rather than diploid. The diploid state is restored when the nuclei of a sperm cell and an egg cell fuse at fertilisation.

Although eggs and sperm are both produced by meiosis, gamete production varies between the two sexes in two important ways. Firstly, in energetic terms, individual sperm are very much cheaper to produce than individual eggs. Essentially, a sperm cell is little more than a haploid cell nucleus with a tail. The nucleus contains the chromosomes and the tail is used for locomotion, being powered by a few specialised organelles, the mitochondria. By contrast, an egg is huge; containing a haploid nucleus surrounded by a reservoir of cytoplasm in which lie organelles, such as numerous mitochondria and ribosomes, and the nutrients to resource the development of the fertilised egg, or zygote. The result is that in most organisms vastly more sperm are produced than eggs.

Secondly, although both eggs and sperm result from meiosis, while meiosis of a single diploid cell leads to four sperm cells in males, in females only one of the haploid products will become a female gamete. The other three products give rise to the yolk of the egg. Partly as a result of selection acting on which of the four meiotic products in females becomes the female gamete, the rate of genetic mutation in females is lower than that in males.

Mating in Adalia 2-punctata

Mating behaviour has been most closely studied in *Adalia 2-punctata*. I shall begin, therefore, by focusing on the mating behaviour in this species before considering variations in other species.

Adalia 2-punctata are highly promiscuous. They mate every day during the breeding season if the weather is appropriate. Mating generally lasts from 1.5 hours to nine hours, depending partly on temperature. Occasionally, if a pair is still *in copula* when the temperature drops below a threshold level of about 11°C, the pair may remain 'locked' until the temperature rises the following day. For comparison, mating duration for *Coccinella 7-punctata* has been recorded as 54 minutes with a range of 41–62 minutes (Rana and Kakker, 2000).

Courtship

These stark facts do not show the complexity of copulation. Unlike many other insects, *A. 2-punctata* does not appear to show any long-distance attraction between the sexes. One sex is not attracted to the other by means of air-borne pheromones or sound. Rather, males seem to find females more or less by bumping into them, although there is probably some short-range visual attraction. Although sexual selection theory (Darwin, 1871) predicts that males should compete among themselves for access to and possession of females, male *A. 2-punctata* make no attempt to defend a territory and rarely fight one another over a female in the wild. Here, where there is no obvious territory holding, theory predicts that males should compete through their searching behaviour. This will be different from female searching behaviour, which will be largely geared to finding food and oviposition sites. Indeed, significant differences between the sexes have been reported in *A. 2-punctata* adults during the breeding season (Hemptinne et al., 1996). While females respond to high prey densities by adopting area-restricted searching, males do not. The males do display area-restricted searching behaviour, but only in response to female-specific contact pheromones. In addition, Honek (1985) observed males flying and walking more frequently between plants than females. Brakefield (1984a) noted that females changed host plant in response to deteriorating nutritional conditions for oviposition, and males followed after a lag. Finally, Hodek (1973) noted that, in general, antennal length is slightly longer in male than in female ladybirds. These observations suggest that male *A. 2-punctata* compete for access to females in the classical Darwinian sense.

When a male ladybird meets another ladybird, he shows little obvious courtship behaviour. A male ladybird's normal reaction when he bumps into an object that is roughly ladybird sized is to climb on top of it, touching it with his antennae. This behaviour is seen whether the object bumped into is a ladybird or not. For example, males will climb on top of a piece of plasticine of the right size and shape. If it is not a ladybird, the male soon climbs off again. However, if the object is a ladybird, it may be male or female, conspecific or a different species, or dead or alive. If both are males, the encounter is broken off rapidly. This is because female ladybirds produce a cocktail of chemicals from pores on their elytra that elicit a mating response from males. The chemicals in this cocktail probably have different functions, some being female-specific and some species-specific. Male ladybirds do not produce the female-specific chemicals and so a mounting male will be able to determine the sex of the ladybird beneath him.

If the male has mounted a female of another species, the species-specific chemicals will be absent, and again the male will be able to recognise that the ladybird beneath him is not a suitable partner. In captivity, if the mounted ladybird is a female of the correct species, but is dead, a male may continue to attempt copulation for several hours. This suggests that the chemicals secreted from elytral pores endure for some time.

If a male has mounted a living conspecific female, he will then move so that he is orientated appropriately for mating. The usual reaction of the female is to attempt to prevent him from mating.

Female Rejection Behaviour

It may seem peculiar that mature female *A. 2-punctata*, carrying fully developed eggs, should frequently attempt to reject mates; however, that is precisely what they usually do. Female rejection behaviour is of two main types. In type 1, the female pulls her abdomen upwards, retracting it tightly into her elytra. The female's genital opening is thus held out of range of the sipho of a male on top of her, and the male is unable to copulate. Type 2 rejection is more complex, involving a variety of behaviours:

(i) She stretches her hind legs to lift her abdomen, and thus the male, off the substrate, so destabilising him (Fig. 5.1).

(ii) She kicks back at the male's genitalia with her hind legs (surprisingly this behaviour is rarely effective!).

(iii) She runs along, waving her abdomen from side to side, to try to slew the male off to one side (Fig. 5.2), and during such behaviour, the female may stop, and try to get rid of the male by rolling onto her side (Fig. 5.3).

(iv) She climbs to a position where she can drop or jump off the substrate and land on her back with the male beneath her. This is the most effective of the type 2 rejection behaviours, but is only employed by about a quarter of females.

These various behaviours may continue for many minutes, and at any time in the series, the female may stop rejecting and accept the male. Then the male may make stable genital contact and start full copulation. Given that type 1 is more effective at preventing copulation, and is energetically less costly than

Fig. 5.1. Mating behaviour of *Adalia 2-punctata*.

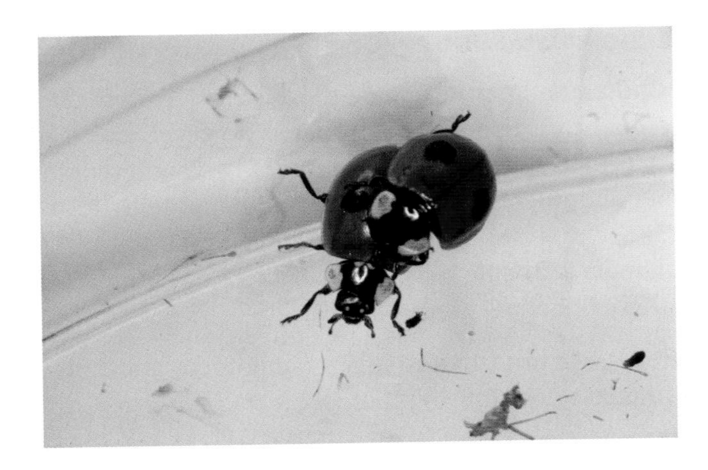

Fig. 5.2. Mating behaviour of *Adalia 2-punctata*.

Fig. 5.3. Mating behaviour of *Adalia 2-punctata*.

type 2, three related questions need addressing. Firstly, do type 1 and type 2 rejection behaviours serve the same purpose? Secondly, in type 2 rejection behaviour, which ladybird, the male or female, ultimately controls whether copulation occurs? Thirdly, what is the function of type 2 rejection behaviour? Haddrill (2001) found that the two basic types of rejection are used by *A. 2-punctata* females in different circumstances. Type 1 rejection is used largely by females that are young, or poorly fed, or do not have mature eggs in their ovaries. Reproductively mature females predominantly exhibit type 2 behaviours. The two types of rejection thus appear to serve different functions.

Majerus (1994a) proposed two hypotheses to explain type 2 rejection behaviours. The first hypothesis proposes that rejection behaviour can be used to prevent unwanted copulation. The suggestion is that at some times mating is disadvantageous to a female, even if she is reproductively mature. This may be when she has very recently mated, or when her spermatheca is full, or if she is hungry. Here, if mating takes place in spite of the female's rejection behaviour, it is a result of the male's behaviour. The alternative hypothesis proposes that type 2 female rejection is a means of testing a male's quality. A male that can overcome type 2 rejection behaviours is a fit male that is likely to carry 'good genes'. It is therefore in the interests of a reproductively mature female to eventually allow such a male to mate with her.

The difference in these two hypotheses highlights the conflicts of interest between males and females over reproduction in most organisms. At the core of this conflict lies the relative size of the gametes produced by males and females and the energetic costs of each. Both males and females will employ reproductive strategies that maximise their own reproductive success. For males, this is usually achieved by mating with as many females as possible. For females, however, the prime motive must be to ensure that sperm carrying good genes fertilise the costly eggs they produce. Thus, while males compete with one another for access to and acceptance by females, females should be choosy.

Haddrill (2001) showed that type 2 rejection behaviours shown by female *A. 2-punctata* function as a means of assessing males, and that females have considerable control over their mating frequency. Moreover, she showed that the degree of rejection shown by females did not change over a series of matings, suggesting that females can maintain this control throughout their reproductive lives. Female ladybirds often consume rejected spermatophores, which can be considered as constituting a nuptial gift. However, it has recently been shown that starved females resisted mating more frequently than less starved females, suggesting that copulation is energetically costly and the nuptial gift does not compensate sufficiently (Perry et al., 2009).

Copulation

Once a female has accepted a male, copulation will occur. In *A. 2-punctata*, copulation consists of from one to three insemination cycles. During each cycle, a sperm package, or spermatophore, is made by the male within the bursa copulatrix of the female and released. This means that males have the ability to multiply ejaculate within a single copulation.

Each insemination cycle consists of two phases. In the first phase, the male engages in a series of three to nine vigorous twisting movements, followed by a pause of approximately 30 seconds and another bout of twisting. This alternation of twisting and rests lasts for about 30–45 minutes following sipho insertion. The male then shifts his position slightly; dropping and appearing to curve the tip of his abdomen further under the female (Fig. 5.4), and engages in an alternation of gentle rocking motions for a few seconds and remaining still for about 30 seconds. This second phase in the insemination cycle also lasts for about 45 minutes. At the end of the second phase, the female ejects the husk of the spermatophore (Fig. 5.5). When ejected, this is gelatinous, but it rapidly hardens in the air. Tests in which males and females were weighed before and after mating show that males produce the materials for the construction of the spermatophore (Ransford, 1997).

Fig. 5.4. Mating behaviour of *Adalia 2-punctata*.

Fig. 5.5. Husk of spermatophore.

If the first mating of a female is interrupted during the twisting phase, she will not lay fertile eggs. However, if interruption occurs once the rocking phase has begun, fertile eggs do result. This suggests that sperm start to be transferred into the spermatophore once it has been formed in the female's bursa copulatrix, on the cusp between the twisting and rocking phases. This indirect estimate of the timing of sperm transfer has been verified by dissection and microscopic examination, using methods developed for *A. 2-punctata* by Ransford (1997). The bursa copulatrix is a muscular organ. During the rocking phase, sperm is transferred from the spermatophore into the spermatheca by muscular contractions of the bursa. First transfer begins soon after the rocking phase is established, and increases rapidly, so that on average over half the sperm are transferred from the bursa to the spermatheca in the first 20 minutes. The precise rate at which this occurs is, however, very variable between pairs. Indeed, estimation of the capacity of the spermatophore is problematic, because sperm continues to be deposited in the spermatophore for some time after contractions shifting sperm into the spermatheca have begun.

At the end of the second phase, some males embark on post-copulatory behaviours. However, other males revert to twisting phase behaviour, without removing their sipho from the female. They then complete a second insemination cycle, similar to the first. A small proportion of males even undertake a third cycle.

A male transfers up to 13 000 (average 10 000) sperm per insemination cycle. Not all of these are transferred to the female's spermatheca, some being left as a residue in the spermatophore when the female ejects it. The number of sperm left in the husk of a first insemination cycle spermatophore can be as many as 2000. This number may be higher for second or third insemination cycle spermatophores because the spermatheca of a female *A. 2-punctata* can hold no more than 18 000 (average 15 000) sperm. This means that although one insemination cycle will not be sufficient to completely fill the spermatheca, two cycles will fill it to overflowing, and with three cycles there will be a considerable excess of sperm transferred over that which the female can store.

The female that ejects the spermatophore husk frequently eats it. However, occasionally the male eats it, if he has completed copulating. Even if this spermatophore is entirely empty of sperm, the husk itself is nutritious.

Post-copulatory Behaviour

Once a male has completed one, two or three insemination cycles, he disengages his sipho from the female, but usually remains upon the female. Then he will typically rotate on the female, sometimes changing direction. During these rotations, males trail their parameres along, and sometimes curve them under the edge of the female's elytra. It has been suggested that males are smearing the female with an anti-aphrodisiac chemical to reduce the likelihood of her mating again in the near future, and so increase their paternity (Hemptinne et al., 1996).

The Mating Behaviour of Adalia 2-punctata *in the Wild*

Some female insects that have been shown to be promiscuous in space-limited captivity arenas are thought to mate with just a single male in the wild (Ridley, 1988). Consequently, it is essential to check that mating behaviour observed in the laboratory is consistent with behaviour exhibited in the wild. Observations of mating behaviour of *A. 2-punctata* in the field have shown that all the elements of copulation observed in captivity occur in the wild (Brakefield, 1984a; Majerus, 1994a; Ransford, 1997; Haddrill, 2001).

Southwood (1977) showed that it is possible to estimate the level of promiscuity by studying the proportion of the population mating at any one time. For example, Brakefield reported figures of between 5.5% and 28.3% of Dutch *A. 2-punctata* populations mating concurrently. Figures recorded for populations in Cambridge in different years were: 3.2%–46.5% (1983) (Majerus, unpubl. data), 0%–27% (1994) (Corry, 1995) and 12.8%–30.57% (1998) (Haddrill, 2001). Using such measures taken throughout the day, estimates of the mean mating time, mean longevity, changes in the population sex ratio throughout the season and appropriate temperature data, a rough approximation of the number of times an average female *A. 2-punctata* may mate in the wild gives a figure of 27.7 times.

Courtship and Copulation in Other Ladybirds

In most species of coccinellids studied, there appears to be little or no long distance attraction through pheromones. However, contact pheromones are important in both species and sex recognition of potential partners. In some species, pheromones are given off by females even before they eclose. Thus, in *Illeis galbula*, males that encounter a female pupa will remain by it, guarding it until the female emerges, when the male will copulate with her (Richards, 1980a).

Romani et al. (2004) have described the morphology of what may be a sex pheromone gland in some *Chilocorus* and *Scymnus* species. The glands are situated on the eversible ovipositor segments in females. Unfortunately, as yet no determination has been made as to the nature of the pheromones produced.

The details of the mating behaviours of a small number of other coccinellids have been reported. In *Harmonia axyridis*, Obata (1988) reported the occurrence of a so-called 'latent period' at the start of copulation, which lasts for about 40 minutes. This is followed by a transition period of bouts of thrusting twists, interspersed with bouts of body shaking, and then the regular repeated bouts of violent body shaking by the male as described by Obata and Johki (1991). A latent period has also been reported in other species of *Harmonia*, *Henosepilachna 28-maculata* and in some *Chilocorus* species (Fisher, 1959; Katakura, 1985). Body shaking, whether more like the twisting or the rocking shown by *A. 2-punctata*, has been reported in many other species of coccinellid, including *C. 7-punctata*, *Coccinella magnifica*, *Coccinella miranda* (Fig. 5.6), *Hippodamia convergens*, *Hippodamia 13-punctata*, *Cycloneda polita*, *Micraspis frenata*, *Halyzia 16-guttata*, *Oenopia conglobata*, *Harmonia 4-punctata*, *Anatis ocellata*, *Myzia oblongoguttata* and *Semiadalia 11-notata* (Obata, 1988; Majerus, pers. obs.). This body shaking is probably a feature of the majority of coccinellids, but is absent from *Cheilomenes 6-maculatus* and *Propylea japonica*. The vigour of shaking varies between species.

Mating duration also varies. As may be expected of a cold-blooded organism, temperature has a strong influence on copulation times; the warmer it is, the shorter the copulation. In *A. 2-punctata*, variation in the number of spermatophores passed also has a strong effect on the length of copulation. Thus, at 21°C, the average length of single insemination cycle matings was 107 minutes; with two cycles it was 223 minutes, and with three cycles, 301 minutes. Multiple ejaculations within a single copulation may also account for the variability in copulation times of *H. 28-maculata*, which range from about one to three hours. Katakura (1985) found that more than a single sperm mass

Fig. 5.6. Mating behaviour of *Coccinella miranda*.

was found in the reproductive tract of some females after mating, concluding that 'ejaculation may occur more than once'.

Within the Coccinellidae, sperm transfer may occur with or without the aid of a spermatophore. Spermatophores have been observed from *Chilocorus* spp. (Fisher, 1959), *C. 7-punctata* (Davey, 1959), *H. axyridis* (Obata and Hidaka, 1987), *Cryptolaemus montrouzieri* (Kaufmann, 1996), *A. bipunctata* (Majerus 1994a), *Adalia 10-punctata, H. 4-punctata, A. ocellata* and *Chilocorus renipustulatus* (Majerus, pers. obs.). In *A. 2-punctata* and *A. 10-punctata*, females, or more rarely males, consume the husk of the spermatophore when the female ejects it. In *H. axyridis*, the spermatophore is also ejected and consumed, although here, because ejection occurs about half an hour after copulation, consumption is invariably by the female, the male having moved on. The same is true of *A. ocellata*, but here spermatophore ejection is delayed even longer than observed for *H. axyridis*. In *C. 7-punctata*, although the spermatophore is ejected by the female between one and six hours after copulation has ceased, it is not usually eaten by the female (Obata and Johki, 1991). In *P. japonica*, *Illeis koebelei* and *H. 16-guttata*, females eject a gelatinous mass after copulation and eat it, yet no true spermatophore is formed (Obata and Johki, 1991; Majerus, 1994a). This material may be analogous to the gelatinous material, secreted from the male's accessory gland, that is used to form a spermatophore in the bursa copulatrix in *H. 28-maculata*. However, in this species, the gelatinous mass is not ejected by the female, but appears to be absorbed by the female internally and may thus be considered a nuptial gift. The spermatophore of *C. montrouzieri* is much reduced in size compared to others in the Coccinellidae and acts merely as a conduit for sperm during its passage from the sipho lodged in the female's bursa into the spermatheca. In some ladybirds, such as *C. 6-maculatus* and *Epilachna admirabilis*, no spermatophore is made, sperm being transferred directly to the female within a few minutes of sipho insertion (Obata and Johki, 1991; Katakura et al., 1994).

The need or lack of need to manufacture a spermatophore also appears to influence copulation duration. Thus, for example, in *C. 6-maculatus*, which does not have a spermatophore, virgin females became fertile following very short copulations of just a few minutes duration (Obata and Johki, 1991). This contrasts with the situation in *A. 2-punctata, H. axyridis, H. 28-maculata* and *Chilocorus* spp. in which sperm transfer only begins 30 or more minutes after genital contact.

Finally, there is variation in the post-copulatory behaviour of males, both within and among species. In *A. 2-punctata*, the rotation of males on females after withdrawing their sipho lasts on average about three minutes. However, some males habitually perform this type of behaviour for longer or shorter periods than others. Indeed, almost a third of males observed in controlled tests with this species dismounted immediately after withdrawal. Conversely, one male rotated for over 10 minutes on each of five females on consecutive days. Similar behaviour was observed in *H. 4-punctata*, *M. frenata* and *C. polita* (Majerus, pers. obs.). In *H. axyridis* and *A. ocellata*, this behaviour has also been observed, but it lasts for considerably longer (mean for *H. axyridis* = 10 minutes; mean for *A. ocellata* = 38 minutes). On the other hand, most male *O. conglobata* dismount from females immediately after withdrawal, and even in the few that did remain on the female, they did so for only a short period (maximum = 3 minutes).

Even in the relatively few species of ladybird that have been closely scrutinised, it is apparent that there is considerable diversity in mechanisms of sperm transfer. Chapman (1969) argued that effecting sperm transfer via a spermatophore is the ancestral strategy in insects, and that direct sperm transfer is derived. If this is the case, *C. 6-maculatus* and *H. 28-maculata* may be considered advanced members of their respective tribes. Certainly, study of the insemination mechanisms of more species of coccinellids would be worthwhile and may have some phylogenetic use. It would also be interesting to attempt to assess the breadth of sperm transfer diversity. Two species that deserve study are *Aiolocaria mirabilis*, in which males may remain mounted upon females for several days (Iwata, 1932), and *Stethorus pusillus*, in which females do not have a spermatheca (Putman, 1955; Moter, 1959).

Several interesting evolutionary questions arise from the pattern of copulation seen in *A. 2-punctata*. Firstly, the high level of promiscuity in female *A. 2-punctata* needs explanation. Secondly, the same is true of the multiple insemination cycles exhibited by males. Thirdly, the extreme promiscuity of *A. 2-punctata* has led to exploitation of this promiscuity by sexually transmitted disease organisms. Finally, other disease organisms that differentially affect the sexes of this and other ladybirds exert significant evolutionary pressures on the mating behaviours as a result of biases in population sex ratios. Here I will discuss the first two of these subjects, leaving a consideration of sexually transmitted diseases of coccinellids and sex ratio distorters to Chapter 7.

Female Promiscuity

Natural selection should favour individuals with a level of promiscuity that maximises their reproductive output, in terms of both number and fitness of progeny. This optimal promiscuity level will depend upon the costs and benefits of mating, which will differ between the sexes. The costs of mating for a female *A. 2-punctata* may be high. Mating lasts for several hours, during which time the female has to carry the male. For this time she cannot oviposit, and, although she may feed, her rate and efficiency of aphid capture is reduced. In addition, increased promiscuity raises the chances of contraction of diseases transmitted during sex. The benefits of re-mating are that carrying more sperm reduces the likelihood of eggs being unfertilised, and there may be fitness benefits to mating with more than one male.

A female *A. 2-punctata* may mate over 25 times in the wild. Theory predicts that as males produce many energetically cheap sperm, in the absence of paternal care of young, they will gain greatest reproductive success by mating with as many females as possible. However, as females produce relatively few energy-rich eggs, they should be protective of them and only mate with the fittest males. Furthermore, as there are costs to mating, females should mate only frequently enough to ensure all eggs are fertilised. Yet, female *A. 2-punctata* mate approximately ten times more often than necessary (Majerus, 1994b).

Five hypotheses have been proposed to explain the promiscuity of female *A. 2-punctata*. In the first, a female gains a direct fitness benefit from each mate in the form of a nuptial gift. This theory may be dismissed because Ransford (1997) has shown that the nutritive value of the spermatophore of *A. 2-punctata* is very small.

A second hypothesis argues that multiple mating is a consequence of females encountering a sequence of males, and mating with any male that is of higher genetic quality than any preceding male (Petrie and Kempenaers, 1998). This 'trade-up' strategy seems unreasonable to account for the extreme level of promiscuity in *A. 2-punctata* and the regularity of copulation.

A third hypothesis that does not depend on a female's ability to gauge a male's quality is known as the 'bet-hedging hypothesis' (Yasui, 1998). In the case of *A. 2-punctata*, it is pertinent to remember that both larvae and adults of this species feed on a variety of aphid species, and that the population densities of aphids are highly variable and unpredictable. Thus, the progeny of females one year, which may have developed on nettle aphids on *Urtica dioica*, may reproduce on lime trees (*Tilia* spp.), or legumes or grain crops or roses, each infested with their own species of aphids, the following year, depending on which plants are host to high aphid populations. In this situation, the best long-term strategy may be to produce as genetically diverse progeny as possible, so that at least some progeny have high fitness in whatever conditions they face. A female may increase the genetic variation of her progeny by mating with many males, thus hedging her evolutionary bets.

A fourth hypothesis is that females mate with many different males to promote sperm competition. Females are capable of storing sperm from several males in their spermathecae. Thus, if sperm from several males are released for fertilisation simultaneously, the fittest sperm should win the race to fertilise the majority of eggs. One problem with this hypothesis is that it is open to cheating, males only needing to produce fit, fast-swimming sperm, which may not be a true indication of their total fitness. However, there is evidence to suggest that males with fit sperm also have high overall viability (Yasui, 1997). Furthermore, patterns of paternity from multiply mated female *A. 2-punctata* suggest that post-copulatory mechanisms, such as sperm competition or cryptic female choice, may be important in this species (de Jong et al., 1993; Ransford, 1997; Haddrill, 2001).

The final hypothesis stems from the growing body of evidence that the nuclear genome of a particular sperm may vary in its compatibility with the nuclear and cytoplasmic genomes of different female eggs. It has thus been suggested the female promiscuity may help minimise genetic incompatibility (Jenions, 1997). Variation in genetic compatibility between sperm from different males and females may be the product of interaction among many aspects of the heritable systems of sexually reproducing organisms. In ladybirds, factors that may contribute to incompatibility include the existence of co-adapted gene complexes, super-genes, epistatic genes, inbreeding depression, the presence of transposable elements (Haddrill, 2001), the presence of heritable intracellular symbionts and nuclear suppressers of these. Importantly, here the fitness of a zygote depends on the genetic compatibility of its mother and father (Zeh and Zeh, 1996). Thus, while in the other hypotheses, a high-quality ladybird will confer this high quality on his or her offspring regardless of their other parent, in the genetic incompatibility hypothesis, a high-quality ladybird will only pass this quality to its offspring if it mates with a compatible partner (Tregenza and Wedell, 1998).

Studies of the reproductive output of females that were mated once to a single male, or repeatedly to the same male (repeated mating) or to a succession of different males (multiple mating), and analysis of the paternity of offspring produced led Haddrill (2001) to conclude that in *A. 2-punctata*, the greatest gain from multiple mating results from a reduction in genetic incompatibility. She showed that multiply mated females had increased fertility, offspring viability and reproductive output compared to both singly mated females and those mated several times to the same male. Similar results have subsequently been obtained with *Propylea dissecta* (Omkar and Mishra, 2005). Finally, using micro-satellite DNA fingerprinting, Haddrill (2001) showed that wild clutches of *A. 2-punctata* had high levels of multiple paternity, with up to six males contributing to some egg clutches. Wild females stored sperm from several males at once and sperm mixing was extensive, with little stratification of sperm in the spermatheca. Furthermore, behavioural and molecular genetic studies have demonstrated that the number of mates was similar to the number of fathers (Haddrill et al., 2008), suggesting that females do not influence the paternity of their eggs after mating. However, there was a positive relationship between mating duration and paternity, which is likely to be a consequence of the high number of sperm transferred within multiple spermatophores (Haddrill et al., 2008).

Haddrill (2001) made a key finding in showing that sperm precedence values were not consistent for individual males when mated with different females. Rather, the sperm precedence values depended on both the male and female in a pair. The fact that different males sired different proportions of offspring with different females is consistent with the expectation that compatibility will vary between pairs (Zeh and Zeh, 1996). Owing to differences in the compatibilities of male and female genomes, sperm precedence values from a particular male should vary between females, as Haddrill found. This will be especially so if post-copulatory mechanisms can act to favour compatible males, or even compatible sperm, in the fertilisation game (Zeh and Zeh, 1997). This may be either through interactions at the gamete level to reduce the likelihood of eggs being fertilised by sperm of low compatibility, or through various female-controlled mechanisms to retain, transfer or use compatible sperm (Eberhard, 1996).

The genome of *A. 2-punctata* includes a variety of features that may lead to incompatibility. While developing molecular genetic marker systems in this ladybird, Haddrill et al. (2002) found that its genome is remarkably variable. She argues that this might be a response to incompatibility resulting from homozygosity. When fertilisation involves two gametes that are genotypically similar, the result may be a zygote of low fitness, as is known to occur in the form of extreme inbreeding depression in *A. 2-punctata* (Lusis, 1947a; Majerus, 1994a). The existence of many multiple-allelic genetic loci will reduce the level of homozygosity, thereby reducing incompatibility. Interestingly, this means that a common genotype may be selected against simply because it is more likely to cause incompatibility through homozygosity than a rare genotype would be. Thus, what may be termed homozygosity incompatibility will lead to negative frequency dependent selection, which may help explain the maintenance of colour pattern polymorphisms in *A. 2-punctata* and many other coccinellids (see Chapter 9).

Multiple Ejaculation within a Single Copulation: A Waste of Sperm?

Male reproductive behaviour also has a feature that is difficult to explain. This is the finding that some males manufacture three spermatophores during a single copulation, and supply each of the three with sperm. On the surface, such behaviour appears maladaptive. The spermatheca of *A. 2-punctata* has an average capacity of around 15 000 sperm (Ransford, 1997; Arnaud et al., 2003).

A male spermatophore contains on average around 10 000 sperm. Two cycles of insemination are therefore sufficient to more than fill the spermatheca of an average female. A male that passes three spermatophores thus appears to be 'wasting' considerable amounts of sperm. The question is why do these males remain *in copula* after two insemination cycles rather than dismounting and moving on to seek another female. One possibility is that this is a crude form of sperm competition, males passing excessive quantities of sperm to dilute or flush out sperm from previous males.

Various models of sperm competition have been proposed. These may be split into lottery models and displacement models. In lottery models, sperm are used for fertilisation either in direct proportion to their abundance in the female (ideal lotteries), or in proportions biased by the order of transfer (biased lotteries) (Parker et al., 1990). In displacement models, sperm displace one another either during sperm transfer (instantaneous mixing during displacement: IMDD), or after sperm transfer (no mixing until after displacement is complete: NMUSDC). The second of these mechanisms is considered both more advanced and more efficient than the first, because the sperm of the last male to mate only displaces sperm from previous matings. In the IMDD model, each sperm from the last male to mate may displace either a sperm from a previous mating, or one of its own sperm.

Ransford (1997) analysed paternity of egg clutches laid by females that were collected as non-virgins from the wild, and were subsequently mated in captivity with males homozygous for one or other of two rare allozyme variants. He showed that the IMDD model operates in *A. 2-punctata*. Ueno (1994), working on *H. axyridis*, also concluded that a displacement mechanism of sperm competition was in operation, but was unable to distinguish between the IMDD and NMUSDC models.

It appears therefore that multiple inseminations within a single copulation in *A. 2-punctata* provide males with fertilisation benefits. These males do 'waste' some of their sperm because the female's spermatheca cannot contain all the sperm from three spermatophores. However, because of the way new sperm and sperm already in the spermatheca are mixed, the sperm transferred to a female in second and third spermatophores continue to increase the donor's level of paternity in subsequent eggs from that female.

One question remains: why do some males pass one spermatophore, others two and still others three? The answer to this question might depend on a number of factors, including the male's own nutritional condition, the male's genotype with respect to genes involved in male reproductive physiology, the male's recent mating history, assessment of female condition in respect of both egg load and sperm load, population density and population sex ratio. The only factors shown to have an influence on the number of sperm transferred in *A. 2-punctata* are the male's recent mating history, the male's genotype and the population sex ratio. The population sex ratio in *A. 2-punctata* is influenced by the age of the population, the number of generations represented in the population by reproductively mature individuals and the prevalence in the population of heritable intracellular sex ratio-distorting symbionts. The effect of these factors will be discussed in Chapter 9.

Sexual Selection: The Case of the Choosy Female

Studies on ladybirds have made significant contributions to several branches of biology. Releases of the Australian ladybird *Rodolia cardinalis* into citrus groves in California laid the foundations of biological pest control. Work on male-killing bacteria in ladybirds is having considerable bearing on appraisals of the conflicting interests of genetic elements in the nucleus and the cytoplasm. Furthermore, work on melanic polymorphism in *A. 2-punctata* during the 1980s played a significant part in verifying, for the first time, a Darwinian theory initially proposed over 120 years earlier in *The Origin of Species* (Darwin, 1859).

Darwin based his theory of evolution on the existence of variations between individuals within a species and 'the struggle for existence', such that the fittest survive, while the less fit perish. This he called 'natural selection'. However, he realised that his theory of evolution would be criticised if there were characteristics of animals, or plants, which could not be explained by his theory. Particularly problematic were the elaborate ornaments of male birds such as peacocks, pheasants and birds of paradise, the antlers of male deer, and the great size of bull seals, sea lions and walruses, compared to their females. If survival were the only factor involved, there appeared to be no good reason for these features, for females of these species survived perfectly well without colourful plumes, antlers or great size.

Darwin's solution to this conundrum was sexual selection, which he defined with precision: 'this [sexual selection] depends not on a struggle for existence, but on a struggle between males for possession of the females; the result is not death to the unsuccessful competitor, but few or no offspring'. Darwin saw that two types of sexual selection would exist. Males would compete with each other for females, and females would choose between males.

Since the publication of *The Origin of Species*, many examples of natural selection have been observed, and understanding of the laws of inheritance has given natural selection theory a sound genetic basis. The theory of sexual selection through male competition (Fig. 5.7), Darwin's 'Law of Battle', has also been readily accepted. However, sexual selection through female choice has been much more controversial. Initial opposition to the notion of female

Fig. 5.7. Two *Exochomus 4-pustulatus* males competing to mate with a single female.

choice undoubtedly owed something to chauvinistic attitudes among the male-dominated Victorian scientific fraternity. More objectively, critics of female choice theory argued that as Darwin sought to use this theory to explain elaborate secondary sexual characteristics, such as peacock tails, he was saying that females must have a highly developed aesthetic sense. Females chose the most beautiful males. The theory was also criticised because Darwin did not put forward any explanation of how female mating preferences could evolve. He took their existence as a premise.

The first mechanism for the evolution of female choice was put forward by Fisher (1930). He suggested that a new, genetically controlled, male character, which was spreading through a population because it conferred on its bearer some slight natural selective advantage, might become the object of a genetically controlled female mating preference. Females that exercise this preference produce better offspring, because they have paired with fitter males. Those males that possess the character picked out by choosy females now gain a reproductive advantage, because they are more likely to mate than males lacking this trait. Given that the preference is genetic, it is selected, along with the advantageous, and preferred, male character. The female preference and preferred character each increases the advantage accruing to the other, so that they advance together, in what Fisher called the 'runaway process of sexual selection'. This process is ultimately halted when the preferred male character becomes fixed in the population, or becomes so exaggerated that it not only loses its original natural selective advantage, but actually becomes disadvantageous to its bearer.

Other theories for sexual selection have subsequently been proposed, including Zahavi's (1975) handicap principle, Hamilton and Zuk's (1982) condition dependent handicaps and the sensory exploitation hypothesis (Majerus et al., 1996). The handicap principle proposes that females should choose mates on the basis of traits that are costly to produce. Only males with good genes would be able to develop these expensive traits or 'handicaps'. Such traits would thus be incorruptible indicators of high genetic male quality (Zahavi, 1975). This hypothesis is problematic because of the lack of a genetic association between the handicap and good genes.

However, Hamilton and Zuk (1982) extended this idea by proposing that females may select males on the basis of traits the development of which depends on the condition of the male. Here, for example, a condition dependent trait may only be fully developed by males that have genes that confer resistance to parasites. These males, by dint of the development of the trait, gain more matings and pass on not only genes for the trait, but also those for parasite resistance.

The sensory exploitation hypothesis proposes that female mating preferences may have arisen in a context other than mate choice. Organisms have sensory biases resulting purely from the nature of the sensory systems that they have to receive information and process it. Female choice may thus arise not because the partner chosen is fitter than the average, but merely as a by-product of biases in the sensory biology of the organism in some other ecological or behavioural context. Hence, selection may act on males to exploit the pre-existing sensory 'tastes' of females (Majerus et al., 1996).

Sexual Selection in Adalia 2-punctata

Adalia 2-punctata has a variety of characteristics that make it suitable for investigations of female choice. It exhibits easily scored, natural, genetically controlled, colour pattern polymorphism. The frequencies of the various colour forms vary considerably from one population to another. The literature contains some reports that mating between different forms of *A. 2-punctata* is not random (Lusis, 1961; O'Donald and Muggleton, 1979). The sexes can be distinguished fairly easily. Females are highly promiscuous so that many mating decisions can be recorded for individual ladybirds, and pairs remain *in copula* for a long time and are easy to find in the field (Fig. 1.4).

During a nine-year project, a number of aims relating to sexual selection in *A. 2-punctata* were addressed:

(i) To test whether mating was non-random in natural populations.

(ii) To compare results, qualitatively and quantitatively, under laboratory and field conditions.

(iii) If mating was found to be non-random, to determine whether this was due to male competition or female choice.

(iv) If female choice was involved, to show whether there was a genetic basis to female mating preference.

(v) To analyse the genetic system determining female choice.

In 1981, a large population of *A. 2-punctata* in the north-west of England was surveyed. The data showed that while predominantly black forms (melanics) comprised only 34% of non-mating males, they made up 49% of mating males. However, among females, melanics comprised 34% of the non-mating and 36% of the mating population (Majerus et al., 1982a, b). This excess of melanic males among mating pairs could have been due to male competition, perhaps because black males may have warmed up more quickly in the morning, and hence mated more readily than red males, or it may reflect female choice. The excess of melanic males in mating pairs, compared to expectation based on random mating, was equivalent to 20% of females choosing to mate with black males, assuming, of course, that female choice was responsible for the excess.

Using the most common red form, f. *bipunctata*, and the most common black form, f. *quadrimaculata*, similar results were obtained in the laboratory using two experimental regimens, one with replacement, the other without (Majerus et al., 1982a). Under both regimens, the representation of melanic males among mating pairs was higher than expected to a level consistent to about 23% of females mating preferentially with melanic males. Therefore, mating in the wild is not random, and qualitatively and quantitatively similar results are obtained in the laboratory.

The question of whether the non-random mating had a genetic basis was addressed by use of artificial selection of the sort animal and crop breeders use to improve milk yield, corn ear yield and so on. Two populations, each of 100 male and 100 female ladybirds in a ratio of 7:3 *bipunctata*:*quadrimaculata*, were set up in cages. From one of the cages, all mating pairs seen were removed, but only females that mated with *quadrimaculata* males were put into a new cage to lay eggs for the next generation of this 'selected' line. This procedure was repeated for three generations, the 7:3 ratio of the forms being restored each generation. The second 'control' population was treated in the same way, except that all mating females, regardless of the forms of their mates, were put into a new cage to provide offspring for the next generation. In the selected line, the proportion of females that mated with melanic males increased rapidly, from less than 25% to over 50% by the fourth generation. However, there was no significant change in the control line. The excess of melanics among mating males therefore has a considerable genetic basis, irrespective of whether the non-random mating results from male competition, female choice or a combination of the two.

Three further mating tests were used to differentiate between the mechanisms of male competition and female choice. In one, males from the progeny of the fourth generation of the selected line, and females from the original unselected stock, were allowed to mate. In the second cage, female progeny from the fourth generation of the selected line were allowed to mate with males from the unselected stock. The third cage contained both males and females from the unselected stock, as a control, to check that the experimental conditions still produced excess melanic males in mating pairs at the levels seen previously with this stock. In each experiment, a 7:3 ratio of *bipunctata*: *quadrimaculata* was used. The results showed that when selected females were mated with males from the original stock, the level of preference is characteristic of the selected line, that is, around 50%. In the second cage, when selected males were used with original stock females, the level of preference was characteristic of the original stock, as it was in the third, control experiment (Majerus et al., 1982a, b). This shows that females, not males, set the level of excess melanic males among mating pairs. This indicates a case of females having a genetically controlled preference to mate with males of a certain genotype.

Further experimental analysis showed that the mating preference is controlled by the dominant allele of a single gene. Females that carry this allele express the preference; those homozygous for the alternative allele mate randomly (O'Donald and Majerus, 1985; Majerus et al., 1986). Other experiments also showed that the preference was for males of specific genotypes, not simply for predominantly black males (O'Donald and Majerus, 1992).

The female preference for males with certain melanic genotypes is a genetic polymorphism, some females having the preference, others lacking it. This polymorphism is variable both in time and space. For example, the frequency of the preference allele in the original population studied was approximately 0.13, whereas the frequency in an East Anglian population was zero (O'Donald et al., 1984; O'Donald and Majerus, 1984, 1985). Moreover, the frequency of the mating preference allele at the original site declined rapidly in the mid-1980s, this decline being correlated to a sharp decline in the frequency of melanic forms in the population (Kearns et al., 1990).

Mating Preferences in Other Ladybirds

Mating preferences are not confined to *A. 2-punctata*, having also been reported from *H. axyridis* (Osawa and Nishida, 1992) and *P. dissecta* (Omkar and Mishra, 2005). For example, in field and captive studies, both female and male *H. axyridis* have mating preferences, but to different degrees (Osawa and Nishida, 1992). In the spring, females preferred to mate with non-melanic males, while in the summer, their preference was for melanic forms. Using manipulation experiments, the patterns of mating were shown to be largely the result of female choice, but that in the summer generation there was also some degree of male choice. The experiments showed that the females use the colour patterns of males in expressing their mate preference, although chemical and physiological traits were also implicated. The use of a suitor's colour pattern in making mating decisions in *H. axyridis* is in contrast to the position in *A. 2-punctata*, where females carrying the preference allele express their preference for melanic males even in complete darkness (O'Donald and Majerus, 1992).

Implications of Female Choice in Ladybirds

The demonstration of genetically controlled female mating preferences in *A. 2-punctata* has a number of implications. Firstly, it verifies Darwin's theory of sexual selection by female choice. Secondly, it has a bearing on the evolution and maintenance of colour pattern polymorphisms in this species and possibly others. Thirdly, it raises questions as to how widespread female choice may be.

How Widespread is Female Choice?

Darwin (1859) raised the theory of sexual selection by female choice to explain secondary sexual characteristics in sexually dimorphic species. Indeed, in the most quoted examples of the evolution of secondary sexual characteristics by female choice, such as the breeding plumage of peacocks, pheasants, birds of paradise and ruffs, the males alone have and display the highly ornate and elaborate traits; their females have rather dowdy, presumably cryptic, plumage. In some of these examples, such as peacocks, males are monomorphic for their breeding colours. In others, such as the ruff, polymorphism exists in the males. However, females are monomorphic in all these cases. *Adalia 2-punctata* and *H. axyridis* are therefore exceptions, for neither shows obvious sexual dimorphism, and both females as well as males are polymorphic for preferred and unpreferred forms. This led to the suggestion that sexual selection by female choice is far more widespread than had previously been supposed (Majerus, 1986). It may explain not only the evolution of elaborate display characters in males, but also the evolutionary stability of simple genetic polymorphisms, which are not limited to one sex. Female choice has been implicated in the maintenance of non-sex-limited polymorphisms in the scarlet tiger moth, *Callimorpha dominula*, the arctic skua, *Stercorarius parasiticus*, the snow goose *Anser coerulescens* and the four-spotted milkweed beetle *Tetraopes tetrophthalmus*. A genetic basis for female choice has not been demonstrated in any of these. However, the simple genetic control of mate preference in *A. 2-punctata* enables us to envisage the appearance, and rapid evolution, of a negative frequency dependent mechanism that may maintain such polymorphisms.

Alternative Behavioural Strategies

Further, that the choice a female ladybird may make when courted by a male – to accept or reject him – is controlled by a single dominant gene, had importance in the context of the sociobiological interpretation of animal behaviour. Much animal behaviour is based on simple choices between two alternatives; for example, whether to be a 'hawk' or 'dove' in a conflict situation, whether to eat or reject a potential food item, or whether to accept or reject a mate. Many theoretical sociobiological models of these sorts of alternative behavioural strategies have assumed that the alternative behaviours represent simple genetic polymorphisms. These simple genetic models of complex behaviours have been challenged as being overly simplistic. However, in the first case where the genetic basis of a complex alternative behavioural strategy was analysed, the female choice behaviour in *A. 2-punctata*, the behaviour was controlled by a very simple genetic system.

Mating Preferences and Speciation

Finally, the work on mating preferences in *A. 2-punctata* may have profound implications for the process of speciation. Mayr (1963) defined sexually reproducing species as: 'groups of actually or potentially interbreeding natural populations, which are reproductively isolated from other groups'. This can be stated differently: individuals of a species share a common gene pool, which is not shared by individuals of other species. The reproductive isolation between species means that each species' gene pool will evolve independently. The biological properties of organisms that prevent interbreeding are called reproductive isolation mechanisms (RIMs). Broadly speaking, RIMs may be split into those that act before the formation of the zygote (pre-zygotic), and those that act after zygote formation (post-zygotic). There are a variety of both groups of RIMs (Dobzhansky, 1951), but all forestall gene exchange between populations.

Against this background, the speciation process may be considered. Envisage what is thought to be one of the most common types of speciation, the division of one species into two, following a period when genetic exchange between two populations of the original species has been prevented by a geographic barrier. These isolated populations will diverge genetically for a number of reasons. New beneficial mutations and genetic combinations may arise in one population, but not the other. Selectively neutral genes, which increase or decrease in frequency by random genetic drift, may be lost from one population, while becoming fixed in the other. Natural selection will act on the populations differently, causing each to become adapted to its own local conditions. This is the first stage of speciation. In theory, the two populations will diverge genetically, given a long enough period of geographic isolation, even in the absence of natural selection. If the two populations come into contact after the genetic constitutions of the two populations have diverged significantly, then hybrid zygotes are likely to be produced. These may show some degree of post-zygotic reproductive isolation, in the form of hybrid inviability or hybrid sterility, if the differences in the genetic constitutions of individuals from the two populations cause problems of chromosome pairing, replication and division during meiosis.

Once some degree of post-zygotic reproductive isolation exists, the second stage of speciation will inevitably involve the promotion of pre-zygotic RIMs by natural selection. This is because any mechanism that avoids the wastage of gametes and resources that occurs if unfit hybrid zygotes are formed will be advantageous.

The discoveries of female mating preferences in *A. 2-punctata* are relevant to this second stage. As female mating preferences maintain traits that are not sex limited, mating preferences need not be confined to females discriminating between males. Darwin used female choice to explain sexual dimorphism, leaving plumage in non-sexually dimorphic birds to be explained by natural selection. The evolution of species-specific, sex-limited traits, such as bird songs, mate-attracting pheromones in many insects, and many types of courtship behaviour, may also be the outcome of selection favouring the evolution of pre-zygotic RIMs. Furthermore, the assumption that such evolution is limited to sex-limited traits may be unjustified. The elaborate plumage and courtship displays in birds in which the sexes differ little, such as great crested grebes and crown cranes, may be the result of the evolution of genetically controlled mating preferences that are expressed in both sexes (Majerus 1986).

Consider a theoretical case in which two populations have diverged so far that there is some degree of post-zygotic reproductive isolation, and assume that there is now some contact between the two populations. Then allow a mutation to arise in one of the populations, causing individuals to recognise some trait exhibited by members of their own, but not the other population, and subsequently mate with individuals bearing this trait. This mutant gene will spread, because it will rarely, if ever, be present in offspring of pairings between the populations, because these have reduced fertility or viability. Furthermore, the advantage to a gene promoting within-population, rather than between-population, pairing would be maintained until the mating preference gene had spread through the whole population, for the disadvantage of pairing with a mate from the other population would always be present. In this way, the evolution of inherent mating preferences would lead to species recognition systems.

Hybridisation in Ladybirds

Mayr's (1963) definition of the biological species implies that matings occur within a species, leading to viable and fertile offspring. The exceptions to the rule – hybrid matings between members of different species – have long fascinated scientists and others.

Hybrid matings among different species of ladybird are rare in the wild. Majerus (1997a) reported only 18 hybrid matings observed in the field. Most of the hybrid matings were between congeneric species (Fig. 5.8), although three were between species from different genera. Sixteen of the 18 pairs were retained and 13 of the females produced eggs. Of these, seven produced progeny that could not be distinguished from their mothers' species. In these cases, it is assumed that the female had mated previously with a conspecific male, that was the father of the progeny. The eggs from a further three hybridisations were all infertile. In two other cases, eggs showed signs of embryonic development but none hatched. Finally, one pairing, between a male *A. 10-punctata* and a female *A. 2-punctata*, produced 196 eggs, of which

Fig. 5.8. Hybrid mating between *Coccinella 5-punctata* (female) and *Coccinella 11-punctata* (male).

five hatched. All of these were reared to adulthood. Examination of larval and adult traits of these showed them to be true hybrids (Fig. 5.9). All were found to be sterile, the two males having underdeveloped testes, and the females having a unique infundibulum, similar to that described previously for laboratory produced hybrids between these species (Ireland et al., 1986).

In captivity, it is fairly easy to induce interspecific hybrid pairings in ladybirds. For example, there are many reports that if virgin *A. 10-punctata* or *A. 2-punctata* females are isolated from opposite sex conspecific individuals for a period of two or more weeks and are kept well fed, they will readily mate with males of the other species (Fig. 5.10) (Lusis, unpubl. Data; Ireland et al., 1986; O'Donald and Majerus, 1992). Using this technique, Majerus (1997b)

Fig. 5.9. Hybrid *Adalia 2-punctata* x *Adalia 10-punctata*.

Fig. 5.10. *Adalia 2-punctata* male mating with *Adalia 10-punctata* female.

attempted 53 interspecific hybridisations involving 16 species combinations (Fig. 5.11). In all combinations, except one, copulations were observed.

Thus, the rejection behaviour that female ladybirds of most species show towards non-conspecific males can be broken down easily, presumably because females have a strong urge to mate. Indeed, in one case a hybrid mating between an *A. 2-punctata* female and a male chrysomelid beetle was observed (Fig. 5.12). Interestingly, there was some indication that sympatric species mated less readily than those that do not have overlapping distributions. Thus, the pairs of species that hybridised most easily were *A. ocellata* and *Anatis labiculata* from England and the United States, respectively, and *H. 4-punctata* and *H. axyridis* from England and Japan, respectively. The reason for this may be that part of the female's mate recognition system involves a

Fig. 5.11. Interspecific mating of *Myzia oblongoguttata* and *Coccinella 7-punctata*.

Fig. 5.12. Interspecific mating between an *Adalia 2-punctata* female and a male chrysomelid beetle.

rejection response to males of other species that they encounter reasonably frequently in evolutionary time. This part of the recognition system is only likely to extend to sympatric species that share similar habitat requirements.

Unpublished work by Lusis on hybridisation between *A. 2-punctata* and *A. 10-punctata* (Zakharov, pers. comm.) is interesting in this context. When Lusis attempted to cross European *A. 2-punctata* with European *A. 10-punctata*, they failed to mate. However, European *A. 10-punctata* readily mated with *A. 2-punctata* from a mid-Asian region lacking *A. 10-punctata*. That European *A. 10-punctata* and mid-Asian *A. 2-punctata* mated readily, irrespective of which species was male or female, suggests that the male ladybird does play some role in species recognition during courtship. Furthermore, the breakdown of species recognition when *A. 2-punctata* are drawn from a region from which *A. 10-punctata* is absent indicates that species recognition traits are not constant over a species' geographic range. Perhaps *A. 2-punctata* from mid-Asia have not evolved traits by which they may recognise, or be recognised by, *A. 10-punctata*, because they have not had to. Alternatively, if there is a cost to species recognition mating preferences, the mid-Asian *A. 2-punctata* may have lost such preferences in the absence of *A. 10-punctata* because they would gain no selective benefit by retaining them. As yet there is no evidence to differentiate between these alternatives.

The one pairing that did not produce a copulation involved female *C. magnifica* that were placed with male *C. 7-punctata*. This failure is of particular interest because the reciprocal pairing did produce matings, although no eggs resulted. *Coccinella magnifica* is myrmecophilous. It is morphologically similar to *C. 7-punctata*, and often occurs with it. Speculation on the evolutionary divergence of these two species has concentrated on the ecological questions of why *C. magnifica* lives close to ants, and why it does not live elsewhere. Less consideration has been given to how gene flow between the two species was arrested. The fact that female *C. magnifica* would not accept male *C. 7-punctata*, even when females had been prevented from mating for a considerable period, suggests that the mate recognition system of these females is stronger than that of most.

When May Hybrid Matings Occur?

The ease with which females can be induced to mate with non-conspecifics suggests that hybrid matings may occur whenever females do not encounter conspecific males very frequently. This may occur in three situations. Firstly, when a species is at very low density, males and females may simply not come across each other very often. Secondly, in univoltine ladybirds there is often a period towards the end of the reproductive season when the sex ratio of adults in mating condition becomes strongly female biased, because of the greater longevity of females. Thirdly, the sex ratios of some ladybird populations are highly female biased due to the occurrence of male-killing bacteria.

One important consequence of the natural occurrence of interspecific hybrid matings is that they will allow the transmission of sexually transmitted diseases between species. For example, interspecific hybridisation may be implicated in the finding that the sexually transmitted mite *Coccipolipus hippodamiae* infests a wide range of coccinellid species.

Egg-laying

The final act of reproduction in ladybirds is egg-laying. There is no evidence of parental care in the Coccinellidae other than the provisioning of eggs by the female, and ovipositing in appropriate places.

Egg-laying strategies in ladybirds are affected by many factors. Mating promotes oviposition (Sem'Yanov, 1970), and increased female promiscuity increases fertility rate (Haddrill, 2001). Female size, female condition, ovariole number, food type and prey size all affect the number of eggs laid at a time. Food availability, the age and structure of prey colonies, the presence of other ladybirds and other competitors for prey in various stages of development, as well as time of year and climatic conditions, affect where and when eggs are laid.

Egg Size and Clutch Size

Females of most insects invest considerable resources in their eggs, and ladybirds are no exception. In aphidophagous ladybirds, the size of egg is partly related to female size.

The questions: how many eggs does a female lay during her lifetime? and what determines the size of the eggs she lays? are interrelated. Comparing eight aphid-feeding species, Stewart et al. (1991a) found that ladybird species of different sizes committed approximately the same energetic investment to reproduction. They found that egg size, multiplied by egg number, is proportional to adult weight. For species of similar size, the options are to lay either many small eggs or few large eggs. On the positive side for laying many small eggs, more progeny can be produced. Furthermore, because newly emerged larvae eat any unhatched eggs, rapid embryo development and early hatching are selected for, which favours smaller eggs. On the negative size, small eggs will produce small neonate larvae, which may encounter problems in subduing prey, and therefore starve to death. Various authors have shown that the efficiency of larvae in capturing aphids is largely a function of size (Dixon, 1959; Mills, 1979), and Stewart et al. (1991a) suggested that the ability to catch aphid prey constrains the lower limit to larval size at hatching, and thus egg size. A further consideration is the length of time that appropriate aphids are likely to be available to developing larvae. Species that lay small eggs, relative to adult size, have a longer larval development period than species that lay large eggs. If larval development time for these species with small eggs is longer than the normal period of abundance of their aphid prey, the larvae may run short of food before they can pupate (Majerus, 1994a).

Although larger ladybirds tend to lay larger eggs, the correlation is not direct for the relative size of an egg compared to the weight of the female that laid it increases in smaller species. Indeed, in very small species, the reproductive investment in a day may be higher than that for larger species. Hence, while in *C. 6-maculatus*, an average egg comprises 0.57% of female body weight, the figure in *Coccinula sinensis* is 1.75% (Majerus and Majerus, 2000). As Stewart (1991a) suggested, the relatively large size of the eggs of small aphidophagous coccinellids, such as *C. sinensis*, is necessary to ensure that developing embryos have sufficient nutrients and space to attain a minimum size that will give newly dispersing larvae a reasonable chance of catching prey. The large relative size of eggs in small ladybirds is likely to place a constraint on clutch size in these species, and on the number of eggs laid per day. Thus, while a female *C. 6-maculatus*, with a maximum reported clutch size of 37 eggs may lose up to 21% of her body weight in a single clutch, in *C. sinensis* (maximum clutch size = 16) this figure increases to 28%.

The majority of, but not all, ladybirds lay their eggs in groups. Most numerous among the species that lay eggs singly or in batches of just two or three are many of the coccid feeders. These often lay eggs under the scales of their prey. In addition, some polyphagous species, such as *Rhyzobius litura*, lay single eggs in leaf cavities (Ricci and Stella, 1988). A few specialist aphid-feeding species also lay eggs singly. For example, the myrmecophilous ladybird *Platynaspis luteorubra* positions individual eggs horizontally, usually on the underside of leaves, close to a leaf vein in the vicinity of aphid colonies. Unusually for a coccinellid, the eggs are flat-bottomed and convex.

In ladybirds that lay their eggs in batches, the number of eggs in clutches varies both within and between species. Baungaard and Hämäläinen (1984) suggested that the number of ovarioles in a single ovary limits maximum clutch size. This supposed limit assumes that only the terminal, and so most fully developed egg in each ovariole can be laid in a particular batch of eggs and that all eggs in a clutch originate from the same ovary. These authors have proposed that records of *A. 2-punctata* 'super-batches', containing up to 85 eggs, probably result from a second oviposition close to an earlier laid batch.

This view is endorsed by Stewart et al. (1991a), who found that the average number of eggs per clutch is usually approximately half the ovariole number. Stewart et al. (1991b) concur that ovariole number sets an upper limit to clutch size. However, observation of clutches of eggs, both in captivity and in the wild, with numbers considerably in excess of the maximum-recorded ovariole numbers, suggests that this assumption is invalid. The maximum recorded number of eggs in a single *A. 2-punctata* batch is 55 (Hurst and Sloggett, pers. comm.) while clutches of over 100 *C. 7-punctata* eggs have been observed (Majerus, 1994a). It seems likely that although eggs in a clutch usually emanate from a single ovary, occasionally clutches include eggs from both ovaries. An alternative explanation is that two or more eggs within individual ovarioles may be 'ripe' and so be included in the same clutch.

Variation in clutch size in a species is affected by the size of the female, the food available to her, her age and whether she is affected by disease. A direct correlation has been demonstrated between female size and ovariole number in *C. 7-punctata* (Dixon and Guo, 1993). As the same authors also showed a direct relationship between ovariole number and clutch size, the correlation between female size and clutch size necessarily follows. Food availability also affects clutch size to a very marked degree. Female *C. 7-punctata* fed excess aphids laid clutches almost double the size of those fed a third of the amount (Dixon and Guo, 1993). Interestingly, in these experiments, the size of eggs did not vary with food availability.

Food type may also influence clutch size. For example, clutch size in aphidophagous species is higher than for mycophagous species of similar size. For instance, the mycophagous *H. 16-guttata* has an average wild clutch size of 6.25 eggs (range 1–24) (Majerus 1994a). This compares with an average of 15.69 (range 5–39) for the slightly smaller *A. 2-punctata* on *Tilia* trees (Majerus, 1994a). The relatively small average clutch size of *H. 16-guttata* may be either a direct consequence of slower nutrient intake by females, or may be adaptive if the number of larvae that can be supported by a particular area of mildew is limited.

In general, older females lay fewer eggs per day, and in smaller clutches, than younger females. This makes sense given that the reproductive output of a female will depend not only on the rate that she can produce eggs, but also on the number of days that she can reproduce. If there is a reasonable likelihood that reproductive output may be curtailed by death, a female will increase her reproduction by laying eggs early in adult life, even if this reduces her longevity. This will lead to a trade-off between the number of eggs laid per day and the longevity of the female. The result of this trade-off is seen in the gradual reduction in the number of eggs a female lays per day, as she gets older (Dixon, 2000). Finally, the reproductive output of a female may also be adversely affected if she contracts a disease. This reduction may be manifested in two ways. Firstly, some diseases reduce both fecundity and fertility, as in the case of the sexually transmitted mite *C. hippodamiae* (Hurst et al., 1995) and fungi of the genus *Laboulbeniales* (Majerus, pers. obs.). Other diseases appear to drain resources from the female so that the decline in eggs laid per day through life accelerates, as in the case of the male-killing *Spiroplasma* in *H. axyridis* (Majerus, 2003).

How Eggs are Laid

Much research effort has been dedicated to ladybird oviposition behaviour. Ricci and Stella (1988) have stressed the importance of the setae (stiff bristles or hairs) on the female styli and tenth abdominal segment, in positioning eggs. Species that lay eggs in batches under leaves, such as *C. 7-punctata*, *H. 13-punctata* and *Anisosticta bitriangularis*, have few short setae, each of which is rooted in its own socket, those nearest the vagina being the longest. The setae on the styli help to ensure eggs are laid upright, with those on the tenth abdominal segment being used to position the eggs side by side. The female genitalia, which are extruded between the upper and lower plates of the eighth abdominal segment during oviposition, may be retracted between each egg deposition (e.g. *Anisosticta 19-punctata*), or they may remain exposed while a complete clutch is laid (e.g. *H. 13-punctata*). In species that lay eggs in crevices, or under coccids, setae tend to be grouped together (e.g. *Chilocorus 2-pustulatus*). In *R. litura*, the setae are of different lengths, longer setae being used to investigate leaf cavities, while the shorter setae are used to guide the egg into position, once a suitable site has been found.

Why Lay Eggs in Clutches?

Perhaps the most obvious reason for laying a number of eggs together is that in doing so a female saves time searching for suitable oviposition sites. However, there are other reasons. Ladybird eggs, in addition to being provisioned with nutrients for embryo development, are also supplied with defensive chemicals. The egg-laying behaviour of most aphid-feeding species, laying batches of brightly coloured eggs in exposed positions (Fig. 2.17), is likely to enhance the impact of the bright, presumably warning colouration. The eggs of other insects containing chemical defences, such as the white butterfly *Pieris brassicae*, are also yellow and laid in batches in exposed positions.

Several studies have shown that the defensive chemicals contained in ladybird eggs may reduce egg predation by larvae of other species of ladybird (Agarwala and Dixon, 1992; Petersen, 1992; Ware et al., 2008). In tests,

A. 2-punctata eggs were relatively unattractive to larvae of *C. 7-punctata*, *A. 10-punctata* and *Coccinella 11-punctata*, but not to conspecific larvae, which ate them readily. The eggs of the other species tested were also less attractive to larvae of other species, than to conspecific larvae, but appeared not as well defended as *A. 2-punctata* eggs. Larvae that tasted one or two eggs of a batch of another species' eggs, then often left the rest of the batch alone (Petersen, 1992). More interesting still is the observation that eggs dipped in hexane for two minutes, which removes any chemicals on their surfaces, lose their aversive qualities. This suggests that surface chemicals protect the eggs.

The demonstration that single eggs are less aversive than batches of eggs, and that the repulsive effect increases with clutch size, indicates that the higher the dosage of these chemicals, the greater their repellent effect (Agarwala and Dixon, 1993). Perhaps the most surprising aspect of this story is that young larvae, once they have dispersed from their egg clutches, show lower levels of cannibalism towards other egg clutches laid by their own mother, than towards those laid by other females (Agarwala and Dixon, 1993). It appears that larvae can recognise sibling eggs. Given the evidence that a cocktail of surface chemicals protects eggs, one might speculate that the exact mix of chemicals varies from mother to mother, and larvae avoid eggs bearing their mother's scent signature, because of their own genetic make-up. Alternatively, they may learn the mix of the cocktail on their own, and therefore their siblings' eggshells, when they consume these just after hatching. This latter possibility is particularly intriguing, for it would mean that recognition of sibling eggs is chemically imprinted on the larvae soon after hatching. Furthermore, it seems the more likely mechanism because, due to the mixed paternity of many clutches (Haddrill, 2001), the larvae that hatch from a clutch are often only half-siblings, thus an explanation based on genetic similarity is less likely than one that is mediated only by the mother.

One deduction from this research on inter- and intraspecific egg consumption is that laying eggs in batches will reduce levels of egg predation by non-conspecific larvae.The potential for egg cannibalism by newly hatched siblings, afforded by laying in batches, must also be considered. From the mother's perspective, it might be considered preferable to lay eggs singly to minimise the chances of this type of cannibalism. However, for the mother, sibling egg cannibalism is a double-edged sword. Although some of her offspring perish, others may have an increased chance of survival because they obtain an easily procured and nutritious meal before dispersing to find aphid prey.

Laying eggs in batches may also facilitate cooperative feeding by siblings. The small size of neonate larvae and the difficulty they have in subduing prey has been noted. As the larvae disperse from their clutch, if one secures a hold on an aphid, the aphid releases alarm pheromones as a warning to other

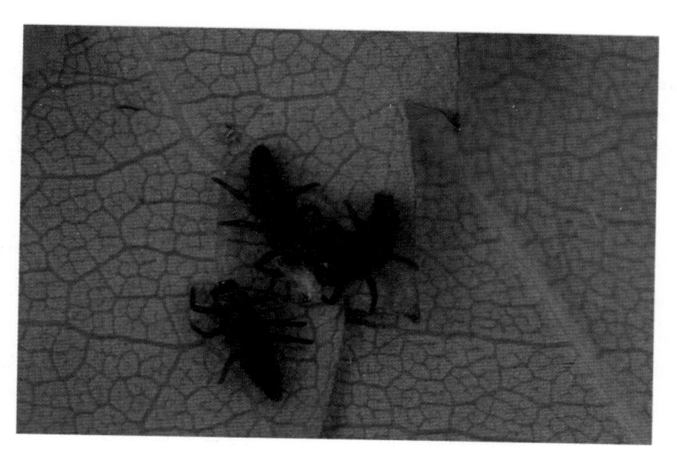

Fig. 5.13. Early instar ladybird larvae feeding on aphid.

aphids in the colony. Other dispersing ladybird larvae respond to the aphid alarm pheromones, moving towards the aphid and showing area-restricted searching behaviour (Hemptinne et al., 2000a). Upon reaching the aphid, they help to subdue and kill it and share in the prey (Fig. 5.13).

Most of the ladybirds that lay eggs singly or in groups of two or three give their eggs dual protection, in the form of chemical defence (Pasteels et al., 1973) and a concealed position under the ovisac or body of scale insects. Upon hatching, the larva eats the contents of the coccid ovisac and then its body. Finding a first meal is thus not a problem for these larvae. Here the advantages of laying eggs in groups are outweighed by the benefits of laying eggs in concealed positions very close to sedentary prey.

The suggestion that females of *H. axyridis* 'purposefully' lay infertile or inviable eggs as nurse or trophic eggs (Perry and Roitberg, 2005) is probably unfounded, with the inviable eggs that they observed being the result of inbreeding depression or male-killer infection.

Oviposition Sites

It is easy to understand why the coccid-feeding ladybirds lay eggs where they do. In laying eggs very close to or on scale insects, they provide slow moving and easily captured prey for their offspring. However, the oviposition sites used by aphid-feeding species are more difficult to comprehend, partly because the information on where these species lay eggs is somewhat contradictory. Many accounts of ladybird natural history state that females lay eggs close to aphid colonies. Banks (1955) suggested that this is because well-fed females oviposit; starved females do not. As females that find heavily aphid-infested plants become well-fed, the likelihood is that they will oviposit on these plants as well. This contention is reinforced by the finding of Evans and Dixon (1986) that female *C. 7-punctata* are more active in the absence of aphids than when aphids are present. However, these authors also found that *C. 7-punctata* females were reluctant to oviposit in the absence of aphids or aphid cues, such as aphid pheromones and honeydew. Furthermore, oviposition may be inhibited even in the presence of high aphid densities. For example, female *C. 7-punctata* are reluctant to lay eggs in the presence of conspecific females, but not males (Petersen, 1992).

Work on oviposition in *A. 2-punctata*, by a group led by Hemptinne, showed that there is an optimum period for ladybird oviposition in the life of an aphid colony, and that, in the field, *A. 2-punctata* do oviposit at this time. Hemptinne et al. (1992) first showed that *A. 2-punctata* oviposition is inhibited in the laboratory by the presence of conspecific fourth instar larvae. Subsequent work has shown that the optimum period occurs once the aphid colony is well established, but before it reaches peak size, in line with theoretical prediction (Hemptinne and Dixon, 1991). Hemptinne et al. (1992) argued that an aphid colony at such a stage of development provides females with nutrients for egg maturation, provides young larvae with a high density of young aphid prey that they will be able to subdue, and will persist long enough for the larvae to complete development.

There are several cues that *A. 2-punctata* could use to assess prey patch quality: the age structure of the aphid colony, the developmental stage of the plant, or the presence of other predators attracted to the colony. Experiments to test the influence of these cues have shown that neither aphid colony age structure nor plant development either inhibit or promote oviposition (Hemptinne et al., 2000b). However, the presence of chemical 'tracks' laid down by both conspecific and heterospecific larvae inhibit oviposition in *A. 2-punctata, A. 10-punctata* and *C. 7-punctata*, with the level of inhibition being greater between species that breed in the same habitat (Magro et al., 2005). In *A. 2-punctata*, the larval tracks consist of a cocktail of semiochemicals, mainly hydrocarbons (Hemptinne et al., 2001). The presence of these chemicals causes both conspecific and heterospecific females to stop laying eggs, become restless, and, in the field, move to other plants (Doumbia et al., 1998; Hemptinne et al., 2001). Ruzicka (2003) has shown that in *Ceratomegilla 11-notata*, semiochemicals in larval tracks are detected by the maxillary palps. Interestingly, while some studies have shown that both conspecific and heterospecific larval tracks deter oviposition, others show that conspecific tracks have a greater deterrent effect. For example, larval tracks of both *Cycloneda limbifer* and *C. 11-notata* deter oviposition by *C. limbifer* females (Ruzicka and Zemek, 2002). However, *H. axyridis* females, while deterred from oviposition by conspecific tracks, were not deterred by *C. 7-punctata* tracks (Yasuda et al., 2001). This may be because *H. axyridis* readily eats *C. 7-punctata* larvae.

The Effect of Reproduction

The reproductive behaviour of *A. 2-punctata* has been closely examined. The strategies of both males and females may be interpreted in the context of the fundamental differences between the sexes. Males compete with one another for paternity, through prolonged copulations and passing excessive amounts of sperm. Females are careful about the males that they mate with, employing various testing behaviours to assess the fitness of suitors, and some have genetic preferences to mate with certain types of males. In addition, females are exceptionally promiscuous, a response to high levels of genetic incompatibility in this species. As our knowledge of the reproductive behaviour of *A. 2-punctata* has grown, it has become increasingly apparent that it is intricately interwoven with almost every other aspect of the ecology and behaviour of this species.

As reproduction is the ultimate aim of life, and is an adult function, we may expect the developmental and survival strategies of immature stages to maximise the chances of reaching reproductive maturity in good condition. We know that food intake of female larvae influences her fecundity, and that both type and quantity of adult food intake influence gonad maturation in both sexes. The ecology of prey influences when *A. 2-punctata* reproduces, where they reproduce and how often they move from one host plant to another. The cannibalistic habits of ladybirds, and in particular neonate larvae, also affect oviposition behaviour and may have affected the population dynamics of deleterious genes, promoting genetic incompatibility, which has led to increased female promiscuity. This promiscuity has in turn led to the species being prone to invasion by sexually transmitted diseases. Sibling egg consumption has also promoted invasion by sex ratio distorting bacteria, the actions of which influence population sex ratios that then impose evolutionary pressures on the reproductive behaviour of both male and female *A. 2-punctata*.

The reproductive behaviour of *A. 2-punctata* has been more closely studied than that of any other ladybird. Examination of less well-researched ladybirds indicates that there is considerable variation among species in virtually every aspect of reproductive behaviour. This variation gives scope for valuable comparative analyses to be conducted in this family of beetles, using *A. 2-punctata* as the model for comparison. In this way, our understanding of the determinants of promiscuity, mating duration, mate guarding, sperm competition, female mate choice, reproductive investment, clutch size and oviposition behaviour may be increased both in this group and more generally.

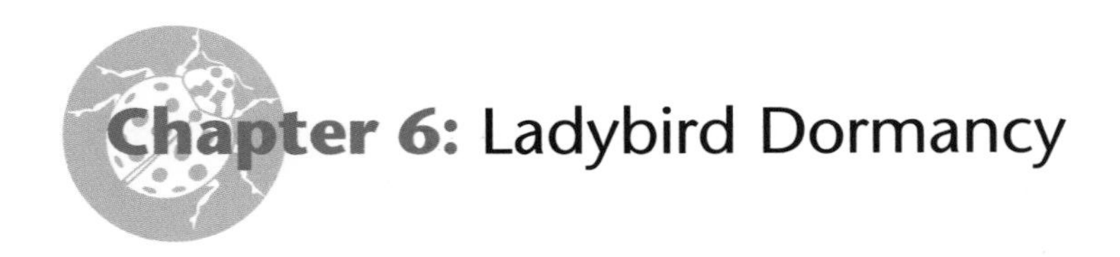

Chapter 6: Ladybird Dormancy

Ladybirds in Unfavourable Conditions

In the wet tropics, conditions are constantly favourable for ladybirds, which can be active and indulge in sex and reproduction all year round. However, in seasonal climates, at least one period of the year is unfavourable, and the drive to procreate must be suppressed by the need to survive until good times return.

To many people in the northern hemisphere, the sighting of the first ladybirds of the year, usually in March or April, means that spring will not be long in coming. We tend to associate ladybirds with hot sunny weather. In some languages the names of ladybirds reflect this: in Estonian ladybirds are Kirilind, meaning little cow of summer; in Flemish they are Sonne Kever and in German Sonnërkafer, both translating as sun beetle; and in the Czech Republic, they are called Slunecko – meaning little sun. The 'great ladybird years' reinforce the summer association because ladybird population explosions tend to occur in long hot summers. Yet, in many parts of the world ladybirds live for almost a year, and can be found as adults in any month. That relatively few are seen in the winter is a reflection of their inactivity during unfavourable conditions.

Not only is cold unfavourable for ladybirds, but in some parts of the world heat or very dry weather are the killers, and there ladybirds have a dormant aestivation period. In a few species, such as *Hippodamia convergens* in northern California, dormancy commences in midsummer and continues, with just a short autumnal period of activity, into and through the winter. Thus, for many ladybirds these unfavourable times span more than half of their adult lives. By the time favourable conditions return and ladybirds resume their feeding and reproductive activities, most will be at least 'middle-aged', having spent from four to nine months of their adult lives in a dormant or semi-dormant state, often secreted away in some sheltered spot, unseen by human eyes.

In arctic, subarctic, temperate and Mediterranean regions of the world, the winter is an unfavourable period for ladybirds. Low temperatures and scarcity of food make activity energetically costly. Consequently, in autumn as food becomes scarce, days become shorter, and temperatures drop, ladybirds must decide what to do. One option is to migrate to warmer climes, as many birds do. The obvious alternative is to find a shelter from the worst of the weather and conserve energy until temperatures rise and food becomes available again. Most coccinellids adopt this second alternative.

Hodek (2012) comprehensively reviewed work on the dormancy behaviour of ladybirds. Areas of study include physiological and metabolic changes prior to and during dormancy, factors that initiate dormancy, choice of dormancy sites and movement to and from these sites, factors that cause dormancy to be broken, and levels and causes of mortality during dormancy. Despite the large amount of effort to understand dormancy in coccinellids, few general conclusions can be drawn from this research. This is due to considerable variation in ladybird dormancy behaviour, both among and within species. For example, some species of coccinellid overwinter on their preferred host plants. Thus species such as *Myzia pullata*, *Anisosticta bitriangularis* and *Harmonia 4-punctata* do not indulge in any significant movement from foraging sites to overwintering sites. Conversely, other species, such as *H. convergens*, *Harmonia axyridis*, *Semiadalia 11-notata* and *Micraspis frenata*, migrate long distances from feeding and breeding sites to dormancy sites. A third group, including species such as *Coccinella 7-punctata*, *Adalia 2-punctata* and *Halyzia 16-guttata*, shows highly variable behaviour, both between populations and among individuals in the same population. This chapter considers the ways that adult ladybirds prepare for and endure unfavourable periods.

Dormancy, Diapause and Quiescence

The most common response of ladybirds to poor conditions is to become dormant. I will use dormancy as an umbrella term to cover any significant slowing or stopping of ladybirds' reproductive, feeding and locomotory activity, even if this is only for a short period.

Dormancy may be subdivided into three classes. Quiescence is a response to sudden, often irregular and unpredictable periods of unfavourable weather. The ladybirds stop feeding and become inactive, but do not regress their gonads and they become active again as soon as favourable conditions return.

Ladybirds may also have a fixed period of dormancy (known as oligopause). They stop feeding, become inactive and can survive for considerable periods on fat reserves. However, they are capable of becoming active and feeding intermittently if conditions improve for short periods within the unfavourable season, and they resume activity immediately when conditions improve at the end of an unfavourable season.

The third category, diapause (including summer or dry season diapause), is the most extreme. Here the response is to long-term, predictable periods of unfavourable environmental conditions. The ladybirds have a definite preparatory period, in which, typically, they either do not mature gonads, or if gonads are already fully developed, these are largely reabsorbed. Nutrients taken in during this preparatory period are used to build up fat stores. They then become inactive, and do not become active or feed during diapause. The breaking of diapause does not occur immediately upon the return of favourable conditions, but often depends on a complex series of events.

Pre-dormancy Behaviour

To Sleep or Not to Sleep?

Newly eclosed ladybirds have a choice to make, either to mature gonads and reproduce or to lay down fat reserves and thereafter enter some form of dormancy. In some species, or some populations of some species, this decision has already been made by past evolution. In these ladybirds, a dormant period is essential before reproduction. In others, some individuals require diapause while some will reproduce if conditions are favourable. Thus, in *C. 7-punctata* from northern England, almost 90% of females require a period of dormancy at low temperature before reproducing, while the remaining 10% mate and begin ovipositing shortly after eclosion if plentiful supplies of aphids are available (Majerus, 1994a). For the majority, the requirement of a period of dormancy before reproduction has been programmed into their genes. Similar results have been obtained for *C. 7-punctata* from other parts of the world (Hodek, 1962). However, in some locations, all females have a requirement for a dormant period before reproducing, as in the Ukraine (Dobrzhanskii, 1922a) and Swedish Lapland (Majerus, unpubl. data), while in southern Spain, only about a quarter of individuals had such a requirement (Hodek and Okuda, 1993). In still other species including, for example, *Propylea japonica, H. axyridis, Cheilomenes 6-maculatus* and *Adalia 10-punctata*, there is no genetic prerequisite for dormancy before reproduction, and whether an individual reproduces or becomes dormant depends on various internal and external cues.

That a requirement for diapause before reproduction can be coded within the genes of a ladybird was determined as long ago as 1922, in two seminal papers by the great Russian-American evolutionary geneticist Theodius Dobrzhanskii (1922a, b). He showed that in the Ukraine, *C. 7-punctata* has two generations per year. Offspring from eggs laid by overwintered adults complete their development in midsummer and adults start to mate and oviposit within two weeks. Females from this second generation eclose from mid-July through to September and do not mature ovaries even when food is abundant. These females only mature ovaries after overwintering. Dobrzhanskii attempted to alter the strategies of females from the different generations by experiment, but failed to do so, concluding that the species had a genetically fixed alternation of strategies. This conclusion received strong support from observations during 1921, in which climatic conditions early in the year were exceptionally harsh. The result was that overwintered females only began to oviposit in early July, so that the resulting first generation adults did not eclose until August and September. In spite of the lateness of the season, these adults matured ovaries, mated and laid eggs, with the vast majority of the resulting larvae dying due to lack of food and frost in September and October (Dobrzhanskii, 1922b).

For ladybirds in which a pre-reproductive dormancy is not obligatory, various factors have been suggested to play a role in the induction of diapause, and these factors vary both within and among species. They include changes in day length, decrease in temperature, change in humidity, physiological ageing of host plants, and the nature, availability and condition of food for adults and/or larvae (reviewed in Hodek, 2012). In some species, a single factor is of paramount importance. For example, in *A. 2-punctata* less than 14 hours of daylight induces diapause, while day lengths longer than 14 hours inhibit diapause (Obrycki et al., 1983; Iperti and Prudent, 1986). Alternatively, in *Coccinella 9-notata*, day length, temperature and food availability are all involved in the induction of two diapause periods each year: one a winter diapause, the other a summer diapause (McMullen, 1967). Some of the factors involved are quite subtle. In *S. 11-notata*, beetles feeding on aphids taking sap from old plants that are deteriorating in condition begin to regress their ovaries and lay down fat reserves (Rolley et al., 1974). The same is true of *H. convergens* feeding on the sexual forms of aphids that develop in the fall compared to those feeding on parthenogenetic aphids that are characteristic of the spring and summer (Wipperfürth et al., 1987).

In some species, two or more factors acting in concert affect diapause induction. Thus, in *Chilocorus 2-pustulatus* in Central Asia, both temperature and photoperiod have an effect, but the critical value of one that induces diapause depends upon the value of the other (Zaslavskii and Bogdanova, 1965). Similarly, diapause induction in *Stethorus picipes* is dependent on a combination of both temperature and day length, with the availability of their preferred prey, spider mites, also being influential (McMurtry et al., 1974).

In many species, induction of dormancy appears to be a very plastic response. Combinations of temperature, photoperiod and food availability are most commonly involved. A good example is the Australian ladybird *Coccinella repanda*, in which both summer and winter dormancy sometimes occur. Dormancy may be initiated by long day length, short day length, high temperature, lack of food or a combination of these factors. Anderson and Hales (1986) concluded that *C. repanda* exhibits an opportunistic life history strategy that is adaptive, allowing efficient use of fluctuating prey populations. I suspect that this conclusion holds for many other aphidophagous coccinellids, including *C. 7-punctata* (Hodek and Michaud, 2008).

Preparation for Dormancy

In temperate regions, the late summer/fall is a crucial time for ladybirds. Once a ladybird has made the 'decision' to enter diapause or quiescence, rather than to start reproducing, it must prepare itself, as best it can, for the winter, by building up its fat reserves. In some years, young adults will be able to find their normal, preferred food, but often this is unavailable, or in short supply, hence it is in this season that adult ladybirds are most usually seen eating alternative foods.

The ease with which food can be found is important, because the likelihood of surviving the winter depends largely on the insect's energy reserves. This means that if food is difficult to obtain, more energy may be lost than gained. The time when an individual stops trying to feed and seeks a place to pass the winter should thus depend on two factors: firstly, the level of its energy reserves, and secondly, the likelihood of this store increasing rather than decreasing if it remains active. If the ladybird's reserves are at such a low level that the probability of surviving the winter is small, it is likely to be worth continuing to seek food, even if food appears scarce. Because the probability of dying before the spring is high anyway, it is worth the risk, even if, in the event of running down reserves even lower, the ladybird dies sooner than it would have done had it remained dormant. The ultimate aim is, after all, to reproduce, and failing this, in evolutionary terms it matters little whether one dies at the beginning or end of the winter. If the insect has an energy reserve level that would give a 50:50 chance of survival should the severity of the winter be no worse than average, but less than 50:50 if the winter is severe, then the decision whether to continue activity, or drift into dormancy, depends on the energetic costs and benefits of continuing to seek food. If energy intake exceeds expenditure it is worth continuing the search.

Survival through dormancy has been shown to be dependent on the level of reserves that a ladybird builds up prior to dormancy (Hodek and Cerkasov, 1963). Barron and Wilson (1998) demonstrated that in *C. 7-punctata*, body weight influenced whether ladybirds remained active or entered dormancy, and that this was correlated to overwinter survival. They assessed the proportions of *C. 7-punctata* still actively foraging or dormant in late autumn, finding that the active ladybirds were significantly lighter than those that had already become inactive. Furthermore, they showed that if foraging ladybirds were prevented from further feeding, only 28% survived the winter, compared with

91% of those that had already become dormant. Interestingly, Honek (1997) had already shown in a study in which *C. 7-punctata* were collected from overwintering sites, weighed and overwintered in artificial hibernacula, that survival is not dependent on body size or sex. The conclusion from Barron and Wilson's (1998) study is that whether ladybirds continue to forage or become dormant depends on their reserves. This suggests that these ladybirds have some physiological mechanism of quantifying their reserves, and that there is a threshold level of reserves required for overwinter survival.

Hodková (1996), reviewing the physiological mechanisms of diapause in insects, noted that the regulation of diapause often involves several steps. Typically, these are perception of environmental signals by sensory receptors; storage and evaluation of this sensory information, frequently through interaction with an internal photoperiodic clock (Saunders, 1982; Masaki, 1984; Zotov, 2008); transmission of interpreted information to the neuroendocrine system (Hodková, 1992); and reaction of the neuroendocrine system, which controls the production of hormones that cause the suppression of reproductive functions, the accumulation of lipids and glycogen, and reduce metabolic rate. The main elements of the neuroendocrine system are neurosecretory cells in the brain, the corpora cardiaca and the corpora allata (Hagedorn, 1985; Delbecque et al., 1990). Specific sensory information reaching various parts of the neuroendocrine system causes specific 'diapause proteins' to be synthesised, while concurrently, production of reproductive hormones is reduced (Ichimori et al., 1990).

Where to Sleep?

The next question for a ladybird must be where to sleep. For most ladybirds, the answer is simply in a position sheltered from the worst of the weather. However, there is considerable variation in the type of dormancy sites chosen by ladybirds. Some species go underground, others move to the litter layer, or low-growing vegetation. Still others remain high on vegetation, some even in exposed positions, while others secrete themselves away in cracks in wood or under bark. Finally, a few species move to high altitude, selecting a sheltered, low-temperature locale to sleep. In addition, some ladybirds sleep alone, while others form small, or in some cases very large, dormant aggregations (Fig. 6.1).

Many studies of overwintering insects have shown that the overwintering site can be critical to survival. It is no surprise, therefore, that ladybirds use a suite of stimuli when selecting their overwintering sites and that these vary among species. These include responses to light (phototaxis), or humidity (hygrotaxis), or temperature (thermotaxis), or gravity (geotaxis). Many species

Fig. 6.1. Large aggregation of *Tytthaspis 16-punctata*.

also show a thigmotactic response; that is, they react to tactile stimuli, squeezing into small spaces, such as hollow plant stems, bark crevices, or spaces under stones. Some species move considerable distances to dormancy sites and are often attracted to prominent landscape features. Finally, some show either spatial or temporal intraspecific variation in their choice of dormancy sites.

One of the most researched examples involves *H. convergens* (Hagen, 1962; Johnson, 1969). Hagen (1962) described the movement of this species from the lowlands of the Central Valley in California, to mountains in the Sierra Nevada. Progeny of diapausing adults eclose in May. These take to the air in the morning, reaching considerable altitude with the aid of convectional updrafts. As the beetles get higher, they encounter lower temperatures and at 11–13°C they stop flying, with the result that they descend. As they fall, the temperature increases again, so that after falling about 300 metres, they resume flight. These vertical oscillations are repeated throughout the day. Direction is largely a function of the westerly winds that carry the ladybirds towards the mountains. Once the ladybirds reach the mountains, they may reproduce if aphids are present. However, most of the beetles feed on pollen and nectar to build up their fat reserves. After a period of feeding, the ladybirds embark on a secondary flight, close to the ground, and select summer dormancy sites. There they remain largely inactive until October, when a tertiary flight, generally to lower slopes, takes them to winter dormancy sites. Similar behaviour takes *H. convergens* to high altitude dormancy sites in other parts of America.

Movement away from light has been shown in a number of ladybirds that overwinter in litter, and this negative phototaxis is sometimes allied to a digging response. For example, *Coleomegilla fisilabris* overwinters in woodland litter layers, moving under leaves due to negative phototaxis. If placed on a yielding surface, this ladybird digs (Park, 1930). The same is true of *Anatis ocellata*, which overwinters in the soil (Majerus, 1992). Not all ladybirds are repelled by light as winter approaches. *Hippodamia convergens* is attracted to sunny clearings on high montane slopes, where large aggregations then form (Hagen, 1962). *Hippodamia 15-signata* shows a similar response, although once aggregations have formed in light clearings, this species passes the winter under rocks and litter (Harper and Lilly, 1982).

Many ladybirds 'hide' in very tight holes or crevices when they become dormant, using tactile stimuli to find a sheltered spot. *Coleomegilla maculata*, for instance, overwinters in forest litter. It moves to the litter layer as a result of negative phototactic and possibly geotactic responses, and once there, squeezes under cover (Hodson, 1937).

Coleomegilla maculata and *H. convergens* also show hygrotactic responses; reacting to humidity gradients once they have moved to a suitable general location to overwinter (Hodson, 1937; Hagen, 1962). While these species are positively hygrotactic, other species, such as *S. 11-notata*, are negatively hygrotactic, choosing well-ventilated crevices in which to pass the winter, rather than more sheltered, damper hideaways (Yakhontov, 1960a; Iperti, 1966a). Iperti (1966a) suggests that the choice of drier shelters can be adaptive, reducing mortality due to fungal disease.

Some ladybirds vary geographically in their choice of dormancy sites. A good example is *C. 7-punctata*. Majerus (1994a), describing the overwintering sites of this species recorded during the Cambridge Ladybird Survey, summarised its preferred sites as follows: 'Very diverse. In exposed habitats, any sheltered position close to ground in leaf litter, plant debris, low herbage, dead foliage of standing plants (e.g. thistle, bracken), grass tussocks, crowns of perennials, tight packed foliage of gorse and conifers. Occasionally below the soil surface. Also occasionally in large aggregations among stones on hilltops. In sheltered habitats (e.g. mature deciduous woodland) in exposed positions on tree trunks and branches.' In contrast, on mainland Europe this species is said to always overwinter on the ground: under stones, in litter, in holes in the soil surface, near the base of plants, in grass tussocks; never in cracks of tree bark or walls (Hodek, 2012).

Aphidecta obliterata is one of several ladybirds to show geographic variation in dormancy sites. In south-east Norway, Graf and Kriegl (1968) reported that most *A. obliterata* overwinter in bark crevices low down on trunks. Parry (1980), working in north-east Scotland, found that most sheltered under bark scales of *Picea sitchensis*, with occasional individuals being found in crevices in the bark of *Larix decidua* and *Pinus* species, while a small number remained on the foliage of *Pseudotsuga menziesii*. In southern England, the majority remains among foliage of needled conifers, rather few being found in trunk crevices (Majerus, 1994a). The proportion that remains among foliage, rather than moving to trunks, is probably related to the fact that winter mortality is temperature dependent (Brown and Clark, 1959). However, the degree of exposure may also be critical, as Graf and Kriegl (1968) showed that in Norway, where temperatures as low as –42°C were recorded in January and February, those overwintering in sheltered sites, at the bases of trunks, suffered lower mortality than those in more exposed positions further up the trunk.

The contrasts in overwintering site choices of *C. 7-punctata* and *A. obliterata* in different geographic regions are probably indicative of the relatively mild oceanic climate of Britain, compared to the more continental climate of central and eastern Europe. In Britain, the relatively warm winters allow a greater range of sites to be used than in most other European countries. This pattern is probably mirrored in other parts of the world with mild, maritime climates.

Are Ladybirds Weather-forecasters?

The preferred overwintering sites of ladybirds also vary between years. *Adalia 2-punctata* usually migrates from its feeding sites to overwinter under bark, in bark crevices, in cracks in wood, including fence posts and telegraph poles or, nowadays, most often in or on buildings, finding suitable refuges around window frames, in double-glazing (including supposedly sealed units), in lofts, between roof tiles, in loose masonry and unused chimneys, and even in the corners of cool rooms (Fig. 6.2). However, Brakefield (1984b) noted that in Holland, *A. 2-punctata* sometimes pass the winter in groups, exposed on *Tilia x europaea* trunks. During the winter of 1981/1982, these groups were common and often contained several dozen ladybirds. However, in subsequent years, few *A. 2-punctata* were found in such situations (Brakefield, pers. comm. in Majerus 1994a). Brakefield (pers. comm.) believes that his observation of *A. 2-punctata* staying on trees in 1981/1982 was the result of autumn temperatures being too low for the ladybirds to make flights from the trees to more sheltered sites. Other examples of temporal variations in dormancy sites are more difficult to explain. In most years in Britain, *Chilocorus renipustulatus* overwinters in the litter layer. However, during the very mild winters of 1988/1989 and 1989/1990, this species was often recorded on trunks and branches of *Fraxinus excelsior*, *Salix* spp., *Populus* spp. and *Betula* spp., and

Fig. 6.2. *Adalia 2-punctata* overwintering indoors.

sometimes the ladybirds were active, or even feeding (Majerus, 1992). Here it is possible that the ladybirds pass the winter in a state of quiescence, and remain in exposed positions until driven lower by severe temperatures. The occurrences of large groups of *C. 7-punctata* overwintering several metres above the ground during the latter winter are more difficult to explain, because there was no evidence that the ladybirds became active or fed during the winter. However, it is the observations of *H. 16-guttata* that are the most difficult to fathom.

In some years, the majority of individuals of *H. 16-guttata* overwinter in the litter layer. In other years, most remain on the trees, generally overwintering below knots or branch joints on slender, smooth-barked trunks, or on the underside of small branches (Fig. 6.3). In still other years, roughly similar proportions of the population remain on the trees or move to the litter layer. The individuals that remain on the trees invariably select a position sheltered from the prevailing south-westerly wind (Fowles, 1990).

After the variation in overwintering sites chosen by this species in different years was first noted (1989/1990), the overwintering sites were monitored annually at three locations in Britain. The proportion of ladybirds that remain on the trees each year is positively correlated to the summed minimum daily temperature for November to February inclusive. The amazing feature of this case is that the vast majority of ladybirds select their overwintering sites by early October, and thereafter hardly move. Certainly, the ladybirds observed still high on trees in late October do not move to lower positions later in the winter, nor do those at ground level move up the trees if the winter turns out mild (Majerus, 1994a).

In 1993, on the basis of four years of data, I wrote: 'It is almost as though ladybirds can predict the severity of the winter several months in advance. I do not even like to speculate on the proposition that these ladybirds are better

Fig. 6.3. *Halyzia 16-guttata* overwintering on a tree.

long-range-weather-forecasters than our best meteorologists, preferring to leave the mystery unsolved' (Majerus, 1994a). Subsequently, with 16 years of data, the correlation between the overwintering sites chosen by *H. 16-guttata* and the severity of the winter is much stronger, but the mechanistic mystery is no nearer solution.

The case of *H. 16-guttata* has attracted the attention of chaos and catastrophe theorists. In meteorology, difficulties in predicting weather more than a week or so in advance were, for nearly four decades, put down to the so-called 'butterfly effect'. The butterfly effect, formulated in the 1960s, proposes that the flap of a butterfly's wings in Brazil might stir up a tornado in Texas. Even with the development of supercomputers in the 1990s, accurate weather prediction more than a week ahead seemed impossible because chaos cannot be predicted. However, some meteorologists continued to seek long- range weather patterns, and were heartened by examples, such as that of *H. 16-guttata*, that seemed to have the ability to judge the weather well in advance. In 2001, researchers at Oxford University and at the European Centre for Medium-Range Weather Forecasting in Reading demonstrated that the butterfly effect had been exaggerated, and that most of the unpredictability it caused were functions of errors in computer models of the atmosphere (Calder, pers. comm.). The sensory information that *H. 16-guttata* uses to decide where to overwinter is still unknown and would be worthy of attention.

Dormancy Aggregations

Many coccinellids congregate into groups that range in size from small to very large to overwinter (Fig. 6.4), or to pass through summer or dry season

Fig. 6.4. Large overwintering aggregation of *Coccinella 11-punctata*.

Fig. 6.5. Group of *Micraspis frenata* in dry season dormancy.

dormancy (Fig. 6.5) (Hagen, 1962; Majerus, 1992, 1994a; Hodek, 1996b, 2012; Honek et al., 2005, 2007).

The mechanisms by which ladybirds are attracted to one another as they form these groups are not well understood. Hodek (2012) noted that the behaviours that may lead to grouping can be either direct or indirect. Responses to environmental stimuli, such as light, gravity, temperature, wind and humidity, may result in large numbers of individuals accumulating at the same place, each responding to extrinsic factors rather than to other ladybirds. Little work has been carried out on the relative importance of environmental stimuli to aggregation formation in most species. Pulliainen (1964) investigated the responses of *Myrrha 18-guttata*, overwintering in pine bark crevices in Finland, to humidity and light. He found that they show a strong negative response to humidity that was only reversed after prolonged desiccation. They were also repelled by short-wave light initially, and by long-wave light after desiccation. These results are important in showing how the physiological status of an organism can alter its behavioural responses, and in indicating that water balance may have a critical role to play in dormancy biology.

Alternatively, there may be a pheromone that causes ladybirds to be attracted to one another at this time. If so, it seems likely that the scent persists for a substantial period as the same sites often are used by ladybirds year after year (Majerus, 1997b). As ladybirds rarely survive from one winter to the next, there must be some mechanism by which ladybirds born in the summer are attracted to these regularly used sites. A pheromone, laid down at a site by the previous winter's tenants, would accomplish this. Hills (1969) suggested that *A. 2-punctata* are attracted to overwintering sites by the smell from the excreta of the previous year's population, or by the few adults that die at such sites each winter. Experiments have shown that washing regularly used overwintering sites of *A. 2-punctata* repeatedly with water during the summer, significantly reduces the use of these sites in the following winter (Majerus, 1997b). These data provide strong circumstantial evidence that some chemical attractant is involved in aggregation formation, at least in *A. 2-punctata*. Further evidence that aggregation semiochemicals may be important was provided, for *H. axyridis*, by Verheggen et al. (2007).

If a pheromonal attractant is involved in group formation, it seems unlikely to be species specific, for clusters often involve ladybirds of two or more species (Fig. 6.6). Moreover, these multispecies aggregations quite frequently involve species from different subfamilies; thus *C. 7-punctata* (Coccinellinae) frequently occur with *Exochomus 4-pustulatus* and *C. renipustulatus* (Chilocorinae) or with *Scymnus suturalis* and *Scymnus nigrinus* (Scymninae), while *Subcoccinella 24-punctata* (Epilachninae) occur with *Psyllobora 22-punctata, Tytthaspis 16-punctata* and *Rhyzobius litura* (Coccinellinae) (Majerus, 1994a).

Why do ladybirds so often form clumps during the winter? If the cues leading to clump formation are environmental and extrinsic to other ladybirds, the answer may be that clump formation is somewhat incidental. Specific responses to particular environmental cues have evolved because they increase the probability of an individual ladybird finding a favourable site to pass the winter. It is not surprising that many individuals should arrive at the

Fig. 6.6. Overwintering cluster comprising *Coccinella 7-punctata, Harmonia 4-punctata* and *Adalia 2-punctata.*

same favourable site. Chemical cues of long duration could also serve to attract numbers of ladybirds to a favourable spot. The lingering scent from ladybirds that occupied a site in previous years may effectively announce that a particular situation has previously been a safe retreat for the winter. However, Hills' (1969) speculation that the corpses of ladybirds could be an attractant seems somewhat illogical in this context, for dead ladybirds are surely not a good indicator of a safe haven.

An alternative hypothesis is that the formation of a group, in itself, is adaptive, increasing the survival chances of the individuals within the group. There are two reasons why this might be so. Firstly, numbers of ladybirds together may be able to regulate their microclimate to some extent. Secondly, group formation may serve to accentuate the bright colour patterns with which so many ladybirds are adorned. The main reasons why toxic or distasteful organisms have bright contrasting colour patterns are to be easily learnt and subsequently remembered by natural enemies. A large group of red and black ladybirds is certainly a striking and memorable image for a potential predator. There are many cases of warningly patterned organisms living communally for parts of their lives. The evolution of these three potential defensive components – toxicity or distastefulness, warning colouration, and grouping behaviour – is interesting, for the adaptive success of each depends to some extent on the others, and it is not intuitively obvious which trait evolved first, or alternatively, how they could all evolve simultaneously.

One other advantage may accrue from grouping behaviour. Ladybirds in groups are likely to be in close proximity to individuals of the opposite sex, and once active in the spring, may mate before dispersing to find food, as noted for some ladybirds (Savoiskaya, 1965; Hodek and Landa, 1971; Solbreck, 1974; Majerus, 1994a), but not others (Majerus, 1994a).

Whether grouping behaviour is incidental or adaptive should probably be judged on a case by case basis. For example, Honek et al. (2005), reporting results from a long-term sampling study in the Czech Republic, concluded that grouping in *C. 7-punctata* was largely incidental, while that in both *S. 11-notata* and *H. variegata* it was adaptive.

Dormancy

The Biochemistry of Dormancy

Many ladybirds that enter dormancy at the start of an unfavourable period die before conditions become favourable again. The survival of ladybirds through the winter is dependent upon two main factors: whether their fat and fluid reserves are sufficient and whether extrinsic factors, such as disease, occur during this vulnerable period.

The role of reserves during dormancy can be interpreted from a consideration of the changes in the chemistry of ladybirds over dormant periods. I have already noted that there are differences in the type of dormancy exhibited by different ladybirds. Those that diapause invariably show a considerable decrease in their metabolic rate, whether compared to that exhibited by beetles arriving at dormancy sites, or the rate that is characteristic of reproducing adults. Estimates of the decrease depend on the activity levels of the non-diapausing ladybirds – reproducing adults have a higher metabolic rate than those preparing for dormancy – and the conditions to which the beetles are exposed prior to and during measurement (Hodek, 1996b). However, reduction to 50% of the summer respiratory rate is common in winter, and can be greater. Thus, in *H. convergens* reduction to less than 20% of reproductive rate have been recorded (Stewart et al., 1967; Lee, 1980).

The beginning of diapause, whether during winter or summer dormancy, is also associated with changes in the production of various hormones (Denlinger et al., 2004) and, in particular, declines in the production of juvenile hormone, which has a stimulatory role in the maturation of ovaries and the development of male accessory glands (Guan and Chen, 1986; Okuda and Chinzei, 1988).

The best researched changes during dormancy concern declines in fat, glycogen and water reserves. Typically, ladybirds enter diapause with a considerable reserve of fat, concentrated in a fat body lying in the abdomen. The size of the fat body varies both within and among species, but in aphidophagous coccinellids is typically 35%–50% of fresh body weight. In coccid feeders, such as *Chilocorus rubidus* and *C. renipustulatus*, the figure is only around 25%. The decline in fat content over the dormant period is also greater for aphid feeders than coccid feeders. Thus, in the former, most reports are of between

50% and 75% of reserves being consumed (*C. 7-punctata*, Hodek and Cerkasov, 1961; Hariri, 1966c; *S. 11-notata*, Hodek and Cerkasov, 1963; *A. 2-punctata* and *Propylea 14-punctata*, Hariri, 1966c), while in the coccid feeders, the declines were in the range 33%–50% (*C. rubidus*, Pantyukhov, 1968a; *C. renipustulatus*, Pantyukhov, 1968b). Rate of decrease is not constant, generally being greatest in the early and late stages of diapause, with losses during the central part of diapause being low, or in some cases, negligible (Parry, 1980).

The pattern of accumulation of glycogen prior to diapause, and its use to fuel metabolism during diapause, are similar to those for lipids, but fluctuate more, probably because glycogen is less chemically stable than most fats. Water content is also important to ladybird survival through dormancy. However, in contrast to fats and glycogen, the changes in water content through dormancy are small.

Survival Through Dormancy

Mortality of ladybirds during the winter is often very high. The chance of a ladybird surviving the winter depends on both internal and external factors. The harshness of the winter, particularly in terms of temperature, is important, but many species can withstand long periods of sub-zero temperatures (Berkvens et al., 2010). Obviously cold tolerance is positively correlated with latitude and altitude. Indeed, some species occurring at high latitude or altitude are able to survive sustained periods of several months embedded in ice. In the case of *Hippodamia arctica*, ladybirds frozen in ice and kept at –20°C for three months became active within six hours of drying out once the ice had melted. Within three days the ladybirds were mating, and within six days, females had begun to oviposit (Majerus, unpubl. data). Overwintering site also has some influence on cold tolerance, Novák and Grenarová (1967) finding that *A. 2-punctata* overwintering in bark crevices had much higher resistance to low temperatures than did *C. 7-punctata*, *Coccinella 5-punctata* or *E. 4-pustulatus*, which overwintered in litter. Similarly, in a comparison of cold hardiness of *S. 11-notata* overwintering exposed to air and *C. 7-punctata* overwintering at ground level and insulated by plant material, Nedvěd (1993) found *S. 11-notata* to have the greater cold tolerance. Aspect was found to play a role in the overwintering survival of *H. axyridis* in The Netherlands, with higher survival of those beetles exposed to a southern, rather than northern, aspect (Raak-van den Berg et al., 2012).

Although ladybirds can survive prolonged cold periods in the depths of winter, they cope less well with sudden drops in temperature in the late autumn or in early spring once they have become active. Therefore, hard early or late frosts may lead to very high mortality, particularly if interspersed with warm sunny days. That said, species vary in this respect. For example, *E. 4-pustulatus* is able to reacclimatise to low temperatures after it has become active in the spring, while *C. 7-punctata* and *S. 11-notata* cannot (Nedvěd, 1995). Humidity also affects the winter survival of many species, particularly those that select sites among plant litter. In wet winters, mortality due to fungal infection is high in species such as *A. 10-punctata, P. 22-punctata, Calvia 10-guttata, Calvia 14-guttata* and *P. 14-punctata*. The number of *Vibidia 12-guttata* sampled at overwintering sites in Poland was negatively related to soil humidity (Ceryngier and Godeau, 2013).

The resistance of ladybirds to low temperatures is an evolved trait, and varies both among and within species. In *A. 2-punctata*, samples taken from northern Sweden and Norway (>60°N) were able to survive for longer periods of refrigeration (<3°C) than those taken from England or France (Tinsley and Majerus, unpubl. data). Similarly, *C. 14-guttata* from Swedish Lapland and Edmonton, AB, which has a continental climate with cold winters, were more resistant to refrigeration than those from the maritime climes of England (Majerus, unpubl. data).

Perhaps surprisingly, some species do not show such consistent variances in cold tolerance. Thus, while several species of ladybird (*Coccinula sinensis, C. 6-maculatus, C. rubidus*) from the Tokyo region of Japan show rather little cold tolerance, others, such as *P. japonica* and *H. axyridis*, are more tolerant than might be expected of species adapted to the warm climes of lowland central Honshu. Here it is possible that gene flow from high latitude or high altitude populations of the latter two species into lowland populations prevents precise local adaptation, hence lowland populations retain relatively high cold tolerance.

Waking Up

What causes ladybirds to wake up after a period of dormancy depends on both the type of dormancy and the pervading environmental conditions. For those species that diapause before they reproduce, diapause has to be terminated before the beetles can become active. In ladybirds that do not diapause, the situation is simpler and activity resumes once environmental conditions have become favourable again.

Termination of Diapause

The main factor affecting when diapause will terminate is time itself. Work on *C. 7-punctata*, from various parts of Europe and Asia, has shown that for those beetles requiring diapause, a minimum period of diapause is necessary. Temperature and day length have an influence on this period, but no set of environmental conditions can reduce it to zero (Hodek, 1996b). In *C. 7-punctata* in western and central Europe, the fall and early winter periods are critical. Ladybirds enter diapause at the end of the summer, and as time passes the intensity of diapause reduces, so that by early December, a return to favourable conditions will allow ladybirds to become active. Favourable periods generally do not return in these regions until March or later, therefore, in the latter part of the winter, the dormancy of *C. 7-punctata* should be considered quiescence rather than diapause.

The situation in the Mediterranean region is somewhat different, because here *C. 7-punctata* has two periods of reproduction; one in spring and a second in the autumn, and two dormant periods; one in winter and one in summer. Here, short day length does not inhibit reproduction as it does further north. Indeed, Bodenheimer (1943) found a pattern of reproductivity and dormancy in Israel that suggests that winter dormancy represents quiescence, while summer dormancy involves an aestivation diapause. Similar observations in Spain and Greece suggest that this pattern is a general feature of *C. 7-punctata* in the Mediterranean region (Hodek et al., 1989; Hodek and Okuda, 1993). This pattern is also seen in *S. 11-notata* in central Greece (Katsoyannos et al., 2005) and in *C. 7-punctata brucki* on Honshu, Japan (Sakurai et al., 1981). Further north in Japan, on the island of Hokkaido, *C. 7-punctata brucki* has a dormancy pattern similar to that of central European populations of *C. 7-punctata septempunctata* (Okuda and Hodek, 1994).

Given that environmental conditions vary geographically, it is not surprising that patterns of dormancy, and the factors that influence return to activity after a dormant period, also vary geographically, both within and among species. *Coccinella 9-notata* has two reproductive periods and two dormancy periods each year (McMullen, 1967). However, in contrast to *C. 7-punctata*, both the winter and summer dormancy periods in *C. 9-notata* involve diapause. Here the beetles reproduce when day and night lengths are fairly similar in spring and fall, and diapause is induced and terminated by both short and long day length. Temperature and prey availability are also involved in diapause induction.

In species without a requirement for diapause, but which enter diapause under certain unfavourable conditions, the factors that terminate dormancy are even more variable. In *A. 2-punctata* in New York State, beetles enter diapause in the winter. This diapause lasts for three to five months before improving environmental conditions can terminate it. Conversely, in western Europe, where winter dormancy in this species involves quiescence, activity and reproduction may begin as little as three days after returning beetles to favourable conditions of day length, temperature and food (Majerus, 1994a).

Once diapause has been broken, the cues that cause each species of ladybird to become active are largely unknown. Undoubtedly, both temperature and day length are important. However, most species vary their emergence time from year to year, becoming active earlier following mild winters, and later after harsh ones. Therefore, it seems that they have some ability to discern, from their previous experience, whether food is likely to be available.

In Britain, observations on the variation in the timing that species left their overwintering sites from year to year during the Cambridge Ladybird Survey suggest that temperature, possibly other climatic conditions and day length affect the timing of reactivation. In addition, different species vary in the timing of final emergence from dormancy sites (Majerus, 1994a). Non-predatory species become active in mild spells during the winter. From early in February, some of the predatory species also venture out. *Coccinella 7-punctata* often clambers out of its sheltered position on bright sunny mornings to bask in the sun, only to return to its shelter, usually finding exactly the same spot, before the sun sets. In most years, it is not until the second half of April that this species finally disperses to find food and a mate. Data collected between 1981 and 2005 suggest that this date is becoming earlier, with the dates of first sighting, first mating and first newly eclosed adult having shifted about twelve days earlier over the last quarter of a century (Majerus, unpubl. data.). This change may be the result of higher temperatures due to global warming. *Exochomus 4-pustulatus* also first venture out on sunny days in February. In some years, vast numbers may be seen basking on pine trunks, branches or cones, catching the sun. However, once active and warm, these ladybirds seek food – on the evergreen, needled conifers they often find it, in the form of coccids, adelgids and aphids that begin to reproduce relatively early in the year. If food is available, *E. 4-pustulatus* may begin mating by the end of February (Fig. 6.7), and ovipositing in early March (Majerus, 1994a).

The timing of the final departure from winter quarters depends largely on the nature of the food of a particular species, and on the likelihood that such food will be available in the early months of the year. Populations of aphids and coccids that feed on conifers begin to increase in February and March. Consequently, by the end of March most of the conifer specialist ladybirds have begun to feed and mate. *Aphidecta obliterata* and *Myzia oblongoguttata*, however, are rarely active before the middle of April. Late in March, *C. reni-pustulatus*, which feeds on scale insects on the trunks, branches and twigs of many deciduous trees, will have started to mate and lay eggs.

In late April, several of the ladybirds that habitually feed on the aphids sucking the sap of herbaceous plants become active. These include *P. 14-punctata*, *Coccinella 11-punctata* and *C. 5-punctata*. *Anisosticta 19-punctata* leaves its haven between old reed-mace leaves and stems, and moves to

Fig. 6.7. *Exochomus 4-pustulatus* mating pair.

the young leaf blades that have recently risen through the mud and water. Both *C. 2-pustulatus* and *Coccinella hieroglyphica* begin to roam the *Calluna* heathland again at this time. By early May, those aphid-eating species that are associated with deciduous trees have become active and begun feeding, mating and laying eggs. Only *H. 16-guttata*, which feeds on powdery white mildews on the leaves of deciduous trees, becomes active later. In this species, mating is not seen until mid-June, presumably because sufficient food for ensuing larvae, in the form of the hyphae of these mildews, has not grown on the leaves of deciduous trees until this time.

The first tasks for most species of ladybird when they wake up are to replenish their food and fluid reserves. Only later do they seek out mates. However, some species that overwinter in groups take the opportunity that close proximity to many members of the opposite sex offers, and mate before dispersing from their dormancy sites. This is the case for both *H. convergens* at the end of winter dormancy and *M. frenata* at the end of its dry season dormancy.

Summer Dormancy and Aestivation

Winter is not always an unfavourable time of year for ladybirds. In many parts of the world, hot dry weather conditions are severely unfavourable, as vegetation dries off and withers and sap-sucking insects become scarce. Then, reproduction takes place either during the spring and fall, or during the winter, or in the drier tropics, during the rainy season. In unfavourable hot dry conditions, ladybirds have the choice of dispersing or becoming inactive, and again, the choice made is likely to depend on a number of factors, such as nutritional status and suitability of conditions for dispersal. Summer dormancy is a regular part of the annual cycle for many ladybirds. In some populations of species, such as *C. 7-punctata brucki*, summer dormancy takes the form of a true diapause, with respiration rate being lowered to between a half and a fifth of normal rate (Sakurai, 1969).

In other species, summer dormancy takes the form of quiescence. This is the case for *M. frenata*, which, in Australia, regularly becomes dormant during the dry season. Like many species that have a winter dormancy, this species forms large aggregations. In Queensland, some sites are used year after year, and aggregations run into many millions. The dormancy here does not appear to be a full diapause, for the ladybirds rapidly become active following the coming of the rains. Furthermore, these ladybirds will become active in response to another environmental stimulus.

Many of the dry season dormancy sites of *M. frenata* lie in the wide expanses of eucalyptus grasslands that fringe the desert interior of Australia. This dry habitat is regularly scarred by fires (Fig. 6.8). The flora is well adapted to these fires, which tend to burn rather coolly, and do little more than scorch off the dry dead grass leaves and the lower foliage of the trees. Both the trees and grasses are fire-resistant, and sprout new growth soon after a fire has passed through. Interestingly, the insects that live in this land are also adapted to living with fire. *Micraspis frenata*, when at its summer dormancy sites, provides

Fig. 6.8. Fire in eucalyptus grassland.

a good example. At the first whiff of smoke, be it smoke from a grass fire carried on the breeze, or smoke from a cigarette, these ladybirds, previously unmoving and apparently dormant, become severely agitated, moving away from the smoke. If the smoke persists, engulfing the trees on which the ladybirds reside completely, they take to the air to escape the oncoming fire (Majerus, 2006a).

A dormant period during unfavourable conditions, whether these are dominated by cold, hot or dry weather, is a dangerous time for ladybirds. Lack of food is the prime reason for habitat unfavourability, and should food not return soon enough, ladybirds may starve to death as their reserves are slowly depleted. Furthermore, during these periods, ladybirds may die from flooding, desiccation, fungal attack, fire or many other reasons. However, although mortality during dormancy may be very high for some species of ladybird, particularly in exceptionally harsh winters or arid summers, these are not the only dangers faced by these beetles. The causes of ladybird death, and particularly the enemies of ladybirds, are the subjects of the next chapter.

Chapter 7: Ladybird Death

The Struggle for Survival

It is early summer. On a nettle-bed infested with aphids, *Coccinella 7-punctata* and *Adalia 2-punctata* adults forage, mate and lay eggs through the day, generally unhindered by insectivorous birds hunting for prey close by. Small parasitoid wasps also frequent the nettle patch. These are looking for ladybirds and, when they find one, will lay a single egg inside the ladybird. The female wasps lay into both species of ladybird, but as time will show, only those laying into *C. 7-punctata* benefit from their reproductive success, as eggs laid into *A. 2-punctata* are encapsulated and die. On many of the plants, batches of yellow ladybird eggs may be seen. Some of these are hatching. In the clutches from *C. 7-punctata* nearly all the eggs hatch, but in some of the *A. 2-punctata* clutches, only half hatch, the embryos in the other half being dead, having been assassinated by an enemy within. These observations are indicative of the struggle for survival of ladybirds.

Death is a major component of Darwin's theory of evolution by natural selection. In using a phrase such as 'the survival of the fittest', we imply the death of the less fit. Prey and their predators, and hosts and their parasites, have been involved in a struggle for ascendancy ever since living organisms began to consume one another. Is it, for example, a coincidence that the fastest two land animals are the cheetah and its main prey, Thompson's gazelle? Natural selection favours those prey that have adaptations that make them difficult to detect, or catch, or eat. Hosts may also develop traits that reduce detection by parasitoids and have strategies that reduce rates of infiltration and reproduction by parasites. Of course, predators and parasites are also subject to natural selection. Evolution will therefore promote predator and parasite traits that counteract the devices of their prey or hosts, increasing their sensory efficiencies, improving catching and handling skills, and developing methods to detoxify prey or circumvent host immunity systems. The complexities of the adaptations and counter-adaptations between the eaten and the eater are a testament to arms races that have endured between prey and their predators, and hosts and their parasites, for millions of years.

Ladybirds are prolific breeders. A female *Harmonia axyridis* may lay over 3000 eggs in her lifetime. Very few of these offspring will survive to reproduce themselves. Some starve or die during dormancy or fall to catastrophic environmental events such as fire or flood. Many others fall foul of a diverse array of other organisms, the enemies of ladybirds.

The enemies of ladybirds can be split into three groups: predators, parasitoids and parasites.

Vertebrate Predators

Birds

It is generally acknowledged that ladybirds, at least those with bright colour patterns, are either toxic or distasteful to most vertebrates. The bright colours serve to advertise this fact, and the ladybirds also deter attacks by releasing droplets of a foul-smelling fluid (reflex blood) when disturbed. This wisdom emanates largely from laboratory studies in which ladybirds were offered to a wide range of vertebrate predators, to assess their acceptability as food (Morgan, 1896; Pocock, 1911; Morton Jones, 1932; Frazer and Rothschild, 1960; Meinwald et al., 1968; Pasteels et al., 1973; Marples et al., 1989). Frazer and Rothschild (1960) offered *C. 7-punctata, Coccinella 11-punctata* and *A. 2-punctata* to mice, voles, hedgehogs, bats, lizards, terrapins, a toad and six species of birds, finding that ladybirds were rejected as prey. Yet, there are a considerable number of reports of birds eating ladybirds in the wild (Nechaev and Kuznetsov, 1973; Muggleton, 1978; Kristín, 1986; Majerus and Majerus, 1997b). This contradiction is exemplified by the finding, through gut analysis, that *Adalia 10-punctata* made up much of the winter diet of *Parus major* (Betts, 1955), but that in palatability tests, these birds actively rejected ladybirds (Lane and Rothschild, 1960).

Muggleton (1978) noted the contradictions between laboratory and field records of birds preying upon ladybirds, citing findings in the literature of 19 bird species that rejected ladybirds in captivity, and 121 bird species that ate ladybirds, mainly in natural conditions. This latter list includes six species of bird found to reject ladybirds under laboratory conditions. Muggleton explained the contradiction by stressing differences in the conditions of birds in the laboratory and in the wild: laboratory birds tend to be well fed, while wild birds are often hungry. Captive birds can thus afford to reject some less tasty food items; wild birds cannot.

Clear evidence that some birds eat ladybirds comes from examination of bird droppings. Muggleton (1978) reported that 38 grams of dried house martin, *Delichon urbica*, droppings, collected by Hounsome in 1976 from under a nest in Manchester, consisted almost entirely of the remains of *C. 7-punctata*, *C. 11-punctata*, *A. 2-punctata*, *A. 10-punctata* and *Propylea 14-punctata*, and notes that the young in the nest fledged successfully. In Britain, the summer of 1976 is known for the population boom of ladybirds; ladybirds would not make up such a large proportion of the diet of *D. urbica* in most years. Nevertheless, *D. urbica* obviously can feed on ladybirds (Majerus, 1991b). Yet Marples et al. (1989) demonstrated that *C. 7-punctata* is toxic to blue tit, *Parus caerulus*, nestlings. This contradiction alone shows the necessity of considering different bird species separately. It suggests that *D. urbica* is tolerant of the toxic effects of *C. 7-punctata*. That this should be so is surely common sense. *Delichon urbica* feeds on the wing, catching insects with a gaping beak as it flies. At speed, the bird would have little opportunity to recognise its prey, and the ladybird would have no opportunity to reflex bleed. Over 40 instances of aerial avian predators taking ladybirds were reported during the Cambridge Ladybird Survey. The ladybirds eaten included *C. 7-punctata*, *A. 2-punctata*, *P. 14-punctata* and *Harmonia 4-punctata* (Majerus and Majerus, 1997b). If birds, such as martins, swifts and swallows, were not resistant to the toxic effects of ladybirds, feeding on the wing would be extremely hazardous. The feeding strategies of these birds will surely have imposed strong selection on them to evolve immunity to any toxins within ladybirds. Other birds also feed on ladybirds. A series of observations of the behaviour of wild birds towards ladybirds give a flavour of the circumstances under which birds may accept ladybirds as food.

The first observations involve adult birds feeding on ladybirds during the winter. The records include European robin, *Erithacus rubecula*, starling, *Sturnus vulgaris* and common treecreeper, *Certhia familiaris*, eating *P. 14-punctata*, *Psyllobora 22-punctata* and *Exochomus 4-pustulatus*, respectively, and a pair of magpies, *Pica pica*, feeding on *C. 7-punctata*.

The behaviour of the magpies is worth describing in more detail, for the birds showed considerable signs of discomfort while eating the ladybirds. At least 60 *C. 7-punctata* were taken from overwintering shelters in dead thistles (Majerus and Majerus, 1997b). The birds would eat two or three and then shake their heads, jab their beaks into the ground and wipe them against vegetation. They also made frequent visits to a nearby pond to drink before returning to the thistle patch. This behaviour was seen during a very cold spell of weather, when little other food was available. Close to starvation, the magpies ate ladybirds, despite their unpalatability, just to survive. The visits to the pond were likened to a child having a sweet after taking nasty medicine 'to get rid of the horrid taste'.

A second set of observations suggests that some birds feed ladybirds to their young, but do not eat them themselves. Tree sparrows, *Passer montanus*, are granivorous as adults but feed their young insect prey (Kristín, 1984). A study in Slovakia and Poland demonstrated that ladybirds of all life stages form a large part of the diet of tree sparrow nestlings (Kristín et al., 1995). A pair of *E. rubecula*, which had nested in an old watering-can and could be kept under close observation, were seen to feed seven *P. 14-punctata*, three *P. 22-punctata*, two *Tytthaspis 16-punctata* and one *A. 10-punctata* to their nestlings (Majerus and Majerus, 1997b). This was out of over 900 food items delivered. Of about 350 food items eaten by the adults, none was ladybirds. One plausible explanation of these data is that the taste buds of the nestlings are not fully developed. Thus, while the adult birds will not eat ladybirds because of the taste, their nestlings will. This requires further testing.

Finally, two carrion crows, *Corvus corone*, were observed eating *C. 7-punctata* in exceptional circumstances on the Norfolk coast in August 1994 (Majerus and Majerus, 1997b). Huge swarms of ladybirds had descended onto beaches in the area, and many had been drowned, so that the tide-line was strewn with dead ladybirds. The crows were feeding on the ladybirds in the tide-line. It was not possible to see whether all the ladybirds eaten were dead, but the crows showed no signs of distress as they fed. It was notable that the crows confined their feeding to the tide-line and did not eat any of the millions of ladybirds crawling on the sands. One interpretation is that the crows were feeding exclusively on dead ladybirds and that due to the lack of reflex bleeding, these were more palatable than live ladybirds would have been. It is also possible that immersion in salt water reduces the unpalatability of the cadavers. This observation highlights the problems of conducting palatability experiments with dead prey. Just how similar are a predator's reactions to live ladybirds that can reflex bleed, and to dead ladybirds that cannot?

Other workers have noted birds eating ladybirds in unusual conditions. For example, Karpenko et al. (1969) reported four species of birds (spotted flycatcher, *Muscicapa striata*; wryneck, *Jynx torquilla*; tree sparrow, *P. montanus* and chaffinch, *Fringilla coelebs*) eating large numbers of ladybirds on the shores of islands formed after flooding. Mass accumulations of ladybirds, at densities estimated as 28 000 per square metre, formed on the shores of the islands. It was noted that chicks fed on these ladybirds developed normally.

Kuznetsov (1997), however, notes that reports of birds eating ladybirds are generally rare in the Russian entomological literature. Perhaps the best data comes from Kuznetsov's collaboration with an ornithologist, V. A. Nechaev. They examined the gut contents of 1125 birds of 190 species from the Primorsky Territory. Only a few species of bird had any ladybird remnants in their guts, and in those that did, ladybirds made up only a very small proportion of the diet (Nechaev and Kuznetsov, 1973). Kuznetsov's (1997) conclusion is that: 'On the whole, the Coccinellidae are only occasionally eaten by birds, and mainly in spring when the birds' habitual diet is lacking'.

Although this conclusion is held by most who study ladybirds, the precise extent of bird predation is certainly not clear. Sadly, few reports contain all the information necessary to clarify the matter: species of bird, species of ladybird, likely condition of predator, abundance of ladybirds, likely abundance of other suitable prey and number of ladybirds eaten in a given time. Careful field observations are needed. Some birds can and do eat ladybirds; others do not, or only do so under conditions of hardship.

Other Vertebrates

There is even less evidence concerning predation by other vertebrates. I have only two personal observations in this regard. The first occurred when my son, Nicolas, at the age of 13 months, put an *A. 2-punctata* in his mouth. His reaction was one of immediate disgust. He quickly spat out the ladybird and became quite distressed for several minutes. The ladybird survived. The second involved cane toads in Queensland, Australia. At a coastal motel north of Mackay, these large toads gathered around ground level path lights in the evenings to catch insects attracted to the lights. Noticing that the toads struck at any movement within range of their long tongues, I released a *Harmonia conformis* onto the light cover. After a moment, one of the toads struck at it, taking it into its mouth. Almost immediately, the ladybird was spat out and the toad began to produce considerable quantities of saliva around its mouth. The ladybird was still alive and after a few moments began to move, cleaning itself. A second toad caught it with its tongue, and again the ladybird was rapidly rejected. At this point I rescued the ladybird, and after considerable cleaning, it seemed none the worse for its ordeal. Over the next three nights, I repeated this experiment many times, each time with the same basic result.

Other workers have conducted palatability experiments with a range of vertebrate taxa. Pocock (1911) offered *C. 7-punctata* to a variety of exotic birds and mammals, including monkeys, lemurs, meerkats and mongooses. He concluded that, although some *C. 7-punctata* were eaten, there was no doubt that they were distasteful to the majority of animals, even to some of those that ate them. Frazer and Rothschild (1960) showed that various insectivorous mammals reject some ladybirds, but again the observations were made using captive animals. A more valuable report is of mortality in Bohemian aggregations of *Semiadalia 11-notata*, which, on the basis of faecal examination, was ascribed to a species of shrew. Perhaps the most dramatic accounts of mammalian predation of ladybirds concern reports of grizzly bears, *Ursus arctos horribilis*, in the American Rockies, feeding on *Hippodamia convergens* in the large overwintering aggregations that this species habitually forms (Chapman et al., 1955; Mattson et al., 1991).

I have found no published accounts of wild insectivorous reptiles or amphibians eating ladybirds. However, I have one record of a fish eating ladybirds. The remnants of three *Aphidecta obliterata* were found in the gut of a rainbow trout, *Salmo irideus*, caught in Scotland (Taylor et al., 1996).

Some herbivores also have a significant effect on ladybird populations. In Britain, *A. 2-punctata, C. 7-punctata* and *P. 14-punctata* breed in very large numbers on *Urtica dioica* beds. Cattle are often put to pasture and graze these nettles. In so doing they inadvertently eat huge numbers of ladybirds in all stages of the life cycle.

Invertebrate Predators

Invertebrate enemies of ladybirds include a variety of arthropod predators, which eat ladybirds, and spiders the webs of which can kill ladybirds. Little is known about the palatability of ladybirds to spiders. Some spiders undoubtedly do kill and eat adult ladybirds (Fig. 7.1) (Majerus, 1994a). Furthermore, Kuznetsov (1997) wrote: 'We have numerous cases where various species of spiders attacked larvae and pupae of common coccinellid species'. However, many spiders do *not* eat ladybirds. For example, an article in Countryman Magazine (1984) contains the following passage. 'Mrs Budibent watched a 7-spot caught in the web of a garden spider, which soon came out for its catch. It did not reach the ladybird but sprang back when about 1cm away, perhaps put off by the exudation of pungent drops of blood which act as an irritant.'

Studies of the palatability of *H. convergens* to the spider *Phidippus audax* showed that both larvae and adults were consumed in some laboratory trials (Bailey and Chada, 1968), but refused in others (Young, 1989). Young (1989),

Fig. 7.1. Spider preying on *Adalia 2-punctata*.

however, failed to record any instance of spider predation in Mississippi when *P. audax* and *H. convergens* occurred together on cotton in great numbers. If dead ladybirds found in spiders' webs are carefully examined, they often show no sign of attack (Fig. 7.2). However, it is often unclear whether this is because the spider that made the web was no longer using it when the ladybird blundered in, or because the web-spinner, still in residence, had left the ladybird alone. A number of photographs submitted through the UK Ladybird Survey illustrate the ability of spiders to capture ladybirds, but they do not indicate whether the cocooned ladybird was subsequently consumed. The tendency for *H. axyridis* to enter buildings during autumn seems to increase the risk of exposure to web-building spiders, which commonly occupy window frames.

Most invertebrates that do eat ladybirds are non-specialist predators. Thus, wasps have been recorded stinging and carrying off both *A. 2-punctata* and *C. 7-punctata*. A number of predatory ground beetles, rove beetles, stag beetles, tiger beetles and soldier beetles have been observed killing and eating larval, pupal and adult ladybirds (Majerus, 1998a). In the case of attacks on adult ladybirds, the beetles consumed most of the abdomen, usually attacking from the back or sides. The elytra, thorax and head were rarely consumed.

Earwigs (Dermaptera) eat the eggs of a variety of ladybirds while many species of predatory bugs (Hemiptera) prey on ladybird eggs, larvae, pupae and adults (Fig. 7.3). In southern England in the summer of 1990, adults and nymphs of *Deraeocoris ruber* fed on pupae and occasionally pre-pupae of *C. 7-punctata, A. 2-punctata* and *P. 14-punctata*, almost to the complete exclusion of other prey. They do this by straddling the pupa with their front legs, inserting the rostrum and sucking out the contents of the pupa (Fig. 7.4).

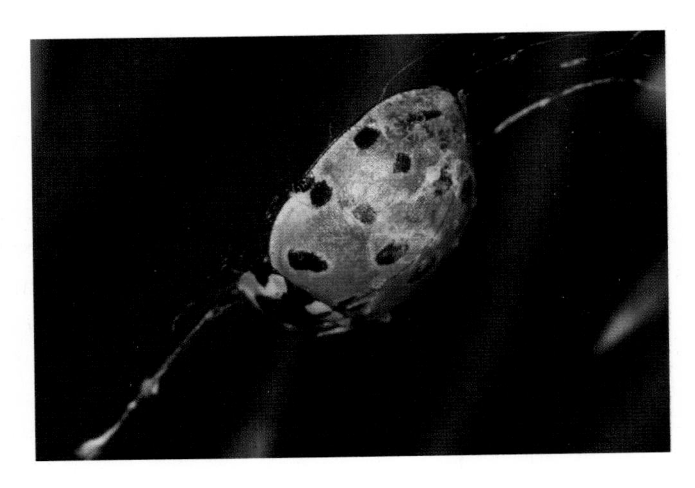

Fig. 7.2. Dead *Anatis ocellata* in spider's web, showing no sign of attack.

Fig. 7.3. Predatory true bug (Hemiptera) feeding on *Coccinella 7-punctata*.

Fig. 7.4. *Deraeocoris ruber* feeding on *Coccinella 7-punctata* pupa.

Kuznetsov (1997) also notes that a number of predatory bugs attack ladybirds, citing members of the families Anthocoridae, Reduviidae and Nabidae as most important in this regard.

Many aphid predators, including larvae of many hoverflies (Diptera: Syrphidae), larvae and adult lacewings (Neuroptera) and predatory ladybirds kill and eat ladybirds. Both interspecific predation and cannibalism are common among the ladybirds. Aphid predators compete with one another for their ephemeral prey. Once the aphid supply on a patch of vegetation has been exhausted, aphid predators may either disperse to seek aphids elsewhere, or seek alternative prey on the same patch. Adult ladybirds and lacewings usually fly to seek food elsewhere, but larvae of both these groups, and hover-fly larvae, are more likely to stay. Generally, eggs, ecdysing larvae, pre-pupae and pupae are most vulnerable to attack. When pairs of active larvae of a variety of species and sizes were placed in Petri dishes without other food, the larger larva usually consumed the smaller (Majerus, 1994a). This observation is echoed by Kuznetsov (1997) who recorded later instar larvae of the lacewings *Chrysoperla carnea, Chrysopa formosa* and *Chrysopa 7-punctata* and of the hoverflies *Episyrphus balteatus* and *Syrphus corollae* eating both eggs and young larvae of ladybirds. Although, in the wild, predation between insects that usually feed on aphids is highest when aphids are scarce, a shortage of aphids is not a prerequisite. As already noted, newly hatched ladybird larvae remain on their egg clutch for some hours after hatching and consume any unhatched eggs as a matter of course. During their initial dispersal, hatchlings also readily eat eggs from other nearby clutches, particularly if these are of the same species. This non-sibling egg predation imposes significant mortality, particularly when eggs are laid some distance from aphid colonies, as when *A. 2-punctata, A. 10-punctata* and *Calvia 14-guttata* oviposit on fruit trees. Here aphid colonies usually reside in curled leaves at the ends of branches. Ladybirds oviposit on leaves some little distance down the branch, and in their initial search for aphids, the young larvae often come upon egg clutches before aphids.

Very little is known about the palatability of coccinellids to invertebrate predators apart from ladybirds. Although some predators do eat ladybirds, others reject them. Lane and Rothschild (1960) showed that the wasp *Vespa germanica* sometimes actively rejects ladybirds, although I have seen *C. 7-punctata* being killed and carried off by this wasp (Majerus, 1994a).

Perhaps the most dramatic instance of rejection of a ladybird was recorded by Thornhill (1976). He studied courtship behaviour in the hanging fly *Hylobittacus apicalis*. Here, males bring females a nuptial gift of a prey item. The size of the prey item is important, with the size of the nuptial gift being strongly correlated to the length of time a female allows the male to copulate. In Thornhill's data, the only prey item that did not reflect this correlation was a ladybird, with the female offered this dainty quickly ending the copulation with the male which had offered it.

One final and perhaps surprising group of invertebrates are known to kill ladybirds. These are the soldier caste of some species of social aphid. For example, gall-forming aphids of the genera *Pseudoregma* and *Ceratovacuna* have sterile soldiers that aggressively defend their gall and are efficient in killing at least small coccinellid larvae as well as other predators and parasitoids that threaten (Aoki et al., 1981; Arakaki, 1992). In response to this aggression, females of the ladybird *Pseudoscymnus kurohime*, which lays its eggs close to such galls, and the larvae of which specialise on these social aphids, protects its eggs with a covering of undigested aphid (Arakaki, 1988).

Ants

The insects that most frequently come into conflict with predatory ladybirds are ants that attend aphids and coccids, the principal prey of many ladybirds.

Insects that are associated with ants are myrmecophilous (from the Greek *myrmex*, meaning ant). Myrmecophilous aphids (Fig. 7.5) and coccids show behavioural and structural modifications to life with ants. When an ant encounters such an insect, it usually strokes it with its antennae. This induces the aphid or coccid to suppress its usual defensive behaviour of kicking out, running away, dropping off the plant, or clamping down. Instead, it raises its abdomen and exudes droplets of honeydew, which the ants then imbibe. Ants gain food from the association, for honeydew is rich in carbohydrates and also contains amino acids, amides, proteins, minerals and B-vitamins (Way, 1963; Carroll and Janzen, 1973; Hölldobler and Wilson, 1990). Under some conditions, the attending ants may also gain protein, by preying on the aphids or coccids.

The benefits of the association to the aphids or coccids are primarily protection from natural enemies and improved hygiene (through the removal of caste skins, dead aphids and honeydew) direct increases in development rate, adult body size, fecundity and reproductive rate and protection from predators, parasitoids and parasites (reviewed in Majerus et al., 2007). Additional benefits may accrue because ants have been reported to transport aphids to new plants when old ones begin to wilt or die, and ants may reduce competition by removing non-myrmecophilous aphids, and other phytophagous insects.

Ants and ladybirds may interact for three different reasons. Firstly, ants that tend aphids and coccids for honeydew will be in competition with

Fig. 7.5. Ants with myrmecophilous aphids.

aphidophagous or coccidophagous ladybirds for food. Secondly, ladybirds may feed directly on ants, although only one species of ladybird is known to specialise on ants (Harris, 1921). Thirdly, ants may prey on ladybirds. These interactions may be either competitive or non-competitive.

Competition between Ants and Ladybirds

Many observations and experiments have demonstrated that Hemiptera-tending ants are more aggressive to ladybird adults and larvae near tended colonies than elsewhere. Adults are usually chased from aphid colonies, while larvae, with their softer bodies, may be picked up and carried away from the colony, or dropped off the plant, or killed (reviewed in Majerus et al., 2007).

Comparison of the densities of ladybirds on ant-tended and untended colonies of aphids and coccids have shown that ants suppress ladybird numbers. This is the case in studies in which the hemipteran colonies being compared are tended or untended by ants (Bradley, 1973; Mariau and Julia, 1977; Itioka and Inoue, 1996; Völkl and Vohland, 1996). Exclusion of ladybirds from hemipteran colonies by ants is beneficial to both tended aphids and coccids (Banks, 1962; Bradley, 1973; Mariau and Julia, 1977; Reimer et al., 1993).

Most non-myrmecophilous ladybirds only feed on ant-tended Hemiptera when untended Hemiptera are scarce (Sloggett and Majerus, 2000b). This means that as aphidophagous coccinellids usually breed when aphids are common, adults can find untended prey. Consequently, the immature stages of most coccinellids rarely come into conflict with ants tending aphids. Only in occasional years, when aphids are scarce, are larval stages likely to attack ant-tended aphids and thus come under selection for adaptations enabling them to feed on such prey (Sloggett, 1998).

Conversely, adult coccinellids often come into conflict with aphid-tending ants, particularly in late summer when building up their reserves for overwintering. At this time, they feed on a variety of foods and some also attack ant-tended aphids (Majerus, 1994a; Sloggett, 1998; Sloggett and Majerus, 2000b). The tolerance of adult ladybirds to ants then becomes critical, for those with little tolerance have to eat non-homopteran food at this period. Few studies allow judgement of the relative tolerances of different ladybirds to ants; most studies either targeting single species of ladybird (Bradley, 1973; Itioke and Inoue, 1996), or clumping all coccinellids together (Banks and Macaulay, 1967; Bristow, 1984). However, two studies suggest that there is considerable variation in the ant tolerance among ladybirds. De Bach et al. (1951) observed that 66% of *Rhyzobius lophanthae* were found with the ant *Iridomyrmex humilis* tending the coccid *Aonidiella aurantii*, on citrus, while only 15% of a *Chilocorus* species occurred in the same situation.

A second study concerned the numbers of six species of ladybird and two types of aphid in the presence or absence of ants in a pine wood (Sloggett, 1998; Sloggett and Majerus, 2000b). The ladybirds were four conifer specialists, *Myrrha 18-guttata*, *Anatis ocellata*, *Myzia oblongoguttata* and *H. 4-punctata*, the habitat generalist *C. 7-punctata*, and the myrmecophile *C. magnifica*. The aphids were *Schizolachnus pineti*, which is not ant-tended, and two *Cinara* species, which are tended by ants. The study site had two adjacent sections, one with a number of nests of *Formica rufa* and the other ant free. Results showed that *M. 18-guttata* and *A. ocellata* had very little tolerance of ants, these species only being found in the ant area once the ants had disappeared in September. *Harmonia 4-punctata* and *C. 7-punctata* had low ant tolerance, for despite being occasionally found in the ant plot, they were much less abundant here than in the ant-free plot. *Myzia oblongoguttata* was found more in the ant-free plot than in the ant plot in early summer, when aphids were abundant. However, once aphids became scarce, it was found equally in both areas. Crucially, it increased in abundance in the ant area once aphids became scarce, suggesting that it moved into the ant area to feed on ant-tended *Cinara* aphids, on which it specialises (Majerus, 1993). This specialisation on a limited number of aphid species, most of which elicit ant-attendance, must have imposed selection pressure for *M. oblongoguttata* to be tolerant to ant attacks. The last species of ladybird, *C. magnifica*, was more abundant in the ant area than in the area lacking ants throughout the study, confirming its myrmecophile status. Larvae of five ladybirds were found on the pines, those of *C. 7-punctata* not being found. No larvae of *M. 18-guttata* were found in the ant area. Larvae of *A. ocellata*, *H. 4-punctata*, and *M. oblongoguttata* were much more common in the ant-free area than in the ant area. Larvae of *C. magnifica* were confined to the ant area.

From these results, Sloggett and Majerus (2000b), produced a hierarchy of ant tolerance for these six ladybirds: *M. 18-guttata* + *A. ocellata* < *H. 4-punctata* < *C. 7-punctata* < *M. oblongoguttata* < *C. magnifica*. They concluded that *C. magnifica* is a true myrmecophile; that *M. oblongoguttata* has some defence against ants both as an adult and larva, and that *C. 7-punctata* are able to coexist with *F. rufa* at moderate levels when aphids are scarce, but does not breed in the presence of *F. rufa*.

Non-competitive Interactions between Ants and Ladybirds

Non-competitive interactions include all those away from ant-attended prey colonies, plus predation of ladybirds by ants (or vice versa). These interactions can affect habitat preferences and the distributions of ladybirds within a habitat. Away from aphid or scale insect colonies, ants that meet ladybirds either attack or ignore them, depending on the species of ant. Several ants that attack ladybirds close to tended homopteran colonies, including *Lasius niger* (El-Ziady and Kennedy, 1956; Banks, 1962), *Formica fusca* (Rathcke et al., 1967), *I. humilis* (Dechene, 1970) and *Myrmica ruginodis* (Jiggins et al., 1993), ignore ladybirds elsewhere. Conversely, some ants that prey on insects attack ladybirds as a matter of course, and exclude many species of ladybird from their forage range. Methods of attack include biting (Fig. 7.6), picking up (Fig. 7.7) or squirting formic acid (Fig. 7.8). Occasionally ants will carry coccinellids back to their nests (Fig. 7.9).

Fig. 7.6. Ants attacking *Coccinella 7-punctata* by biting.

Fig. 7.7. Ant attacking *Coccinella 7-punctata* by picking up.

Fig. 7.8. Ant attacking *Coccinella 7-punctata* by squirting formic acid.

Fig. 7.9. Wood ants carrying *Anatis ocellata* to their nest.

Ladybird Defences Against Ants

The level of tolerance of ladybirds for ants depends on the defensive capabilities of the ladybirds. Ladybirds use an array of behavioural, physical and chemical defensive strategies when faced with ant aggression (Pasteels et al., 1973; Richards, 1980b, 1985; Majerus et al., 2007).

Behavioural Defences

Most ladybirds show some defensive behaviour when attacked by ants. Larvae run away, or drop to the ground, while adults can also use these tactics or may fly, or 'clamp down', pulling their legs away under the body, their heads in close to the thorax, and pressing down tightly to the substrate. Many coccinellids, particularly the Chilocorinae, use this clamping down behaviour. Members of the Chilocorinae have a flat ventral surface and a lip around the edge of their elytra, so that the contact made with a flat substrate is very tight, thereby preventing the ants from gaining access to the ladybird's less well protected ventral surface. Some members of the Coccinellinae do not clamp down completely, but adopt a rolling motion, dropping the side being attacked to the substrate (Jiggins et al., 1993; Sloggett, 1998).

Eisner and Eisner (1992) have suggested that the flicking behaviour of ladybird pre-pupae and pupae is an anti-ant defence, with the joints between abdominal segments of the pupae acting as 'gin-traps' that damage ant mouthparts, legs and antennae. However, a more likely explanation of this behaviour is that it acts to reduce oviposition by pupal parasitoids, such as scuttle flies (Majerus et al., 2007).

It is not clear whether any of these behavioural defences evolved as specific responses to ant attacks, or whether they are general anti-predator/parasitoid devices, although Sloggett (1998) argues that some of these behaviours are more extensively developed in species that encounter ants frequently.

Physical Defences

Some ladybird larvae are covered by a network of wax filaments (Fig. 2.21), which Pope (1979) proposed was an adaptation against ant attack. The wax covering may serve as a defence in several ways. Firstly, it may be difficult for ants to bite into. Secondly, as some of the waxes are sticky, it may cause ants to break off attacks to clean their mouthparts. Völkl and Vohland (1996) assessed the defensive efficiencies of wax coverings of two species of *Scymnus* experimentally. They showed that normal larvae (waxy) of *Scymnus nigrinus* and *Scymnus interruptus* were less likely to be killed by attacks from *Formica polyctena* and *L. niger*, respectively, than were larvae from which the wax covering had been removed. Thirdly, the wax may also provide a form of mimicry, eloquently demonstrated by Eisner et al. (1978) with *Chrysopa slossonae*, lacewing larvae, which disguise themselves as their woolly aphid prey, as protection from ants, by coating themselves in wax covering removed from the woolly aphids. Some ladybirds with wax coverings feed on mealy aphids, and look very similar to their prey. Consequently, mealybug-tending ants may ignore such coccinellid larvae because they do not recognise them as a threat to the mealybugs. For example, *Cryptolaemus montrouzieri* larvae were ignored by *Pheidole megacephala* when on colonies of waxy mealybugs tended by this ant, yet were attacked by the same ant when on tended colonies of the waxless

Coccus viridis (Bach, 1991). The wax coverings of coccinellids are secreted by the larvae themselves.

Some ladybird pupae also have wax coverings. *Scymnodes lividigaster* larvae smear wax onto the surface of the substrate before pupation (Richards, 1980b) and the pupa is both wax-covered and spiny. Richards (1980b) has suggested that both the wax smear and the pupal covering act to deter aphid-tending ants, although it is not clear whether any deterrent effect is due to the physical barrier posed by the pupal covering, the texture of the wax, its chemical composition, its colour, or a combination of these (Sloggett, 1998).

Pupae of *Rodatus major* have a very dense wax covering. This may provide both a physical defensive barrier against ant aggression and have a mimetic function. The species feeds mainly on eggs of the coccid *Monophlebulus pilosior*, and Richards (1985) has proposed that the wax covering gives *R. major* larvae a strong resemblance to the ovisac of this coccid.

The main physical defence of an adult ladybird against ants is its hard dorsal surface. Coupled with the clamping and rolling behaviours described previously, this provides a stout barrier to injury.

Chemical Defences

Coccinellids are well known for their bright colour patterns, advertising unpalatability. This unpalatability is largely chemical in nature. At the centre of coccinellid chemical defence lie a diverse array of alkaloids (Daloze et al., 1995) and pyrazines (Moore et al., 1990). Indeed, about 50 different alkaloids have been identified in ladybirds (Laurent et al., 2005). The alkaloids are produced in the haemolymph and distributed throughout the ladybird's body. Pyrazines, in contrast to alkaloids, are volatile and are thought to be involved in signalling the unpalatability of ladybirds and, as such, are absent from cryptically coloured ladybirds (Moore et al., 1990; Daloze et al., 1995).

Coccinellids can reflex bleed, which means they secrete a foul-smelling, distasteful fluid from the tibiofemoral joints of adults or the dorsal surface of larvae and pupae. The reflex blood of many ladybirds is thought to be distasteful to ants (Happ and Eisner, 1961; Pasteels et al., 1973; Sloggett, 1998). Moreover, it appears that ants contaminated by reflex blood have reduced mobility caused by the reflex blood as it dries out (Happ and Eisner, 1961; Bradley, 1973; Bhatkar, 1982). How much adult ladybirds reflex bleed when attacked by ants varies among species and circumstances. Some studies found that ladybirds rarely reflex bleed, even under sustained attack by ants (Jiggins et al., 1993; Marples, 1993), while others have noted ladybirds reflex bleeding readily when attacked (Banks, 1962; Bhatkar, 1982). Majerus (1994a) argued that reflex bleeding is a last defence against ants, used mainly when other defences, including fleeing, have been unsuccessful. Reflex blood is energetically costly (de Jong et al., 1991; Holloway et al., 1991, 1993), and thus is only deployed when the ladybird is in severe jeopardy and other strategies have failed.

Sloggett (1998) has argued that the reluctance of adult coccinellids to reflex bleed, except as a last resort, indicates that reflex bleeding did not evolve initially as a defence against ants. He notes that predatory ladybirds usually encounter aphid-tending ants when untended aphids are rare. At such times, ladybirds are likely to have low energy reserves and hence reflex bleeding would incur a high cost. It is notable that the phytophagous coccinellid *Epilachna varivestis* reflex bleeds readily when attacked by ants (Happ and Eisner, 1961). This species synthesises a complex array of defensive alkaloids (Eisner et al., 1986; Attygalle et al., 1993; Proksch et al., 1993; Radford et al., 1997; Shi et al., 1997). Sloggett suggests that this complexity may result from its plant diet, because, unlike hemipteran predators, it will rarely be food limited and so can devote more resources to its chemical defence.

Ladybird larvae reflex bleed much more readily than adults when attacked by ants (El-Ziady and Kennedy, 1956; Happ and Eisner, 1961; Bradley, 1973; Sloggett, 1998). Owing to their softer exoskeleton, larvae are at greater risk of suffering injury than are adults. Moreover, as larvae usually occur at times of prey abundance, they are less resource limited than are adult ladybirds (Sloggett, 1998).

Ant aggression probably played little role in the initial evolution of reflex bleeding in coccinellids. However, it may have a role in shaping the exact balance of defensive traits of ladybirds to their various enemies. In species of coccinellid that frequently encounter ants, relatively more resources may be devoted to defences against ants than in species that rarely interact with ants. In addition, ants may reduce the density of potential coccinellid predators and parasitoids occurring within ant forage ranges, producing enemy-free space (Jeffries and Lawton, 1984). If so, ladybird species that commonly co-occur with ants, including myrmecophilous species, may invest fewer resources in defences against potential predators and parasites that are excluded by ants, than do ladybirds that rarely occur with ants (Sloggett, 1998).

Coccinellid eggs and some pupae also have chemical defences. The eggs of some coccinellids are coated with defensive chemicals that are repellent to some predators including ants (Godeau, 1997; Sloggett, 1998). In addition, many aphidophagous ladybirds that lay eggs in batches contain defensive chemicals in the soma. These chemicals, once tasted by predators, including ants, deter further attack, so that while one egg may be destroyed, the rest of the clutch is then left alone (Dixon, 2000). Some coccinellid pupae (e.g. Chilocorinae) can reflex bleed and the reflex blood probably has some deterrent effect against ants. Pupae of *E. varivestis* have a covering of hairs. These hairs are glandular, each hair bearing a droplet of alkaloid at its end. This droplet is repellent to the ant *Leptothorax longispinosus* (Attygalle et al., 1993).

Myrmecophilous Ladybirds

A few coccinellids frequently live in association with ants and have been suggested to be myrmecophilous (Berti et al., 1983; Sloggett, 1998). These are listed in Table 7.1. In a few taxa, the case for myrmecophily is highly speculative. In the tribe Monocorynini, suggested myrmecophily is based only on antennal morphology, and there are no records of any of these ladybirds living with ants (Sloggett, 1998). In *Scymnus fenderi* and *Scymnus formicarius*, the possibility of myrmecophily derives from a few observations of adults and pupae with ants. The myrmecophily of *Hyperaspis acanthicola* is also questionable as it is based on reports of larvae living in hollow spines of acacia trees abandoned by ants. In all these cases, observations to confirm or refute myrmecophily are needed. Experimental work to determine whether these coccinellids gain benefit from any associations found with ants would be even more valuable.

The myrmecophily of the other seven taxa is more certain, but in some, little is known of how the ladybirds interact with ants. In *Brachiacantha 4-punctata*, *Brachiacantha ursina*, *Hyperaspis reppensis* and *Ortalia pallens*, myrmecophily appears to be limited to the larvae. Larvae of these species have been found in ants' nests feeding on ant-tended coccids or ant-tended fulgorids and, in the case of *O. pallens*, on the host ants. The predation of ants by *O. pallens* is probably the result of a dietary shift from homopteran prey once the species had become myrmecophilous (Sloggett, 1998). Both larvae and pupae of *Thalassa saginata* develop in *Dolichoderus bidens* nests (Berti et al., 1983). The larvae mimic cuticular lipids of the ants' brood, although whether these larvae feed on the brood or some other diet is unknown (Orivel et al., 2004).

In *C. magnifica* and *Platynaspis luteorubra*, myrmecophilous adaptations are found in larvae and some other life-history stages. Here myrmecophily has been well studied and findings shed light on the ecology and evolution of myrmecophily.

Coccinella magnifica

The myrmecophily of *C. magnifica* (Fig. 3.12) is well established (Sloggett et al., 2002 and references therein). *Coccinella magnifica* is distributed across Europe and Asia, with its local distribution resulting from its association with ants.

Table 7.1 Coccinellids that have been Suggested as being Myrmecophilous (Adapted from Sloggett, 1998 and Ceryngier et al., 2012)

Coccinellid	Associated ant(s)	Evidence for myrmecophily	References
Subfam.: Coccidulinae			
Rodatus major	*Iridomyr* ex sp.	Larvae feeding inside Coccoidea ovisac; larvae covered in wax, which at pre-pupa resembles Coccoidea ovisac. Larvae exhibit defensive behaviours	Richards, 1985
Azya orbigera	*Azteca instabilis*	Waxy filaments on larvae prevent ant attack	Liere and Perfecto, 2008
Subfam.: Scymninae			
Scymnus fenderi	*Pogonomyrmex subnitidus*	One adult recorded from *P. subnitidus* nest. Ant is gramnivorous and does not tend Homoptera. Myrmecophily unproven	MacKay, 1983; Hölldobler and Wilson, 1990
Scymnus formicarius	*Formica rufa*	Little known. Adults apparently found with ants	Wasmann, 1894
Scymnus interruptus	*Lasius niger*	Potentially chemical and behavioural protection	Godeau et al., 2009
Scymnus nigrinus	*Formica polyctena*	Wax covering of larvae repels ants	Volkl and Vohland, 1996
Scymnus (Pullus) posticalis	*Lasius niger*, *Pristomyrmex pungens*	Larvae ignored by ants	Kaneko, 2002

Brachiacantha 4-punctata	*Lasius umbratus Formica subpolita (=F. camponoticeps)*	Waxy larvae prey upon tended coccids and adelgids within ant nests. Closely related species are probably also myrmecophilous. Other ant hosts are probable	Mann, 1911; Wheeler, 1911; Gordon, 1985; Montgomery and Goodrich, 2002
Brachiacantha ursina	*Lasius* spp.	Probably the same behaviour as *B. quadripunctata*	Smith, 1886; Montgomery and Goodrich, 2002
Hyperaspis reppensis	*Tapinoma nigerrimum*	Larvae feed on ant-tended fulgorids in ants' nests. Adults are attacked by ants	Silvestri, 1903
Hyperaspis acanthicola	*Pseudomyrmex ferruginea*	Larvae found in hollow spines of *Acacia* spp. abandoned by ants. Myrmecophily unproven	Chapin, 1966
Hyperaspis conviva	*Formica obscuripes*	Wax cover of larvae	Bradley, 1973
Ortalia pallens	*Pheidole punctulata*	Myrmecophilous larvae feed on ants. Adult habits unknown	Harris, 1921
Thalassa saginata	*Hypoclinea bidens*	Pupae found with ants. Chemical mimicry demonstrated for larvae, pupae and adults. Production of chemical attractant. Myrmecophily probable. Diet unknown, hypothesis of ant brood predation by larvae and/or adults	Berti et al., 1983; Corbara et al., 1999; Orivel et al., 2004
Subfam.: Ortaliinae			
Apolinus lividigaster	*Crematogaster* sp., *Paratrechina* sp.	Long projections and wax cover on body of larvae and pupae, defensive behaviour of larvae, wax around pupae	Richards, 1980b

Table 7.1 *(cont.)*

Coccinellid	Associated ant(s)	Evidence for myrmecophily	References
Subfam.: Chilocorinae			
Platynaspis luteorubra	*Lasius niger* *Myrmica rugulosa* *Tetramorium caespitum*	Multiply recorded with a variety of ant species. Larvae and pupae show myrmecophilous morphology	Pontin, 1959; Majerus, 1994a; Völkl, 1995
Phymatosternus lewisii	*Pristomyrex pungens*	Coccid-like larvae	Kaneko, 2007
Subfam.: Coccinellidae			
Coccinella magnifica	*Formica rufa* group	All stages found with ants	Wasmann, 1912; Donisthorpe, 1919–1920; Majerus, 1989; Sloggett, 1998; Sloggett et al., 1998; Sloggett and Majerus, 2003

Fig. 7.10. Wood ant nest.

In north-western Europe, where it has been most studied, it is restricted to habitats in which ants of the *F. rufa* group forage.

Donisthorpe (1919–1920) placed *C. magnifica* and *C. 7-punctata* on *F. rufa* nests (Fig. 7.10), and found that the former were only slightly attacked, while the latter were 'vigorously assailed'. Pontin (1959) and Majerus (1989) have made similar observations. Indeed, Majerus (1994a) notes that when *C. magnifica* and other species of ladybird were released onto *F. rufa* nests, all the other species were strongly attacked by the ants, but *C. magnifica* was largely ignored. Both larvae and adult *C. magnifica* were treated alike by the ants, appearing to share the same immunity. Donisthorpe (1919–1920) observed that adult *C. magnifica* reflex bled freely when attacked by ants. However, others have failed to make similar observations and have argued that the immunity of *C. magnifica* to ant attacks is not a consequence of reflex bleeding (Majerus, 1989; Jiggins et al., 1993).

Several hypotheses have been proposed to explain the low levels of aggression of ants towards *C. magnifica*. *Coccinella magnifica* may secrete a pheromone that placates ants by advertising distastefulness or toxicity (Majerus, 1989). Alternatively it may secrete chemicals that mimic the ants' own scent, or the odour of aphids (Majerus, 1989). A third hypothesis is that *C. magnifica* produces a chemical that is toxic to ants (Donisthorpe 1919–1920) or is at least a deterrent (Sloggett, 1998).

To test these various hypotheses, detailed field and laboratory studies of the interactions of *C. magnifica* and other ladybirds that occur with *F. rufa* in conifer and mixed woodland were undertaken (Sloggett *et al.*, 1998; Sloggett and Majerus, 2003). In tests in which *C. magnifica* and *C. 7-punctata* were introduced onto *F. rufa* trails, *C. magnifica* was attacked occasionally, but very much less than *C. 7-punctata*. This observation has special importance as Godeau et al. (2003) have shown that *C. magnifica* follows ant trails to locate aphid colonies. Moreover, when Sloggett and Majerus (2003) introduced the two *Coccinella* species onto ant-tended aphid colonies, *C. magnifica* remained on the colony longer and were more successful in feeding on aphids than *C. 7-punctata*. Although both species were attacked, the degree of aggression towards *C. magnifica* was much less than towards *C. 7-punctata*. When attacked, *C. 7-punctata* dropped off plants or flew away more often than *C. magnifica*. *Coccinella 7-punctata* adults occasionally reflex bled while *C. magnifica* did not. Larvae of both species reflex bled when attacked by ants. However, *C. magnifica* larvae produced less reflex fluid than *C. 7-punctata*, yet the fluid that was produced had a greater deterrent effect on the ants (Sloggett and Majerus, 2003). *Coccinella magnifica* larvae were also often found in situations that minimised ant aggression towards them, feeding on aphids dislodged from colonies. Finally, it is notable that none of the defensive behaviours of *C. magnifica* to *F. rufa* were unique to *C. magnifica*. All also occurred in *C. 7-punctata*, but the two species used the various behaviours in the repertoire to differing degrees, suggesting that *C. magnifica*'s defences against ants may have evolved by gradual changes of *C. 7-punctata* behaviours (Majerus et al., 2007).

Sloggett (1998) also investigated the relative chemistries of *C. magnifica* and *C. 7-punctata*. He showed that dead *C. 7-punctata* were more often attacked on ant trails than were *C. magnifica*, irrespective of whether whole corpses, corpses without elytra or wings, or just elytra were used. From these observations, he deduced that the low level of aggression shown by ants to *C. magnifica* has a chemical rather than a behavioural basis. Analysis of the cuticular lipids of the two species showed little difference between them, and little similarity to the surface lipids of *F. rufa* (G. Lognay, J.J. Sloggett and J.-L. Hemptinne in Sloggett, 1998), suggesting that *C. magnifica*'s immunity to ant attacks is not due to chemical mimicry of the ants. Moreover, the similarity in the cuticular lipids of the two species make it unlikely that *C. magnifica* gains immunity by mimicking some other element of the habitat. Transfer experiments showed that *C. magnifica*'s defence is not specific to a particular *F. rufa* nest, or indeed to just *F. rufa* (Sloggett, 1998). Sloggett concluded that *C. magnifica*'s defence against ants is probably based on repellent chemistry involving alkaloids, or possibly pyrazines. Interestingly, while the predominant alkaloids synthesised by most *Coccinella* species are coccinelline and precoccinelline, those of *C. magnifica* are hippodamine and convergine (Dixon, 2000; Sloggett, 2005), and convergine is more repellent to ants than is coccinelline (Pasteels et al., 1973).

Platynaspis luteorubra

The myrmecophily of *P. luteorubra* is suggested by the morphology of its larvae (Fig. 7.11) and pupae, which have shapes unlike those of most ladybirds, but similar to other myrmecophilous larvae (e.g. some lycaenid butterflies and hoverflies of the genus *Microdon*). In addition, larvae of *P. luteorubra* have been

Fig. 7.11. *Platynaspis luteorubra* larva with aphids and ant.

recorded with various ant species that tend aphids, including *L. niger*, *Myrmica* spp. and *Tetramorium caespitum*, in both underground galleries and on plants (Pontin, 1959; Völkl, 1995; Godeau, 2000).

Völkl (1995) has demonstrated that *P. luteorubra* is a true myrmecophile. Intensive field studies on a variety of plants showed that *P. luteorubra* larvae were found significantly more in ant-tended colonies than in untended colonies. The species has various morphological and behavioural adaptations to life with ants, thereby allowing them access to ant-tended resources. Ants do not recognise larvae of *P. luteorubra* as a threat to aphids. This may be a result of the larva's unusual shape and its slow movements. Völkl (1995) also assumed that the larvae possesses 'camouflage' chemicals. This was confirmed by transfer experiments in Germany. When larvae were moved between colonies of the aphids, *Aphis fabae* and *Metopeurum fuscoviride*, the response of ants towards them varied. Larvae moved to a new colony of conspecific aphids were not attacked, while those moved to a colony of the other species were (Oczenascheck, 1997). Analysis of the cuticular lipids of the larvae showed that these were similar, both in type and quantity, to those of their prey. As the cuticular lipids of the two species of aphid differ, a change in prey altered the cuticular lipids and therefore the efficiency of the larvae's chemical mimicry. This is a very efficient form of scent mimicry because the larvae do not have to manufacture different cocktails of mimetic chemicals when feeding on different prey species.

Völkl (1995) also found that *P. luteorubra* pupae were often attacked by *L. niger*, but that their dense covering of long hairs protected them from injury. Here then, the chemical mimicry of the larvae appears not to be carried across into the pupal stage. This may be because alcohols rather than lipids dominate pupal cuticular compounds.

Adult *P. luteorubra* do not appear to have chemical defences against ants. In Belgium, adult *P. luteorubra* often feed on *A. fabae* tended by *L. niger* on *Cirsium arvense*. These adults are often attacked by *L. niger* and respond either by fleeing or by clamping down on to the substrate (Godeau, 2000).

The myrmecophily of *P. luteorubra* is adaptive. Larvae in ant-tended aphid colonies are more successful in capturing prey than those in unattended colonies, and adults from ant-tended colonies are larger than those from untended colonies (Völkl, 1995).

There are striking differences in the adaptations that the two best-studied myrmecophilous coccinellids have evolved for life with ants. *Platynaspis luteorubra* larvae chemically mimic aphids, in effect sequestering mimetic chemicals from their prey. *Coccinella magnifica* uses ant-repellent chemicals, and physical and behavioural defences. Despite these differences, the main reason for the myrmecophily of both species is the same: to enable them to feed on ant-tended aphids when other aphids are scarce. Comparative work on the myrmecophily of these two species and other related non-myrmecophilous species, such as *C. 7-punctata*, has provided insights into the evolution of habitat preferences, as discussed in Chapter 3.

In summary, many predatory coccinellids encounter hemipteran-tending ants regularly because they both use resources provided by Hemiptera. Ants are therefore an important factor in the ecology of many ladybirds. However, caution should be taken when studying ladybird–ant interactions. Too often, the reactions of ladybirds to ants have been viewed in isolation. In reality, many of the defensive strategies employed by coccinellids against aggressive ants are merely adaptive modifications of general defences. Thus, it is those few species of ladybird that have the closest association with ants, the myrmecophilous ladybirds, which may be most instructive. In these species, the closeness of the association means that many of the behaviours of the coccinellids to ants have changed specifically because of the association. Studies of *C. magnifica* and *P. luteorubra* have already given insights into not only specific interspecies interactions, but also the roles of enemy-free space, resource utilisation, interspecific competition and the evolution of habitat specificity (Majerus et al., 2007). Certainly, other myrmecophilous coccinellids await discovery, particularly from the tropics. If other ladybirds with close associations with ants are identified, then close scrutiny will surely provide novel insights into a range of phenomena.

It is perhaps worth mentioning here, as a tangential note, that there are a very small number of anecdotal reports of ladybirds living with termites. To give one example, Pasteels (pers. comm.) has observed larvae of unidentified coccinellids feeding on termites within termite nests. The larvae attack the thoracic area of the termite (Fig. 7.12) and only detach when ready to ecdyse. For workers in the tropics, hunting for termitophilous coccinellids may prove fruitful.

Parasitoids and Parasites

Ladybirds are parasitised by a range of invertebrates, including fly (Diptera) and wasp (Hymenoptera) parasitoids, mites (Acarina), roundworms (Nematoda), protozoa and various pathogenic fungi, bacteria and viruses.

A parasitoid is defined as an organism (usually insect) that lays its eggs in or on another insect species, with the larva of the parasitoid developing inside its host, ultimately killing it. Parasitoids are represented by species from two orders of insects, the Diptera and the Hymenoptera. They may attack any stage in an insect's life history. In the ladybirds, egg, larval, pupal and adult

Fig. 7.12. Larva attacking thoracic area of termite.

parasitoids are known. That said, egg parasitoids are rare, and are only found in non-predatory ladybirds.

The lack of egg parasitoids of predatory ladybirds is probably the result of the sibling egg cannibalism practised by young hatchlings. As egg parasitoids of other groups usually hatch from their egg hosts after non-parasitised eggs have hatched, parasitoids of ladybird eggs would be devoured by hatchlings before completing their development. This speculation is supported by a few observations of egg parasitoids of phytophagous coccinellids that rarely indulge in sibling egg consumption.

Diptera

Several true flies (Diptera) parasitise ladybirds. The literature contains reports of the fly *Medina* (= *Degeeria*) *luctuosa* (Tachinidae) (Fig. 7.13), which is usually a parasitoid of chrysomelid beetles, parasitising ladybirds. In all, 17 species of coccinellid have been recorded as hosts to this fly. However, Belshaw (1993) argued that these records are probably all of another fly of the same genus, *Medina separata*. Female flies lay a single egg under an elytron of a ladybird, and the resulting larva burrows into the host, where it develops in the body cavity (Belshaw, 1993). Respiration is via a tube attached to the aperture through which the larva entered. In the final instar, the larva eats its host's vital organs, killing it, and exits through the abdominal wall, falling to the ground and pupating in the soil. It overwinters as a larva in the host. Levels of parasitism tend to be low, generally well below 10%, and are not thought to represent a significant mortality factor.

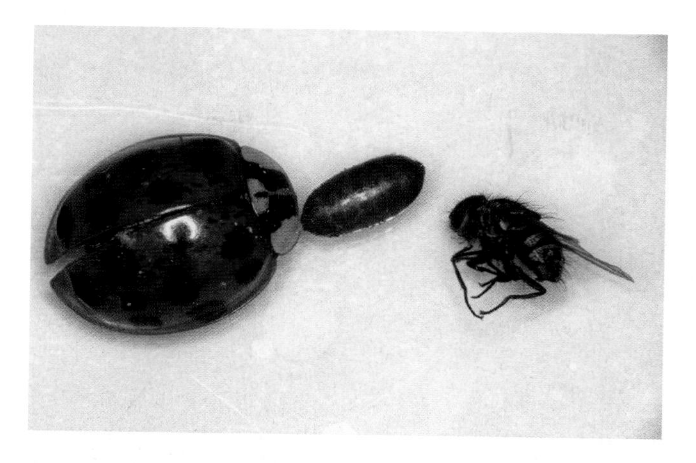

Fig. 7.13. *Harmonia axyridis* with pupa and adult of *Medina luctuosa* (Tachinidae).

Far more important are scuttle-flies of the genus *Phalacrotophora* (Phoridae). These flies are endoparasites of ladybird larvae, pre-pupae and pupae, mainly attacking members of the Coccinellinae and Chilocorinae in Europe and Asia, but not in America (Acorn, pers. comm.). They parasitise many different ladybirds (Disney et al., 1994; Majerus, 1994a; Ceryngier and Hodek, 1996; Ceryngier et al., 2012). The published host lists are unlikely to be exhaustive, and I suspect many species are only excluded because their pupae are rarely examined. These flies can cause significant ladybird mortality. Ford (1979) found 10 out of 12 pupae of *C. 7-punctata* to be parasitised by *Phalacrotophora fasciata*. Disney (1979) cited data from samples in which *Phalacrotophora berolinensis* parasitised 30% of *A. 2-punctata* pupae, and *P. fasciata* parasitised 13% of *A. 2-punctata* and 20% of *C. 7-punctata*.

The life histories and behaviours of the various scuttle-flies that attack ladybirds are similar. Females find a ladybird pre-pupa, tending to select one that is close to pupation (Hurst et al., 1998), and then attract one or more mates, probably by 'calling' pheromonally. The pairs mate, with little obvious courtship, on, or near to, the pre-pupa. Male competition or female mate choice behaviour may be significant, as instances of two or more males attending a female are not uncommon. Indeed cases of 'triplets', consisting of a female, mounted by a male, that was in turn mounted by a second male, have been observed. It is likely that females are polyandrous; a female *P. fasciata*, first observed ovipositing on an *A. 2-punctata* pupa, then moved to an adjacent pre-pupa, where she was mated by another male less than a minute later.

Once males withdraw from females, they dismount almost immediately and move away. The females stay by, or on, the pre-pupa, and usually remain inactive until the host's final larval skin begins to split, when oviposition

Fig. 7.14. *Phalacrotophora fasciata* egg-laying on ladybird pupa.

begins usually on the ventral thoracic region of the newly forming pupa. Eggs are laid on the surface of the exoskeleton, most often between the legs, or internally (Fig. 7.14). Data cited by Disney et al. (1994) suggest that neither *P. berolinensis* nor *P. fasciata* oviposits in active larvae of *A. 2-punctata* or *C. 7-punctata*. However, once larvae have become inactive in the pre-pupal phase, these species may be attacked. These observations can be contrasted with reports of flies of the related genus *Megaselia* that do attack active ladybird larvae (Howard and Landis, 1936; Le Pelley, 1959).

The number of fly larvae that emerge from a single coccinellid pupa is largely dependent on host size. In *A. 10-punctata*, the average is two, with a recorded maximum of six. In *A. 2-punctata*, these figures are three and six, respectively; in *Hippodamia variegata*, three and seven; in *C. 14-guttata*, four and seven; and in *C. 7-punctata* six and 30 (Filatova, 1974; Disney et al., 1994; Majerus, 1994a; Majerus, pers. obs.).

The size of fly puparia and resulting flies is strongly influenced by the numbers of maggots that emerged from a pupa. The size of the puparia and adult flies decrease as the number of developing maggots within a pupa increases (Smith, Disney and Majerus, unpubl. data). The high numbers of flies in some pupae are thought to be the result of several different females laying eggs on, or in, a single pupa. While there are no proven instances of multiple ovipositions by one species of fly into the same ladybird pupa, *A. 2-punctata* pupae have been recorded to yield larvae of both *P. fasciata* and *P. berolinensis*, indicating that multi-parasitism does occur (Disney, 1979; Disney et al., 1994).

The phorids use ladybird pupae as more than just a food source for their offspring. They also feed on them themselves. Martelli (1914) first recorded the adults of a phorid sucking fluid from a ladybird pupa. His observation of *P. fasciata* feeding from a *P. 22-punctata* pupa has been confirmed for *P. fasciata* with several other ladybird species, and for *P. berolinensis* feeding on *A. obliterata* pupae. Disney et al. (1994) described the behaviour of *P. fasciata* feeding on *C. 7-punctata*, and *P. berolinensis* feeding on *A. 2-punctata* pupae. The flies made an incision into the dorsal surface of the host's thorax, or anterior of the abdomen, and then sucked up the fluid issuing from the wound. Feeding lasted for nearly half an hour in one case. The ladybird pupae were still alive at this time, but all died subsequently, with fly maggots emerging from the majority of them, implying that the adult feeding behaviour does not have a strong adverse affect on the development of their progeny.

Coccinellid pre-pupae and pupae are not completely defenceless when attacked by phorids. Firstly, pre-pupae have the ability to reflex bleed, fluid being secreted from the dorsal surface of the abdominal segments. This defence is not very efficient against phorids unless the fluid gets onto their legs, in which case they often leave the pre-pupa to clean themselves, and rarely go back. A second common defence strategy is for pre-pupae and pupae to flick in response to the activities of phorids. This entails raising the anterior end sharply upwards and then dropping back down again (Fig. 7.15 and Fig. 7.16). Such behaviour is sometimes induced by phorids simply walking about on pupae, and invariably when the ovipositor is inserted (Fig. 7.17). In observations of pupal flicking, Disney et al. (1994) recorded that approximately 25% of parasite attacks were deterred by this behaviour, verifying that pupal flicking is an evolved defensive adaptation.

In almost all reports of these flies emerging from coccinellids, the appearance of a scuttle-fly maggot has been associated with the death of the host.

Fig. 7.15. Pupa of *Anatis ocellata*.

Fig. 7.16. Pupal flicking behaviour of *Anatis ocellata*.

Fig. 7.17. *Dinocampus coccinellae* ovipositing in *Coccinella 7-punctata.*

The only exception is of two *A. 2-punctata* pupae surviving parasitism by single *P. fasciata* larvae (Majerus et al., 2000a). In both cases, the larvae that emerged gave rise to abnormally small male flies.

Hymenoptera

The most widespread parasitoid of ladybirds among the Hymenoptera is the braconid wasp *Dinocampus coccinellae.* This wasp is parthenogenetic; viable eggs are laid without fertilisation and virtually all develop into daughters. Less than a dozen males (Fig. 7.18) have been recorded (Geoghegan et al., 1998a; Shaw et al., 1999). It is cosmopolitan in distribution. Ceryngier et al. (2012), reviewing the known host species for this wasp, included 59 species of coccinellid. This list is certainly not exhaustive. Sloggett and Majerus (2000c) reported the first record of *Myrrha 18-guttata* as a host of *D. coccinellae* (Fig. 7.19).

The wasps usually attack adult ladybirds (Geoghegan et al., 1998b), although there are also some records of larvae and pupae being attacked (David and Wilde, 1973; Sunderland, 1978; Obrycki et al., 1985). A female will locate a ladybird, initially approaching it with her ovipositor sheath pointing backwards. She taps the ladybird with her antennae a number of times, often changing antenna. Then, after moving a little back from the ladybird, she approaches again, with her abdomen curled under and forwards, so that the sheath points forwards between her legs. She then taps the ladybird again with her antennae and brings the tip of her ovipositor sheath to slide gently along the surface of the ladybird, gradually circling the ladybird. This antennal tapping and sheath sliding continues for several minutes, with the wasp

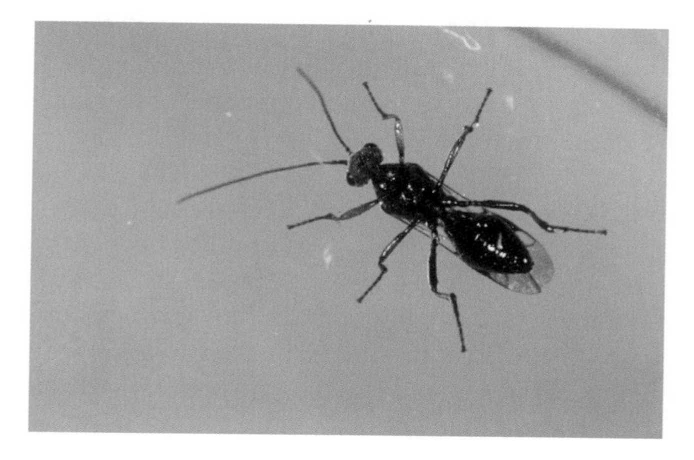

Fig. 7.18. Male *Dinocampus coccinellae*.

Fig. 7.19. *Myrrha 18-guttata* with *Dinocampus coccinellae* pupa.

walking on just her middle and back legs, the front legs being held upwards due to the position of the ovipositor sheath. It is assumed that the sliding of the sheath over the surface of the ladybird enables the wasp to locate a weak point in the hard exoskeleton of the beetle. The sheath and ovipositor is finally inserted with a sharp thrust, or series of thrusts, usually through the joints between head and thorax, or pronotum and elytra, or between abdominal segments (Fig. 7.17). The thrust of the ovipositor may be strong enough to rock the ladybird from side to side. A single, rather elongate egg is laid.

The ladybird usually appears more or less oblivious to the wasp until the moment of oviposition, at which point it reacts by running away.

Once laid, the egg grows in size rapidly for about three days, and hatches after about five days. When the egg hatches, a young larva emerges together with a dozen or so clumps of 30–100 trophic cells. These cells then separate into spherical cells covered with microvilli. They are very active metabolically, and swell by absorbing and storing nutrients from the host. Although a female will only lay one egg in a ladybird, several wasps may lay in the same host. However, the resources in a ladybird appear only sufficient for the development of one wasp. In consequence, the first instar larva has very large curved mandibles and, in cases of super-parasitism, the first larva to hatch kills other eggs or hatching larvae, so that, although up to 47 eggs have been found in a single *C. 7-punctata* (Maeta, 1969), only one will survive to the second instar. The first instar larva feeds on the fat body in the host. Thereafter, the larva feeds on the swollen trophic cells, most of which are consumed during the larva's development. The larva passes through three larval instars (Kadono-Okuda et al., 1995). It has little effect on most of the tissues of its host, but development of the ovaries and testes is regressed, and the structure of the fat body is altered, due to nutrients being diverted elsewhere.

Insights into the behaviour of the wasp larva inside its host have been made by use of technology more normally associated with human brain and body scans. A team in Dundee has used magnetic resonance microscopy (MRM) to study the larva and changes in the internal anatomy and body chemistry of hosts. This technique does not require either the host or the larva to be killed, so interactions between parasitoid and host can be monitored throughout the wasp's development (Chudek et al., 1998). This technique showed that infection of *C. 7-punctata* by *D. coccinellae* causes changes in the water and lipid contents of the ladybird and regression of ovaries similar to those observed when ladybirds are fed on a standard artificial diet rather than aphids. These changes also show broad similarity to changes observed during preparation for dormancy (Geoghegan et al., 2000). Hopefully this technique will lead to further advances. In particular, MRM should allow real-time study of the effects of super-parasitism in this system and other host–parasitoid relationships.

When fully fed, the larva is reputed to attack its host's leg motor neurons, thereby effectively immobilising the ladybird. Certainly the host becomes immobile some half-hour or so before the larva eats its way out, the only movement being occasional jerks or tremblings of the ladybird's body and appendages.

The larva escapes from its host head first, by burrowing through the thinner cuticle between two of the ventral abdominal plates (Fig. 7.20). The larva then

Fig. 7.20. Emergence of *Dinocampus coccinellae* larva from *Coccinella 7-punctata.*

moves around and orientates itself to lie between the ladybird's legs. It then constructs its cocoon, initially producing a small base pad of silk on the substrate and then attaching strands to the legs of the ladybird. Some strands are attached to the substrate, two or three millimetres beyond the ladybird's legs, thereby fastening the ladybird down more firmly. Once the outlying net of loose silk is complete, the larva forms its cocoon with back-and-forth and side-to-side movements (Majerus, 1991c).

The wasp emerges from the more pointed end of the cocoon. First indications are a small puncture. The mandibles become visible, and are used like a pair of scissors to cut around the end of the cocoon, leaving only a small area attached. The end is then pushed back as the wasp exits, leaving the cocoon like a 'pitcher with a hinged cap'. The wasp moves away from the cocoon to expand and dry its wings, and to clean its appendages, before taking flight.

Amazingly, the host usually remains alive until the wasp emerges. Not only does it remain alive, but it also has the ability to reflex bleed if attacked by a predator. The wasp pupa is thus effectively protected, at least from some predators, by the warning colouration and the chemical deterrents in reflex blood of the very insect that is own activities will ultimately have killed. Most ladybirds parasitised by *D. coccinellae* starve to death, or succumb to fungal disease, usually within a week of the emergence of the parasite (Fig. 7.21). Even when adults have an empty cocoon carefully removed from between their legs, and regain at least some use of their legs, they are so weakened that they usually die within a week or two (Sloggett, unpubl. data).

Emerging wasps are fully mature and are able to search for ladybird victims as soon as their wings are dry. Several generations are produced in a year, the number depending on availability of hosts and length of climatically suitable

Fig. 7.21. Dead *Coccinella 11-punctata* with *Dinocampus coccinellae* pupa.

periods for development. *Dinocampus coccinellae* passes the winter as an egg or first instar larva within the ladybird and remains dormant until the host begins to feed in the spring, whereupon the larva becomes active and starts to develop.

Some coccinellids are more heavily parasitised than others. This is undoubtedly a consequence of both the suitability of different coccinellids as hosts and to active host selection by the wasps. For example, in England, *C. 7-punctata, C. 11-punctata* and *H. 4-punctata* typically have prevalence levels of 10%–30%, while species that appear comparable in size, such as *C. magnifica, A. ocellata, C. 14-guttata* and *Coccinella 5-punctata*, have prevalence levels of only a few per cent (Majerus, 1997c). In some years, however, prevalence levels may become highly inflated, reaching as high as 68% (Geoghegan et al., 1997). Interestingly, parasitism of the alien *H. axyridis* within its invaded range is very much lower than that of native species (Comont et al., 2014; Roy et al., 2016).

Levels of parasitisation are higher in samples from overwintering beetles than in active ones (Iperti, 1964; Parker et al., 1977). A number of workers have found that female hosts are more heavily affected than males (Maeta, 1969; Parker et al., 1977; Cartwright et al., 1982). This is a consequence of active host selection by the wasps, which preferentially lay in female ladybirds when offered a choice (Davis, D.S. et al., 2006). This choice is adaptive because female ladybirds, in addition to being slightly larger, on average, than males, also consume more food than males. The wasp also shows adaptive preferences to oviposit into adult rather than pre-imaginal ladybirds (Geoghegan et al., 1998b) and for young rather than old adults (Majerus et al., 2000b).

Interestingly, I have also recorded a *Dinocampus*-like wasp from *T. 16-punctata* (Fig. 7.22). The wasps were morphologically identical to *D. coccinellae,*

Fig. 7.22. *Dinocampus*-like wasp pupa with *Tytthaspis 16-punctata*.

but about half the size of *D. coccinellae* that have developed in *C. 7-punctata* or *H. 4-punctata*. Presumably, it is the same species, its small size simply being a consequence of developing within a very small host.

Several types of chalcid wasp also parasitise coccinellids, including members of the genera *Homalotylus*, *Tetrastichus*, *Oomyzus*, *Aprostocetus*, *Baryscapus* and *Pediobius*. In general, these are small wasps, and unlike *D. coccinellae*, many wasps may develop in a single host.

Homalotylus flaminius attacks members of the tribe Scymnini (Klausnitzer, 1992); *Homalotylus eytelweini* parasitises members of the Coccinellinae and Chilocorinae (Majerus, 1994a), while *Homalotylus platynaspidis* has, as yet, only been reared from *P. luteorubra* (Völkl, 1995).

Eggs are laid in coccinellid larvae, usually while the larva is ecdysing. Larvae parasitised by some species of *Homalotylus* may be recognised because several days after being attacked, they attach to the substrate, swell, darken, and become hard. However, *H. platynaspidis* does not complete its development until its host reaches the pupal, or occasionally the pre-pupal stage (Völkl, 1995).

Where several wasps develop in a single host, each makes its own chamber, using host tissues to divide itself off from others. The wasps emerge from their host as adults, each one making a small round exit hole. The number of wasps that emerge from a host again appears to be determined by host size, with the greatest number having been reported from *Anatis 15-punctata* (Kulman, 1971). *Homalotylus* appear to pass through the winter, or other unfavourable periods, as final instar larvae or pre-pupae. All species appear to be essentially multivoltine, and the prevalence levels increase through the year, but then decline over the winter. Maximum prevalence levels for *Homalotylus* species within a year are variable, and generally lower in temperate or cooler climates. In Mediterranean, subtropical and tropical climates, these wasps may have a significant impact on coccinellid populations. In America, 13% of *Coccinella californica* larvae were found to be parasitised (Frazer and Ives, 1976), while the level in *A. 15-punctata* was twice as high at 26% (Kulman, 1971). In other parts of the world, parasitisation rates by *Homalotylus* may be very much higher. For example, Iperti (1964) found 80% of *C. 7-punctata* larvae infected in southern France. Rubtsov (1954) reported over 90% of *Chilocorus 2-pustulatus* to be affected, while Yinon (1969) noted prevalence levels approaching 100% for *Rodolia cardinalis* in India.

Völkl (1995) reported that infestation rates of *P. luteorubra* by *H. platynaspidis* may be over 50%. Interestingly, Klausnitzer (1969) and Völkl (1995) noted *H. platynaspidis* to be the only known parasitoid of this coccinellid. The close relationship between these two insects may be a consequence of the coccinellid's myrmecophilous habit. Other potential parasitoids may be deterred by the presence of ants. Yet Völkl reported *H. platynaspidis* not to be unduly disturbed by attendant ants, although the prevalence levels of ladybirds in untended aphid colonies was sometimes higher than among tended aphids colonies. Although ants do not harm *H. platynaspidis* directly, they may reduce the rate at which it attacks *P. luteorubra*, simply because the ants' movements around and over an aphid colony may disturb the wasp.

From the Tetrastichinae, *Oomyzus scaposus* has been recorded from the Chilocorini, Coccinellini and Scymnini; *Aprostocetus neglectus* is found in the Chilocorini and Scymnini, while *Tetrastichus epilachnae* appears restricted to the subfamily Epilachninae. *Baryscapus thanasini* and *Oomyzus sempronius* usually parasitise larvae of members of the lacewing genus *Chrysoperla*, or clerid beetle larvae, although *C. 2-pustulatus* may also be infested. Finally, *Oomyzus ovulorum* and *Tetrastichus ovicida* are egg parasitoids of phytophagous ladybirds.

All these species of Tetrastichinae are gregarious parasites, with up to 40 adults emerging from *C. 7-punctata* and *A. ocellata*, and up to 15 from *A. 2-punctata*, *Coccinella 5-punctata* and *Chilocorus* species. Females usually lay eggs in third and fourth instar larvae, less commonly in pupae. The wasp larvae kill their hosts fairly soon after they have hatched and feed in the mummified corpse. All species are multivoltine, and larvae of the last generation in the year pass through the winter in the mummified host, emerging the following spring. All the wasps from one host use the same exit hole, whether they emerge from larvae or pupae (Filatova, 1974). Klausnitzer (1969) noted that all parasites emerging from an individual host are of the same sex, a possibility due to the female's control over progenic sex ratio: fertilised eggs become females, while unfertilised eggs become male. Females are reproductively mature as soon as they emerge, and because development time may be very short, as little as 12 days according to Telenga (1948), up to seven generations per year may be produced.

Finally, the chalcid wasp *Pediobius foveolatus* attacks larvae of the Epilachninae, mainly in tropical regions. The biology of this species is similar to that of other chalcid parasitoids of coccinellids. They preferentially attack third or fourth instar larvae, are multivoltine and many wasps may emerge from the same host. When this occurs, the sex ratio is always strongly female biased, as it often is in cases of hymenopterous super-parasitism (Majerus, 2003). In America, attempts have been made to use this wasp to control the Mexican bean beetle, *E. varivestis*. However, success has been limited due to winter mortality of the wasps (Stevens et al., 1975; Schaefer et al., 1983).

Mites

The parasites of ladybirds can broadly be split into three groups, ectoparasites, endoparasites and microbial pathogens. Of the ectoparasites, only mites (Acarina) have been much studied. The mites that may be found on coccinellids are divisible into those that are parasitic and those that simply use ladybirds as a means of transport. This latter, phoretic group includes species in the families Hemisarcoptidae, Winterschmidtiidae and Acaridae. These mites prey on coccids and similar insects. The hypopus stage does not feed, but attaches itself to the outer, usually ventral surface of an arthropod and hitchhikes to new host plants and prey colonies. Mites of the genus *Hemisarcoptes* are important predators of diaspidid scale insects (Gerson et al., 1990; Izraylevich and Gerson, 1993; Ji et al., 1994), and are generally vectored between prey colonies by adult ladybirds of the genus *Chilocorus* that also prey on these coccids (Houck and O'Connor, 1991).

In three very limited surveys of the mite fauna of ladybird samples from southern England, Holland and Belgium in the early 1990s, four new species of hemisarcoptid mite were discovered (Fain et al., 1995, 1997). This suggests that many new species of these mites that are phoretic on ladybirds await discovery.

Although not parasitic on coccinellids, the phoresy of these mites on coccinellids is important in biological control. As species of *Chilocorus* are used as biological control agents against scale insects, their efficiency in control programmes may be increased by ensuring that released beetles bear hypopi of mites that will also attack the scale insects.

A variety of truly parasitic mites have also been recorded from ladybirds. These include some species, such as *Leptus ignotus* (Prostigmata: Erythraeidae), that parasitise a wide variety of arthropods (Hurst et al., 1997a), and mites of the genus *Coccipolipus* (Podapolipidae) that specialise on coccinellids.

Coccipolipus*: A Sexually Transmitted Disease-causing Agent of Ladybirds*

The mite genus *Coccipolipus* comprises over a dozen species, all of which have been found on coccinellids (Ceryngier and Hodek, 1996). Although most of these species are tropical, the best researched is the widely distributed species *Coccipolipus hippodamiae*. This mite has been recorded from the United States, Russia, central and eastern Europe, and Zaire (McDaniel and Morrill, 1969; Husband, 1981; Majerus, 1994a; Hurst et al., 1995; Webberley et al., 2002; Ceryngier et al., 2012). Work on the mite in Europe has shown that this species may be regarded as the causative agent of a sexually transmitted disease.

Coccipolipid mites live under the elytra of ladybirds, usually with their mouthparts embedded into the elytra (Fig. 7.23), but occasionally into the dorsal surface of the abdomen. They feed on haemolymph. Female *C. hippodamiae* lay eggs that hatch into mobile larvae. These larvae do not feed initially, but migrate to the posterior of their host and wait until the ladybird next mates. They then migrate to their original host's mating partner while the two are *in copula*. Thereafter, the larvae metamorphose into adults, embed their mouthparts into their new host, and begin to feed (Majerus, 1994a; Hurst et al., 1995). The mite is highly detrimental to female ladybirds. Infection leads to a reduction in fecundity of about 25%. More crucial is the effect

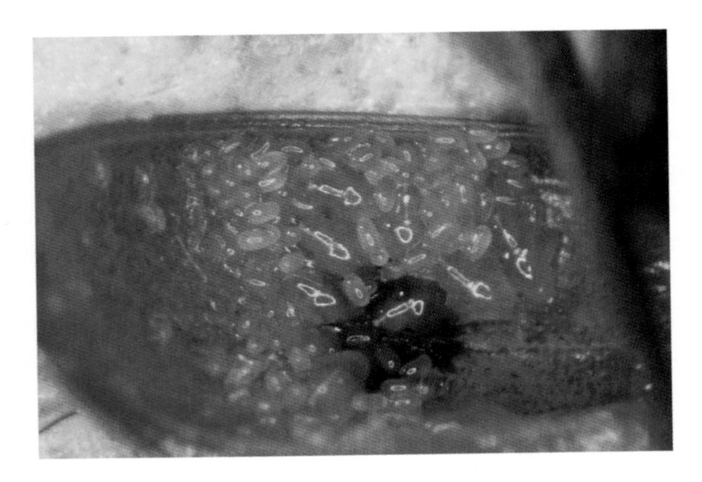

Fig. 7.23. Coccipolipid mites under the elytra of *Adalia 2-punctata*.

infection has on the viability of eggs that are laid. Egg hatch rates begin to decline soon after a female has become infected, and in less than three weeks is close to zero (Majerus, 1994a; Hurst et al., 1995).

Coccipolipus hippodamiae has been recorded from a variety of ladybird species, being particularly common on *A. 2-punctata* and *A. 10-punctata*. The highest prevalence has been recorded from *A. 2-punctata*, in which up to 90% of some populations have been found to be infected by the latter part of the breeding season, a consequence of repeated matings (Webberley et al., 2006a). The mite occurs commonly on *A. 2-punctata* over most of Europe, but is lacking from British populations, coastal areas of north-west continental Europe, and is scarce in most of Scandinavia (Majerus, 1994a; Webberley et al., 2006b). This is because transmission from one generation to the next requires some mating between generations, and in some years in north-west Europe; the old generation has died before the new generation is reproductively mature.

The only ladybird from which *C. hippodamiae* has been recorded in Britain is *C. magnifica* (Sloggett, pers. comm.). The presence of the mite on this species is probably the result of the extended longevity of this ladybird, promoted by its habit of feeding on ant-tended aphids. The long survival of overwintered adults means that some of these adults always survive until the first of the next generation have eclosed, allowing some intergeneration mating each year, and transmission of the mite down host generations (Sloggett, pers. comm.).

The mating behaviour of hosts has a critical effect on the population dynamics of *C. hippodamiae*. The pattern of increase in the prevalence of the mite in *A. 2-punctata* populations through the year is a direct consequence of mating rate and the extent that matings occur between individuals of successive generations. Furthermore, variation in the prevalence of the mite on four species of coccinellids has been directly attributed to variation in the promiscuity of the different hosts (Webberley et al., 2004). The highest prevalence is seen in *A. 2-punctata*, which is more promiscuous than the less commonly infected *A. 10-punctata* and *Oenopia conglobata*. The lowest prevalence was recorded in *C. 14-guttata*. A high proportion of individuals of this species in Europe have a requirement for an overwintering diapause before they mate, with the result that intergeneration matings in this species, although not unknown, are rare. This may also be the reason for low prevalence of *C. hippodamiae* in *C. 7-punctata*.

Despite the unusual severity of this mite on the reproductive output of females of the species it infects, its hosts do not appear to have evolved resistance to it, nor do females show any preference to mate with uninfected rather than infected males (Webberley et al., 2002).

Studies of the epidemiology of *C. hippodamiae* on its coccinellid hosts have greater importance than merely the study of a disease agent of beneficial insects, important though those are. Very few sexually transmitted diseases of invertebrates are known and of those that are, none are as amenable to study as that caused by *C. hippodamiae*. Consequently, this association is of interest because it provides an out-group with which to test models of factors, such as promiscuity, population age structure, generation time and rate of transmission, on the evolution, ecology and epidemiology of sexually transmitted diseases in vertebrates, including humans.

Little work has been carried out on the interactions between other species of *Coccipolipus* and their hosts. However, they are probably also transmitted during host mating, and are truly parasitic. Indeed, Schroder (1982) reports that *Coccipolipus epilachnae* reduces fecundity of *E. varivestis* by two-thirds and increases mortality by 40%. Consequently, *C. epilachnae* has been introduced into the United States to control this coccinellid, which is a pest of soybean.

Worms

Thus far, apart from a single report of a primitive, thread-like worm of the genus *Nematomorpha* attacking *Coccinella transversalis* (Anderson et al., 1986), only nematode worms have been recorded as endoparasites of ladybirds. The parasitic nematodes of ladybirds belong to two families, the Allantonematidae and the Mermithidae. From the former family, *Parasitilenchus coccinellinae* has been reported from *P. 14-punctata, A. 2-punctata, Hippodamia variegata, Oenopia conglobata* and *Ceratomegilla 11-notata* in Europe (Iperti, 1964; Iperti and van Waerebeke, 1968), and *Cheilomenes 6-maculatus* and *Illeis indica* in India. Over 100 adult female worms may infest a single ladybird with up to 10 000 larvae and young adults. These nematodes are not usually fatal, but retard maturation of the ovaries, consume host resources and have a general debilitating effect on their hosts. The method of transmission from one host to the next is unknown, but it may be sexually transmitted. Certainly, *P. coccinellae* is common in the reproductive organs of its host.

The immature stages of several members of the Mermithidae parasitise adult ladybirds, and occasionally larvae (Delucchi, 1953). Identification to species level is difficult from the juveniles found in coccinellids. Mermithids have been recovered from *H. variegata, C. 11-notata, P. 14-punctata* and *C. 7-punctata* (Iperti, 1964; Rhamhalinghan, 1986). Only a single parasite has been found in a host. The parasites cause reduction in weight, reduce respiratory rate, the size of the fat body and reproductive organs, and ultimately cause damage to vital organs resulting in paralysis and death of the host.

Microbial Pathogens

Our knowledge of the Protozoa, fungi, viruses and bacteria that attack ladybirds is generally sparse and fragmentary.

Protozoa

Sporozoans of the suborder Microsporidia and of the family Gregarinidae are known to infest ladybirds (Iperti, 1964; Lipa and Sem'Yanov, 1967; Laudého et al., 1969; Ceryngier et al., 2012). Gregarinids are taken up as spores by either larvae or adult ladybirds with their food. The spores then give rise to eight sporozoites, which live in the alimentary canal, attached to epithelial cells, and destroy the intestinal cells. These then metamorphose into trophozoites that live free in the gut and may indulge in sex to produce a binucleate gametocyst. The nuclei undergo multiple divisions to produce gametes that unite to produce zygotes. These develop into spores that are released from the host's gut to be taken up orally by new hosts. These sporozoans have a general debilitating effect on their hosts, reducing longevity and fecundity (Laudého et al., 1969), and may cause death if the gut becomes blocked by gametocysts.

Fungi

Fungal infection can be a significant mortality factor, particularly where ladybirds form large aggregations to aestivate or to overwinter. Members of the genus *Beauveria* (Ascomycota) are perhaps the most virulent, with mortality rates of up to 34% in some aggregations of *C. 11-notata* in France (Iperti, 1964). Iperti (1966b) found that infection rate in overwintering *C. 11-notata* was related to altitude. Beetles overwintering at high altitude were less likely to suffer from fungal attack than those at lower altitudes.

Beauveria bassiana, a ubiquitous fungus found in soil and on vegetation (Ormond et al., 2010), is the most commonly recorded species, and has been recovered from 17 species of ladybird, most reports being from overwintering hosts. A ladybird that has succumbed to *Beauveria* is recognisable on dissection, because the body cavity is filled with a homogeneous, off-white, cheese-like mass. *Coccinella 7-punctata* that overwinter in aggregations show lower levels of mortality due to *B. bassiana* than do those that overwinter alone (Ormond et al., 2011). Moreover, in choice tests, *C. 7-punctata* adults avoid overwintering near dead *C. 7-punctata* with sporulating fungi (Ormond et al., 2011). Roy et al. (2008b) found that larvae of *H. axyridis*, *C. 7-punctata* and *A. 2-punctata* are all highly susceptible to *B. bassiana*. However, *H. axyridis* adults suffered lower mortality to this fungus than do *C. 7-punctata* or *A. 2-punctata*, although female *H. axyridis* inoculated with the fungus had reduced fecundity. Roy et al. (2008b) speculate that these females diverted resources from egg maturation to immunity systems in response to the fungus. It should be noted that species reported as *B. bassiana* are likely to be one of many from within the species complex of the genus *Beauveria* and further work is required to elucidate the taxonomic resolution.

Other species of pathogenic fungi recorded from coccinellids include *Beauveria tenella, Metarrhizium anisopliae, Isaria* (= *paecilomyces*) *farinosus, Isaria* (= *paecilomyces*) *fumosoroseus* and *Lecanicillium* (Ceryngier and Hodek, 1996; Ceryngier et al., 2012).

Another distinctive group of fungi known to infect coccinellids are species of *Hesperomyces* (Ascomycota: Laboulbeniales), which are of interest because, like the podapolipid mites, they are thought to be primarily sexually transmitted. The fruiting bodies of these fungi are yellow and grow on the surface of the ladybird, being most abundant on the ventral surface of males and the dorsal surface of females (Fig. 7.24). However, recent research indicates that transmission during overwintering, when many species of ladybird aggregate, is also important for the population dynamics of *H. virescens* (Nalepa and Wier, 2007). Levels of infection may be very heavy (Fig. 7.25). Laboulbeniales have been recorded from a variety of ladybird hosts in Mediterranean, subtropical and tropical regions, particularly in humid habitats. In cooler climates, they are generally rare because of low temperature. Indeed, in London, *Hesperomyces virescens* shows a strong reduction in prevalence with distance from the city centre, which is probably the result of temperature gradients across the city associated with urbanisation (Welch et al., 2001).

Fig. 7.24. Fruiting bodies of *Hesperomyces* (Ascomycota: Laboulbeniales) on *Adalia 2-punctata*.

Fig. 7.25. *Adalia 2-punctata* heavily infected by *Hesperomyces*.

Interestingly, in Britain, the distribution of *H. virescens* on *A. 2-punctata*, which was previously confined to the London area, has increased over the last few years, possibly as a result of increased temperatures due to global warming.

One final group of organisms, recently ascribed to the fungal Kingdom, are deserving of attention. Microsporidia have only rarely been reported from coccinellids, but are probably common. Indeed a recent study indicated the high prevalence of microsporidia within *H. axyridis* (Vilcinskas, 2013). Three species of the genus *Nosema* have been described from ladybirds. They have been reported from around the world, infesting a variety of ladybird species, with each *Nosema* species attacking different host tissues or cells. Undoubtedly, the true extent of microsporidial disease in coccinellids is unknown, and it would be valuable if modern molecular genetic and microscopic techniques were used to assess the range of these organisms. This is particularly the case because some microsporidia are intracellular, can be maternally inherited, and cause sex ratio distortion in their hosts via feminisation or late male-killing (Majerus, 1999).

Bacteria and Viruses

With the exception of male-killing bacteria that follow one unusual life-history strategy, virtually nothing is known about the bacterial or viral diseases of coccinellids. Certainly, those who have cultured large numbers of ladybirds will have observed that if reasonable measures are not taken to maintain culture hygiene, by removing old food and frass, more larvae die showing symptoms of enteric disease characteristic of bacterial or viral infections of the gut. Yet, I know of no studies of viral diseases, and only a few investigations of bacterial pathogens (James and Lighthart, 1992; Strong-Gunderson et al., 1992; Giroux et al.; 1994), and these are essentially assays of the bacteria present. In fact, compared to most insects, ladybirds suffer little from these types of disease, even in captivity. However, one group of pathogenic bacteria does affect ladybirds very significantly. These are the male-killing bacteria.

Male-killing Bacteria

Certain types of organisms have become associated with particular fields of science or scientific hypotheses, because they have attributes that make them 'model' organisms for the study of specific phenomena. Among the invertebrates, examples include the Lepidoptera for the study of colour pattern variation and its inheritance, fruit flies of the genus *Drosophila* for many genetic studies, squid for the study of neurology and the nematode, *Caenorhabditis elegans*, for the study of development. The ladybirds have been influential in the study of various phenomena, from pest control, through sexual selection by female choice, and the evolution of aposematism to the impacts of invasive alien species. However, it is in the field of sex ratio biology that they have scope to be most important. Over the last 20 years, the Coccinellidae have become the model system for investigation of sex ratio distortion due to a phenomenon known as male-killing (Majerus and Hurst, 1997).

The Sex Ratio

Female-biased sex ratios in a ladybird were first reported by Lusis (1947b), who observed that some female *A. 2-punctata* produce only or predominantly female progeny. This is unusual because, as Darwin noted, most dioecious organisms (species with separate male and female individuals) produce roughly equal numbers of male and female progeny. In the first edition of *The Descent of Man*, Darwin (1871) gave an explanation of why this should be so, writing: 'Let us now take the case of a species producing from the unknown causes just alluded to, an excess of one sex – we will say of males – these being superfluous and useless, or nearly useless. Could the sexes be equalised through natural selection? We may feel sure, from all characters being variable, that certain pairs would produce a somewhat less excess of males over females than other pairs. The former supposing the actual number of the offspring to remain constant, would necessarily produce more females, and would therefore be more productive. On the doctrine of chances a greater number of the offspring of the more productive pairs would survive; and these would inherit a tendency to procreate fewer males and more females. Thus a tendency towards the equalisation of the sexes would be brought about.' (Darwin, 1871, p. 316.)

Interestingly, Darwin was obviously unconvinced by his own arguments over this issue, for in the second and subsequent editions of *The Descent of Man*, he omitted this explanation, writing instead: 'I formerly thought that when a tendency to produce the two sexes in equal numbers was advantageous to the species, it would follow from natural selection, but I now see that the whole problem is so intricate that it is safer to leave its solution to the future.' (Darwin, 1874, p. 399.)

One explanation of the commonness of the 1:1 sex ratio is that, in most species, sex is determined by the segregation of sex chromosomes during meiosis in the heterogametic sex, and that meiosis generates a similar number of X-bearing and Y-bearing gametes. This is true as far as it goes. However, many organisms do not have sex chromosomes and a variety of mechanisms that result in sex ratio biases exist. Therefore, were it beneficial to produce a sex ratio different from 1:1, such mechanisms would surely become prevalent. An explanation of numerical sexual equality would be more convincing if it were based on selection. Such an explanation exists. The explanation, usually attributed to Sir Ronald Fisher (1930), differs little from that given by Darwin, almost 60 years previously (Edwards, 1998).

The logic of Fisher's explanation is simple and beautiful. Assume that genes acting in the parental generation determine the sex ratio of the next generation. For example, we may envisage genes that either act in the heterogametic sex, by altering the ratio of male and female determining gametes produced, or act in the homogametic sex, by altering the success of the two types of gamete in fertilisation. Suppose then that a particular population contained twice as many females as males. Subsequently, assuming that all females are mated, the males would, on average, produce twice as many offspring as the females, for each would father two families, while each female would only produce one family. A gene, acting in a parent, would be transmitted to more descendants if it could cause that parent to produce the more rare sex, in this case sons. A pair producing many males would contribute more to future generations than pairs producing equal numbers of sons and daughters, or many daughters. Similarly, if there were more males than females in a population, a gene causing parents to produce more daughters than sons would spread. The consequence of this is that, other things being equal, there should be equal investment in male and female offspring, and that the evolutionarily stable sex ratio should be 1:1, as only then is the reproductive value of a son equal to that of a daughter.

The evolutionary stability of sex ratio equality based upon negative frequency dependent selection has been hailed as the most celebrated argument in evolutionary biology (Edwards, 1998). The Darwin/Fisher argument is pivotal to almost every aspect of evolutionary considerations of population biology. It was the first example of an evolutionary stable strategy (ESS) and few of those described since can compete with it in elegance and simplicity. It also beautifully illustrates that selection most often acts on individuals rather than groups. Here it is pertinent because it predicts that sex ratio equality should be the norm and makes those species exhibiting distorted sex ratios of interest to evolutionary biologists.

It is in this context that Lusis' observations of female-biased sex ratios being produced by some pairs of Russian *A. 2-punctata* is significant. Lusis (1947b) noted that the trait was inherited, but only through the female line. That is to say, the probability of an individual producing a biased sex ratio was related to the sex ratio produced by the mother, but not the father. In 1990, I collected a small sample of *A. 2-punctata* from a site in southern England. Five out of 26 pairs produced female-biased sex ratios. By breeding from the daughters and few sons of these five families, Lusis' observation of maternal inheritance of the biased sex ratio trait was confirmed (Hurst et al., 1992). In addition, the proportion of eggs that hatched in these sex ratio families was only about half that of normal families. The inference was that all females were producing roughly equal numbers of male and female eggs, and that something in the fertilised eggs was killing those destined to be males; but what?

The Case of the Feminist Bacterium

Most cells in an individual are composed of two parts, the nucleus, which contains the chromosomes, and the cytoplasm. Although most of the fundamental genetic material, DNA, is concentrated in the nucleus, some DNA also occurs in the cytoplasm, in the mitochondria, and in a variety of intracellular, parasitic microbes such as bacteria, viruses and protozoa. A female gamete (egg) has both nucleus and cytoplasm, but a male gamete (sperm) has virtually no cytoplasm. Consequently, the genetic material in mitochondria and cytoplasmic microbes is normally inherited only through the female line; it is maternally inherited. In experimental science, when faced with a variety of alternative explanations, it is expedient to test those that can be verified or refuted quickly (and cheaply), before resorting to more complicated experiments. One possible explanation was that the females producing mainly daughters were infected with a cytoplasmic bacterium that killed male but not female embryos. Were this so, 'curing' these females, by administering an antibiotic to kill the bacteria, should be possible. But how does one get a ladybird to take medicine? Fortunately, ladybirds have something of a 'sweet tooth'. In the field, they eat honeydew and in the laboratory, they eat golden syrup. When an antibiotic, tetracycline, was mixed with golden syrup, and given to females with the biased sex ratio trait, the females ate it readily. Soon thereafter these females began to produce egg clutches with high hatch rates that subsequently gave rise to as many sons as daughters (Hurst et al., 1992). This test provided strong circumstantial evidence that a bacterium, living in the cytoplasm, was causing the death of male embryos. Subsequently, these findings were confirmed by microscopic and molecular genetic technology. Both light microscopy of haemolymph using the stain DAPI (diamidinophenylindole), which binds to DNA and fluoresces under ultraviolet light, and transmission electron microscopy revealed bacteria to be present in the cytoplasm of haemocytes and other cells of females with the sex ratio trait, but not in normal females or males.

Using DNA technology, the identity of the bacterium was established. In *A. 2-punctata*, the 16S rDNA gene of the bacterium associated with the female-biased sex ratio trait was found to be very similar to bacteria of the genus *Rickettsia*, only varying in 2% of its nucleotides from *Rickettsia typhi*, which causes murine typhus, *Rickettsia prowazekii*, which causes epidemic typhus, and *Rickettsia rickettsia*, the causative agent of Rocky Mountain spotted fever. The male-killer was thus placed in this genus (Werren et al., 1994).

What is Male-killing?

Microbes that live inside the cells of hosts and are inherited in the cytoplasm of eggs have their interests best served if they promote female hosts that pass them to subsequent generations, at the expense of male hosts that do not. Thus, these microorganisms have evolved a variety of extraordinary behaviours to bias the sex ratios of hosts in favour of females. These strategies include the feminisation of genetic males, the induction of asexual reproduction, and male-killing (O'Neill et al., 1997; Majerus, 2003).

Male-killing is the least sophisticated of the mechanisms of sex ratio distortion practised by inherited symbionts, and probably the most common. The basic mechanism involved is for the symbiont to kill male but not female hosts. As a consequence of the coarseness of this approach, male-killers are more taxonomically diverse, both with respect to hosts and symbionts, than other microbe sex ratio distorters. Among arthropods, male-killing has been reported from five orders of insects and two species of mites. In addition, male-killers themselves are taxonomically diverse, with a variety of different bacterial groups and one group of protists being recorded (Majerus, 2003).

The male-killers in ladybirds are termed early male-killers because they cause the death of male hosts early in embryogenesis. Thus far, all those that have been identified are bacteria.

The rationale behind male-killing is simple. By killing male hosts, bacteria increase the fitness of copies of themselves in female hosts. Because of the nature of reproduction in these bacteria, the progeny of an infected mother will carry copies of a symbiont that are identical. It follows that if in killing male hosts the symbiont increases the fitness of the dead males' sisters, a fitness benefit will also accrue to 'exact' copies of the male-killer within these females.

Two classes of benefit from male-killing are possible. Firstly, mating with brothers is prevented, so loss of fitness through inbreeding depression is reduced. An infected female, the brothers of which have been killed by a male-killer, cannot indulge in full-sibling mating. Secondly, the resources that would have been used by male offspring of an infected female may be allocated to female siblings. It is this second type of benefit that occurs in the male-killers of ladybirds. Crucially, the resources made available to females from the death of males must become preferentially available to infected rather than uninfected females. This will be the case when male and female progeny from one mother are found in close proximity. This type of limitation means that the distribution of male-killers is not random in the invertebrates. Groups of species with certain behavioural or ecological similarities will be more prone to early male-killer infection than groups lacking such characteristics, because only in the former will fitness compensation from male-killing be sufficient to allow the invasion or evolution of this strategy. A non-random pattern of male-killer distribution in the invertebrates has been observed (Majerus and Hurst, 1997). Hot spots occur in the milkweed bugs (Hemiptera), nymphalid butterflies, particularly of the genus *Acraea*, and the ladybirds. Of these groups, ladybirds have been most extensively studied (Majerus, 2006).

Early male-killing bacteria have been recorded from 14 species of ladybird and are suspected to occur in four others. Bacteria of five different groups (*Rickettsia, Wolbachia, Spiroplasma,* Flavobacterium and γ-proteobacterium) are known, showing that male-killing has evolved independently many times. The situation in *A. 2-punctata* is complex and unexpected. Following the identification of the *Rickettsia* in English populations of *A. 2-punctata*, samples of *A. 2-punctata* from other countries have shown that male-killing is widely distributed in this species. However, the causal agent of male-killing varies. To date, three other bacteria, a *Spiroplasma* (Hurst et al., 1999a) and two strains of *Wolbachia* (Hurst et al., 1999b) have been found. The *Rickettsia* appears to occur on its own in Britain. However, in other populations, different male-killers coexist, all four sometimes being present together (Majerus et al., 2000c). Theoretical models of the evolution of male-killing suggest that two or more male-killers cannot coexist in a single host population at equilibrium, except in the presence of male-killer suppressers, and even then, the parameter space for coexistence is severely limited (Randerson et al., 2000). The finding of four male-killers in a single population, and the wide geographical distribution of the male-killers in question (Majerus, 2003), therefore suggests that the assumptions upon which these models are based are not applicable to ladybird male-killers.

Given the still small number of species of coccinellid in which male-killing has been sought and identified, it is unlikely that the full taxonomic diversity of male-killers in ladybirds has yet been revealed. However, analysis of the rate at which new genera of male-killing bacteria of insects have been described suggests that the total diversity of early male-killers will not be orders of magnitude greater than that already known (Hurst et al., 2003). Moreover, representatives of all the major bacterial groups containing early male-killers have been recorded in the Coccinellidae.

It is the behavioural traits and the ecologies of aphidophagous ladybirds that confer high resource benefits to daughters of infected females as a result of the death of males. All coccinellids in which male-killers have been found are aphidophagous, oviposit in clutches and exhibit neonate larval cannibalism/consumption of sibling eggs. These features are not independent. Prey ephemerality promotes sibling egg cannibalism, in turn leading to evolutionary advantages to rapid embryonic development (Majerus and Majerus, 1997a). The result is that neonate larvae are very small and have minimal energy reserves, leading to high starvation rates in the period just following dispersal from their egg batches (Banks, 1955, 1956; Wratten, 1976). Here is the context for a high fitness compensation to accrue from male-killing.

In *A. 2-punctata*, the habit of consuming unhatched eggs in their clutch before dispersing provides two advantages to male-killing. The first involves direct resource reallocation to infected females through the consumption of male-killed eggs. The hatch-rate of egg clutches laid by females not infected with male-killers is typically around 90%. This compares to hatch-rates below 50% in clutches from infected females. Thus, in infected clutches, neonate larvae, which will be predominantly female, will have, on average, one male-killed egg to consume (Fig. 7.26), compared to up to about 0.1 unhatched eggs available to neonate larvae from uninfected clutches. The additional resources gained by eating a single egg allow neonate larvae to survive half as long again as those denied such a meal (Majerus, 1994a). Furthermore, larvae from infected clutches are larger when they disperse to seek aphid prey than those from uninfected clutches (Majerus, 1994a). The longer survival time, and

Fig. 7.26. Neonate larvae consuming male-killed eggs.

larger size of dispersing larvae from a male-killed clutch, leads to a greater search area, more rapid first instar larval development and higher likelihood of survival to first ecdysis, particularly at low aphid densities. Similar or greater advantages resulting from resource reallocation via sibling egg consumption have been observed in other aphidophagous ladybirds (Majerus, 2003; Elnagdy et al., 2011).

A second advantage to male-killing resulting from the cannibalistic behaviour of neonate larvae is that slow-developing female larvae in infected clutches are less likely to be attacked and eaten by clutch-mates than similar larvae in uninfected clutches. The reduction in this probability follows from both the fewer larvae that hatch and the greater number of unhatched eggs in a male-killed clutch than in a normal one. Evidence that this reduction in cannibalism allows a greater number of female larvae to hatch in male-killed than in normal clutches has been shown for both *A. 2-punctata* and *Coccinula sinensis* (Hurst, 1993; Majerus, 1994a, 2003).

A third possible advantage to male-killing has yet to be verified by experimental study. Ladybird larvae not only consume unhatched eggs in their clutch before dispersing, but they also eat conspecific eggs and larvae once they have dispersed. In interactions between two larvae, assuming that neither is in the process of ecdysis, the larger larva usually wins and eats the smaller (Majerus, 1994a). Because larvae from male-killed clutches are, on average, larger when they disperse than those from normal clutches, larvae infected with the male-killer will win and thus gain another advantage if clutches of eggs are laid at the same time and close together by infected and uninfected females (Majerus, 2003).

When Suicide is Painless

The advantage and disadvantage of male-killers to their ladybird hosts – death of sons balanced by benefits to daughters through resource reallocation – are not difficult to understand. However, evolution of male-killing should also be viewed from the bacterium's perspective, for male-killers appear to do something very peculiar. In killing their hosts, they commit suicide, and for those who believe in Darwinian evolution – the survival of the fittest – committing suicide seems an odd survival strategy. However, on closer examination, this behaviour has a simple and sound evolutionary rationale.

Bacteria replicate, making exact copies of themselves, in their host cells. Usually, when an infected host cell divides, both daughter cells receive at least one copy of the bacterium. As the bacterium cannot be transmitted through sperm because of its lack of cytoplasm, it would make evolutionary sense for the bacterium to avoid being transmitted into males. However, as the inheritance of these bacteria is through the gametes of female hosts, this is not an option. All female gametes are essentially the same with respect to sex, each carrying an X-chromosome. It is only just before the egg is laid that a sperm fertilises it and the sex of the ensuring zygote is determined. If an X-bearing sperm fertilises the egg, the zygote will be female. If a Y-bearing sperm fertilises the egg, a male zygote results. Because fertilisation occurs just before oviposition, the bacterium has little chance of responding to the sex chromosome carried by the sperm that fertilises the female host gamete. Once in a male egg, the bacterium is in an evolutionary cul-de-sac, whatever happens. It is effectively dead anyway, and loses nothing by killing its host, committing suicide in the process. Furthermore, through resource reallocation and the reduction in cannibalism of female siblings, by killing its host, the bacterium provides benefit to the hosts of clonally identical copies of itself. Here is an extreme and elegant example of kin selection (Hamilton, 1967), with bacteria in male hosts enduring the ultimate sacrifice – death – to the benefit of genetically identical copies of themselves.

The Vertical Transmission Efficiency and Cost of Male-killers

Considering the dynamics of early male-killers, Hurst (1991) showed that three parameters affect the spread of a male-killer in a host population. These are the level of fitness compensation, the vertical transmission efficiency, and any cost that infection imposes on infected females. While the level of fitness compensation through resource reallocation in ladybirds will depend largely on the availability of prey that normal neonate larvae can catch (Majerus, 2006b), the vertical transmission and costs of male-killers vary as a result of interactions between different male-killers and their various hosts.

The vertical transmission of male-killers in ladybirds is never 100% efficient. Values obtained from laboratory cultures vary from 72% for a *Rickettsia* from a Muscovite population of *A. 2-punctata* to 99.98% for a *Spiroplasma* from a Japanese population of *H. axyridis*. Comparative analysis of the various male-killers in different hosts, and the four male-killers in *A. 2-punctata* suggest that bacteria of some genera (e.g. *Spiroplasma*) have a tendency for high transmission, others have a tendency for low transmission (e.g. *Rickettsia*), and still others have a very variable transmission efficiency, as is the case for the Flavobacteria that cause male-killing in five species of ladybirds. The vertical transmission efficiency may be influenced by a range of factors, such as temperature, antibiotics, host activity, host dormancy and host density, all of which may vary in the wild (Hoffman and Turelli, 1997).

The imperfect vertical transmission of male-killers means that in each generation some of the females produced by infected mothers are uninfected. Furthermore, these females will also produce a few sons. These sons may be important in the dynamics of the host population, because these sons of infected mothers will gain the same resource advantage as their mainly infected sisters. Consequently, while selection acting on the male-killer will favour efficient male-killing, the selection acting on ladybird hosts will act to reduce the vertical transmission of the male-killer (Majerus, 2003).

The cost to females of bearing a male-killing bacterium, in addition to the loss of male progeny, has been assessed in five species of ladybird. Each case has shown some negative fitness effects, such as decreased oviposition rates, lower overall fecundity, higher infertility levels, or shorter adult life-span. In addition, in *H. axyridis* infected with *Spiroplasma*, egg hatch rates of infected females, which are initially around 50%, decline as the female ages. The pattern of decline is probably a result of the bacterium causing the death of some female offspring as bacterial density increases with host age (Majerus, 2003).

It is interesting that in each of the ladybirds studied to date, costs rather than benefits are imposed on infected female hosts. Because the bacteria are maternally inherited, it would be beneficial for these symbionts to reduce their detrimental effects on female hosts, becoming harmless, or even beneficial. That this has not happened may indicate that metabolic limitations and the need for the symbiont to proliferate to ensure its transmission into the next host generation limit its potential to evolve mutualism towards female hosts (Majerus, 2003).

How Do Male-killers Kill?

Male-killers and the population sex ratio biases that they cause in their hosts affect host evolution in a number of ways. However, one final question is pertinent to the discussion of male-killers as enemies of ladybirds. It is simply, how do male-killers kill? How does a bacterium determine that it is in a male rather than a female embryo and then only kill if it is in the former? Male-killers are known to occur in a wide variety of taxa, including some major crop pest species, such as the army worm moth, *Spodoptera littoralis*. Were a male-killer to use a specific gene product to kill males, it may be possible to produce this synthetically. If it can be made species specific, such a toxin may be less environmentally damaging than traditional chemical insecticides and less prone to the evolution of resistance against it.

To date, few studies have thrown any light on the mechanism of male-killing. The few strands of evidence we have suggest that sex recognition by the bacteria does not depend upon male-specific genes on the Y chromosome, because male-killers also occur in Lepidoptera in which males are homogametic, and in parasitoid wasps in which males are haploid. Some work on a male-killer in the fruit fly *Drosophila melanogaster* has suggested that female determining genes may counter the killing activity of male-killers. Experiments by Sakaguchi and Poulson (1963) showed that flies with two X-chromosomes survived infection, while those with other chromosome complements died, even if they were phenotypically female. More recently, Veneti et al. (2005) have shown that a male-killing *Spiroplasma* fails to kill male *D. melanogaster* carrying mutations in any of five genes that are involved in male sex determination, suggesting that the bacterium uses the proteins normally encoded by these genes to identify the sex of its host.

The other strand of evidence comes from work on ladybirds and fruit flies where low bacterial density allows some infected males to survive, and high bacterial density causes some females to die (Schulenburg et al., 2001b; Majerus, 2003). This suggests that the bacteria may be inherently pathogenic, and that their pathogenicity is density dependent. If killing occurs only when the bacteria exceeds a certain threshold density, it is feasible that this density is much lower in male than female hosts, thereby producing the sex bias in host death.

There is no reason to suppose that the mechanism of male-killing will be the same for all male-killers and all hosts. While the other reproductive manipulations of detrimental inherited bacteria are practised by just one or two genera of bacteria, male-killers are phylogenetically diverse. This diversity in the bacteria that have evolved male-killing independently may be the result of there being more than one way to kill a male. Indeed, related bacteria that infect other insects have a diverse range of impacts on host fitness including helping protect their hosts from infectious diseases; whether the same is true of ladybird male-killers would be worthy of future investigation.

Arms Races and the Red Queen

The enemies of ladybirds are numerous and varied. They influence the evolution of ladybirds, for those ladybirds of a species that have the best defences against the enemies that are most prevalent and detrimental make the greatest contribution to the next generation. In some, but not all of these interactions, we see evidence of enduring evolutionary arms races of the type that led Van Valen, (1973) to propose the Red Queen hypothesis. This theory proposes that as soon as one of the protagonists evolves a trait against the other, selection will favour the development of some counter-adaptation. Thus, eaters and eaten will be constantly posing new evolutionary problems for each other. As the Red Queen said to Alice in Lewis Carroll's (1962) *Through the Looking Glass and What Alice Found There*: 'Here, you see, it takes all the running you can to keep in the same place.' The aposematism of many coccinellids to many predators, such as birds, that find prey by sight, is one obvious anti-predator device. This has been countered by some birds, such as martins, that have developed immunity to ladybird chemical defences. However, multistep arms races are not obvious in many of the associations between ladybirds and generalist predators. While predators will impose strong selective pressure on their prey, for these will lose their lives should their defences fail, the pressure imposed by ladybirds on these predators will be less, for the predator may simply move on to some other type of prey, should ladybird defences be too strong. It is between ladybirds and some of their parasitoids, parasites and pathogens that co-evolutionary arms races are more likely to occur, and in particular, with those enemies that specialise on ladybirds, almost or completely to the exclusion of other hosts. Thus, we see evidence of arms races in the resistance of *Adalia* ladybirds to attack by *D. coccinellae,* in the flicking behaviour of ladybird pupae to ovipositing scuttle-flies, in the scent mimicry of larvae of *P. luteorubra* living with ants. In the case of ladybird male-killers one ladybird species, *Chilomenes sexmaculata,* which is infected by a male-killing y-proteobacterium has indeed evolved resistance against the male-killer. The resistance operates as a "rescue gene" such that if a female ladybird or the male she mates carries the gene the male eggs are rescued from the effects of the male-killer and survive to adulthood (Majerus and Majerus, 2010).

The Crucial Last Two

A female ladybird lays many eggs. Some may lay over 3000. From the moment the eggs are laid, a female's offspring start to diminish in number. Some of the eggs fail to hatch because they were not fertilised and male-killing bacteria kill some. Others are cannibalised by their own siblings or preyed upon by other larvae or other egg predators. Many of the larvae that do hatch die of starvation, particularly in their first instar. Predators, parasites and pathogens take a further toll on larvae, pre-pupae and pupae. If food runs short for developing larvae, a second peak of mortality due to starvation or cannibalism may hit immature stages. Once the remaining offspring have reached adulthood, they are faced with a new suite of predatory or parasitic enemies. On top of this, dispersal carries risks, and adult ladybirds may also suffer high levels of mortality during dormancy periods. Finally, even if they avoid all these risks, some ladybirds will fail to find a compatible and healthy mate willing to copulate with them. Throughout their lives, inclement weather, incidental death due to the activities of large animals, and the multitude of disruptions and disturbances caused by people, are also a threat to ladybirds.

Given all these dangers, it is perhaps not surprising that, on average, only two of the eggs laid by a female survive to reproductive maturity and produce offspring in their turn. Indeed, with the range of risks to which ladybirds are exposed, it is perhaps more amazing that as many as two survive. But usually they do, and two is enough! Two will maintain the population. In reality, the average number varies somewhat around two: higher than two, and the population increases; lower than two and the population declines. Over time, it is essential that the average of two be maintained; and over eons of time, the average has remained around two. Nevertheless, in recent times, the average for many species has begun to decline, due to the ladybird's greatest enemy. Ironically, that enemy is a professed friend of ladybirds: us!

Chapter 8: Ladybird Colouration

Colour: A First Line of Defence

The bright, eye-catching colour patterns sported by most ladybirds have evolved as an advertisement of unpalatability/toxicity to potential predators. Thus, these colours are a first line of defence, warning that these beetles are foul smelling, bad tasting and, in some cases, poisonous. It is easy to understand that bright contrasting colours, say red and black, or yellow and black, are associated with warning. We ourselves use such colours in traffic lights, on road signs and many other danger signals. However, bright colours do not always mean danger. They can be a means of attraction. Product advertisers use bright colours on packaging to draw the consumers' attention. Many plants have bright red or yellow berries to attract the attention of birds, for it is to the plant's advantage to have its berries eaten. The hard seed at the centre of the berry passes through the gut of the bird undamaged, and will be deposited some distance from the parent plant. The bright colour is thus instrumental in seed dispersal. Furthermore, for some plants, passage of seeds through the crop and stomach of a bird, where they are scarified and acted upon by a range of digestive chemicals, actually increases germination rate. Many flowers use bright colours to attract pollinators, while many animals, including humans, use bright colours as sexual attractants, the exotic plumes of male birds of paradise, peacocks and pheasants being obvious examples. The importance of bright colours in all these contexts is that they are memorable. In experiments involving several ladybird species, Dolenska et al. (2009) found that unspotted ladybirds were attacked by great tits, *Parus major*, more so than were spotted ladybirds, indicating recognition by the birds of the warning colouration of ladybirds.

Chemical Defences

Much research has been carried out on the chemical defence systems that the bright colours of ladybirds advertise. At the centre of coccinellid chemical defence lies a group of related nitrogen-containing molecules, known as alkaloids. The first of these to be identified was N-oxide coccinelline, and its corresponding free base precoccinelline, extracted from *Coccinella 7-punctata* (Tursch et al., 1971). These also occur in many other species of the genus *Coccinella*, and some other genera (Henson et al., 1975). Subsequently, similar compounds have been extracted from other species. N-oxide convergine and its free base hippodamine were found in *Hippodamia convergens* (Tursch et al., 1974), while myrrhine was extracted from *Myrrha 18-guttata* (Tursch et al., 1975). Slightly less similar alkaloids, propyleine (a dehydroprecoccinelline), and adaline (a homotropane alkaloid), were yielded by *Propylea 14-punctata* and *Adalia 2-punctata*, respectively (Tursch et al., 1972, 1973). Other alkaloids have been detected in other species. For example, several predatory ladybirds, including *Harmonia axyridis*, produce an aliphatic diamine, harmonine (Braconnier et al., 1985; Enders and Bartzen, 1991; Sloggett et al., 2011), while several alkaloids have been extracted from the plant-eating ladybird *Epilachna varivestis* (Attygalle et al., 1993; Proksch et al., 1993). Although many coccinellids synthesise more than one alkaloid, in a particular species or genus, one alkaloid usually predominates. Thus, in species of *Chilocorus*, chilocorine A or chilocorine B are most commonly present in quantity, while in *Exochomus 4-pustulatus*, exochomine predominates.

Many alkaloids have now been extracted from coccinellids (Table 8.1). They have been found in members of all the subfamilies of coccinellids that have been investigated. However, in *Aphidecta obliterata*, *Subcoccinella 24-punctata* and some species of the tribes Scymnini and Coccidulini, alkaloids could not be detected (Pasteels et al., 1973). These species are rather dull in colour and most are small. The subfamilies to which they belong all contain species that do synthesise alkaloids. It seems probable that at least some of these species, such as *A. obliterata*, have evolved cryptic colouration, and have foregone the production of expensive defensive chemicals that are of little value to insects that rely on not being detected by predators for their defence. Indeed, in feeding tests, these species did not appear unpalatable to European quail (*Coturnix coturnix*),

Table 8.1 The Alkaloids Extracted from Different Coccinellid Subfamilies

Subfamily	Alkaloids
Chilocorinae	Chilocorine A, chilocorine B, exochomine
Scymninae	Euphococcinine, piperidine
Coccinellinae	Adaline, adalinine, coccinelline, convergine, harmonine, hippocasine, hippocasine N-oxide, hippodamine, myrrhine, n-octylamide, precoccinelline, propyleine
Epilachninae	Epilachnene, euphococcinine, piperidine, pyrrolidines, 2-phenylethylamine, Nx-quinaldyl-L-arginine

although it should be noted that these birds are not normally insectivorous. That said, these ladybirds do still reflex bleed, and it may be that they produce other defensive chemicals aimed at invertebrate predators rather than birds (Majerus, 1994a).

Many coccinellids also contain relatively high concentrations of histamine (Frazer and Rothschild, 1960). Other volatile chemicals, such as quinolenes and pyrazines, have also been extracted from some coccinellids (Rothschild, 1961). Among these is 2-isopropyl-3-methoxy-pyrazine, which is produced by many chemically defended and brightly coloured insects, and is at least partly responsible for the strong smell given off by many coccinellids when disturbed (Al Abassi et al., 1998).

The diversity of alkaloids and other defensive chemicals in the coccinellids, and the variation in concentrations of the various substances present in these cocktails, indicate that these insects were probably some of the first to use combinatorial chemistry in their defence (Schröder et al., 1998).

Many but not all of the defensive chemicals found in ladybirds are synthesised by the ladybirds that bear them (Tursch et al., 1974; Ayr and Browne, 1977; Jones and Blum, 1983). Coccinelline and precoccinelline have been extracted from both the eggs and larvae of *C. 7-punctata*, but could not be detected in their aphid prey (Tursch et al., 1971). Furthermore, by feeding ladybirds with radioactively labelled acetate, Tursch et al. (1975) demonstrated that coccinelline was synthesised internally, via a polyacetate pathway. It is not known whether all quinolenes and pyrazines are synthesised by ladybirds, although some undoubtedly are. However, some coccinellids also have the ability to store and use defensive chemicals from their prey. *Coccinella 11-punctata* and *Hippodamia variegata* both sequester cardiac glycosides from *Aphis nerii* (Rothschild et al., 1973; Rothschild and Reichstein, 1976). *Coccinella 7-punctata* sequesters pyrrolizidine alkaloids when feeding on *Aphis jacobaeae* (Witte et al., 1990). *Hyperaspis 3-furcata* gains a major weapon in its armoury by storing anthraquinone carminic acid from its main prey, the cochineal insects of the genus *Dactylopius*. This chemical appears to be particularly effective in deterring ants (Eisner et al., 1994). In these cases, the defensive chemicals are actually manufactured by the plants on which the aphids feed, so the ladybirds get this element of their defence third hand.

There is great variation in the ways in which ladybirds use their chemical armoury, both among and within species. Experiments carried out on the reflex blood exuded by *C. 7-puncata* and *A. 2-punctata* showed that the defensive chemicals are distributed throughout adult ladybirds, but occur at much higher concentrations in reflex blood than elsewhere (de Jong et al., 1991; Holloway et al., 1991). The blood is exuded from joints between the femur and tibia of the legs (Fig. 8.1). All six legs may reflex bleed simultaneously

Fig. 8.1. Reflex blood exuded from joints between the femur and tibia of the legs.

or independently. The amount of reflex blood and the concentration of defensive chemicals therein vary both among individuals and with time. *Adalia 2-punctata* females produced higher concentrations of alkaloids in their reflex blood than did males (de Jong et al., 1991), possibly reflecting their need to arm their eggs with chemical defences. *Coccinella 7-punctata* produced very little reflex blood when just removed from overwintering sites in early spring. However, the concentration of coccinelline in the blood that was secreted was very high. Fluid loss is a critical factor in winter survival for many ladybirds, and it is probable that the cost of reflex bleeding during the winter is higher than at other times, because the fluid lost in this activity is not easily replenished. That the amount of reflex blood produced by the ladybirds increased rapidly over four days feeding on *Acyrthosiphon pisum*, while the concentration of coccinelline declined, reinforces the suggestion that fluid content is limiting during the winter. Spring feeding allows fluid reserves to be replenished.

Holloway et al. (1991) suggest that overwintering ladybirds, or those that have just emerged from overwintering sites, may not be able to reflex bleed. They propose that the reflex bleeding system is shut down, so as not to waste precious resources required for overwintering. This is not the case for most species, although overwintering ladybirds do tend to be much more difficult to stimulate to reflex bleed, and can be induced to exhaust their reflex blood reserves (Majerus, 1994a). Field tests on 50 individuals each of eight species (*C. 7-punctata, A. 2-punctata, P. 14-punctata, Tytthaspis 16-punctata, Calvia 14-guttata, Harmonia 4-punctata, E. 4-pustulatus* and *Scymnus suturalis*) showed that most reflex bled, both in September, just after they had retreated to overwintering sites, and in December. In March, a significant number of some species failed to produce any discernible reflex blood (up to 38%), and some produced blood from some but not all legs. However, the majority still did reflex bleed, given enough provocation (Majerus, 1994a). These results suggest that ladybirds do not switch off their reflex bleeding system during the winter. It is more likely that ladybirds have a considerable degree of control over the reflex bleeding response. This is presumably a sophisticated automatic response, the action of reflex bleeding being mediated by an intrinsic cost–benefit analysis. Fluid content, external temperature, humidity, rate of fluid loss, recent fluid intake, and the perceived threat from the disturber, are all likely to be involved. Observations on the reluctance of adult ladybirds, and greater willingness of larvae, to reflex bleed when attacked by aphid-tending ants support the idea that the action can be understood via cost–benefit analysis. That ladybirds can reflex bleed from individual legs (Fig. 8.2) also suggests that a relatively sophisticated control system exists. In *H. axyridis*, Grill and Moore (1998) found that larvae that were reflex bled at regular intervals during development, tended to develop more slowly and produced smaller adults than larvae not treated in this way. This supports the contention that the production of defensive chemicals is energetically costly.

Holloway et al. (1991) found that *C. 7-punctata* parasitised by the wasp *Dinocampus coccinellae*, produced remarkably large quantities of reflex blood and they suggested that this could have been the result of the parasite causing widespread internal damage. However, remembering the 'cut-throat' way in which this wasp uses ladybirds to protect itself during its pupal stage, using the warning colouration and reflex bleeding defences of its immobilised ex-host, one wonders whether the wasp has some influence over the production of reflex blood. If it could cause an increase in production for a period prior to and just following its exit from its host, it would thereby maximise the benefit it gained from its host's defences.

Fig. 8.2. *Coccinella transversoguttata richardsoni* reflex bleeding from some but not all legs.

Alkaloids vary in their effects on particular predators. While coccinelline is toxic to particular birds, adaline and propyleine are not, although they may be distasteful (Marples, 1993; Majerus and Majerus, 1997b). A number of authors have suggested that the typical form of *A. 2-punctata* is a Batesian mimic of the more toxic *C. 7-punctata* (Brakefield, 1985a; Marples et al., 1989). In Batesian mimicry, one species, the mimic, which is palatable, gains protection by closely resembling the other species, the model, which possesses some form of inherent protection. The species involved in a Batesian mimetic association play two very different roles, the mimic gaining at the expense of the model without having to pay the energetic cost of bearing the protection. The speculation here is thus that the relatively poorly protected *A. 2-punctata* mimics the more toxic *C. 7-punctata*, tricking predators that have experience of and avoid the latter into leaving the former alone as well.

While this may be so, my own view is that too much emphasis has been placed on birds as predators of ladybirds in this context. Moreover, although adaline may not be toxic to birds, it does have a deterrent effect (Marples, 1993). Furthermore, it may have a greater deterrent effect against other enemies of ladybirds than does coccinelline. To give one possibility, *D. coccinellae* parasitises a wide spectrum of ladybirds but not *A. 2-punctata* nor *A. 10-punctata* to any significant extent. At one time it was thought that these species were too small to allow the development of the parasite (Hodek, 1973; Majerus and Kearns, 1989). However, the recording of this wasp from smaller species, such as *P. 14-punctata*, *Exochomus 4-pustulatus*, *Coccinula 14-pustulata*, *Coccinula sinensis* and *T. 16-punctata*, makes this idea untenable. Both *A. 2-punctata* and *A. 10-punctata* produce adaline, and it may be this chemical that makes the *Adalia* species less suitable as hosts to this wasp.

Returning to the colours of ladybirds, we have a range of closely related beetles, containing closely related chemical defence systems. Three questions may be asked. Firstly, how does true warning colouration evolve? Secondly, why do all species of ladybird not look more or less the same? Thirdly, why is there so much colour pattern variation among individuals in some species of ladybird? For the rest of this chapter, I will discuss the first and second of these two questions, leaving the description of variation within species and its evolutionary causes to Chapter 9.

The Evolution of True Warning Colour Patterns

The evolution of true warning or aposematic colour patterns presents certain problems. The main difficulty is that two very different characteristics, conspicuousness and chemical defence, apparently have to evolve simultaneously to give an efficient defence. As novel characteristics only arise rarely, if both traits have to arise together, the evolution of aposematic colouration would be expected to be very, very rare. Yet, a great wealth of organisms have adopted aposematic colour patterns for their main defence against predators that hunt by sight. A number of evolutionary pathways have been proposed to account for the evolution of aposematic colouration. These can be divided into those in which the chemical defence evolves first, with bright colouration evolving afterwards; those in which the order of evolution of the two traits is reversed; and those in which both parts of the system evolve at the same time.

First Unpalatable, Then Conspicuous

Imagine an inconspicuous, ancestral coccinellid that feeds on sap-sucking insects, such as coccids or aphids, and is threatened with extinction as a result of competition for food from other predators. A female lays some of her eggs in the vicinity of prey that are feeding on a chemically defended plant and sequester the plant's defensive chemicals. The resulting coccinellid larvae feed on these prey, but have to cope with their chemical defence. There is likely to be a metabolic cost to this detoxification, in the form of slower prey processing and development. However, because these chemically defended prey are not attacked by many species of predators, the coccinellid larvae gain an advantage through reduced interspecific competition. If during the detoxification process, some of the defensive chemicals in the prey are stored by the ladybird larvae, they will gain some measure of unpalatability. Furthermore, once an insect has evolved means of containing sequestered toxic chemicals within itself, it becomes more likely to evolve the ability to manufacture defensive chemicals itself. This is because a mutation that causes the manufacture of a toxic alkaloid, for instance, will be disadvantageous in an insect that has no mechanism to cope with the toxic effects of the alkaloid. However, in an insect that can detoxify alkaloids, the mutation may be beneficial if the presence of the alkaloid adds to its defensive capabilities.

As a result, we have a chemically defended coccinellid. Irrespective of whether the defensive chemicals are sequestered or manufactured, we now need only to increase the efficiency of defence. For an insect that is unpalatable due to the chemicals it contains, advertising its distastefulness would seem advantageous. However, this also presents a problem. Suppose a new mutation arose that made our inconspicuous beetle more obvious. This more conspicuous individual would attract the attention of a predator that could have no experience of its obvious colouration, and hence would not know that it was unpalatable. In the ensuing attack, the mutant ladybird might be fatally injured, and the mutation lost. Alternatively, suppose the mutation occurred in the germ-line cells of a female that then laid a clutch of eggs. An individual resulting from one of these eggs that expressed the new mutation would not be alone. Many of its brothers and sisters would have the same mutant phenotype, and they would be close together. Although a naive predator may attack and kill one of these obvious ladybirds, the unpleasantness of the experience would deter it from attacking others that looked the same. Here then, the evolution of aposematic colouration involves kin selection (Fisher, 1930). The loss of one of a family increases the fitness of some other members of the family that express the same mutant allele. The production of progeny within a small area is critical in this scenario, for it leads to a local abundance of conspicuous and chemically defended individuals.

The other traits that are frequently associated with aposematic insects may evolve subsequently. These include slow movement, resting in exposed positions (Fig. 8.3), aggregative behaviour, being resistant to injury and secreting volatile, foul-smelling chemicals. Many ladybirds show all these secondary features.

Conspicuous First, Then Chemical Defence

Intuitively it seems probable that unpalatability will evolve before bright colouration during the evolution of aposematism because a conspicuous and palatable beetle is likely to be seen and eaten. However, there are at least two scenarios that would allow unpalatability to evolve once bright colouration had already evolved. Firstly, defence against predators is not the only function of the colours and patterns of an organism. Bright colour patterns and conspicuous displays are part of the courtship of many animals, and black or dark colouration, which is a feature of many aposematic colour patterns, is also often involved in thermoregulation. Thus, should obvious colouration evolve due to factors other than defence, the risk of predation by hunters that use their vision to find prey would then impose a selection pressure for chemical defence to evolve.

The second scenario starts with Batesian mimicry. One critical feature of a Batesian mimetic association is that the mimic gains greatest protection when it is comparatively rare compared to its model. This is because predators are likely to encounter and learn the distasteful qualities of the model before they come upon a mimic. Now imagine that a Batesian mimic's model species becomes extinct. What is the fate of the Batesian mimic? There appear to be four possibilities. Firstly, the mimic may follow its model into extinction, due to predation pressure resulting from being conspicuous and having no effective defence in the form of an unpalatable model. Secondly, it may evolve similarity to an alternative unpalatable model. Thirdly, it may lose its mimetic

Fig. 8.3. Group of *Coccinella 7-punctata* resting in an exposed position.

warning colouration and revert to an inconspicuous cryptic colour pattern. Finally, and of most relevance here, it might evolve chemical defence itself, thus swapping its Batesian mimicry for an aposematic strategy.

Although I know of no case in which it has been demonstrated that either of these two scenarios has applied, the amount of black patterning on many ladybirds is correlated to both altitude and latitude, suggesting that thermoregulation has some part to play in the evolution of colouration. Furthermore, changing the defensive strategy of the colour pattern occurs as a normal part of the species' life history, as *Anisosticta 19-punctata* changes from being inconspicuous to being aposematic during its adult life. It would be interesting to investigate whether the timing of this change in colouration is correlated to a change in defensive chemistry. A recent study demonstrated the influence of temperature during pupation in determining the size of spots on the non-melanic f. *succinea* of *H. axyridis* (Michie et al., 2010). The spot size of individuals pupating at low temperatures was larger than for those pupating at high temperatures. Furthermore, it has been shown that individuals of *H. axyridis* with small spots are more alkaloid-rich than those with larger spots (Bezzerides et al., 2007). These studies provide some intriguing insights into the adaptability of the alien *H. axyridis* and it would be interesting to expand them to further species.

Bright Colouration and Unpalatable Simultaneously

The simultaneous evolution of bright colouration and defensive chemistry is perhaps the least likely pathway to the evolution of aposematism. Given the rarity of both prey shifts and mutation, it seems unlikely that two discrete traits, one involving a change in pigment chemistry, the other involving unpalatability, would arise exactly together. But unlikely is not the same as impossible, therefore this avenue should not be excluded completely. It is, for instance, not impossible that pigmental and palatability chemistries are sometimes metabolically linked. For example, following a prey shift to a chemically defended prey species, the detoxification process might not only result in the sequestration of some of the prey's defensive chemicals, but also to novel pigments resulting from the detoxification process (Majerus, 2002).

In the evolution of aposematic colour patterns, a pluralistic approach is appropriate. I suspect all the above pathways have been followed during the evolution of some cases of true warning colouration, although that in which bright colouration follows the evolution of chemical defence is probably most common.

Why Are Ladybirds Not All Alike?

To discuss why different species of ladybird do not all look alike, we need first to investigate why, theoretically, we might expect them to. There seems no doubt that the family Coccinellidae had a single ancestral species, and that our current species have evolved from this (Sasaji, 1968; Majerus, 1994a; Dixon, 2000). What this ancestor looked like, in terms of colour and pattern, is not known, but it gave rise to a group of beetles with a very wide range of colours and patterns. Ladybirds come in almost all colours of the rainbow. In some the elytra are unpatterned (Fig. 8.4), but most sport two or three strongly contrasting colours. Perhaps the most familiar combination is of black spots on a bright red background, but this combination may be reversed. Black on

Fig. 8.4. *Coccinella californica.*

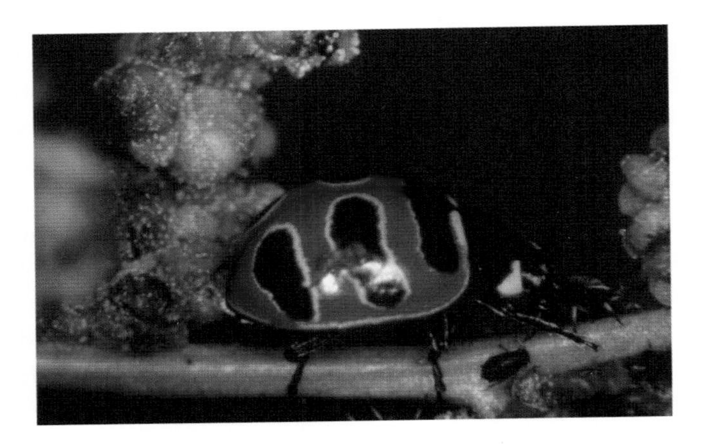

Fig. 8.5. *Coccinella 3-fasciata.*

orange, black on yellow and black on pink are also frequently met. Less common are simple spotty combinations of black on beige or buff, of white on orange or maroon or brown, of yellow on black, of orange on black or brown or blue or dark turquoise. In some, the colour patterns involve stripes rather than, or as well as, spots. Some have even more complex patterns with three colours, as in the superb *Coccinella 3-fasciata* (Fig. 8.5).

The Theory of Müllerian Mimicry

Research to date suggests that the majority of ladybirds have alkaloid-based chemical defence systems. Many authors in the past have suggested that some species are Müllerian mimics of one another. Mimicry is traditionally defined as the resemblance of one species of organism to another for protective or aggressive purposes. The two main classes of mimicry are termed Batesian and Müllerian after the scientists who first proposed them, H.W. Bates (1862) and F. Müller (1878).

In contrast to Batesian mimicry, in which the mimic gains at the expense of its model, Müllerian mimicry involves a much more equitable partnership, for here resemblances occur between two or more protected species. All members of a Müllerian mimicry complex benefit, for the losses due to inexperienced predators learning to avoid protected prey by means of colour pattern are spread across the individuals of many species.

The evolutionary pathway that would lead to the production of a Müllerian mimicry complex is simple. Suppose two differently coloured protected species occur in the same geographic region at the same time, but that species A is a hundred times more abundant than species B. If an inexperienced predator takes ten attempts to learn that each colour pattern should be avoided, the more rare species, species B, suffers proportionately more than species A. Consequently, if a mutation occurs that makes an individual of species B resemble species A, it will gain an advantage by being one of many, rather than one of few. This mutation will increase in frequency in the species B population, so that the appearance of species B converges towards that of species A. Other species may later be drawn into the complex. Theoretically, at least, there is no limit to the number of species that may be involved in a Müllerian mimicry complex.

There is certainly some evidence that birds that have had experience of one species of ladybird, and have learned its distasteful qualities, thereafter avoid not only this species, but also other similarly coloured species (Marples, 1993). There can be little doubt that some groups of ladybird species constitute Müllerian mimicry complexes. However, the theory, taken to its expected conclusion, should lead to all species being similarly coloured and patterned, which they are not. Not only do different species have different colour patterns, but also, as we shall see, many species exhibit a great range of colour and pattern variation within a population.

Different Influences on Ladybird Colour Patterns

All ladybird species do not look alike because the colour pattern of a species is not the product of defence against predation alone. Different ladybirds live in different habitats, under different climatic conditions, feed on different foods, are exposed to different parasites and predators, and contain different defensive chemicals. The evolutionary circumstances of each species, both past and present, must be considered individually.

The evolution of a ladybird's basic colour pattern will depend on a number of factors including its evolutionary origins and past evolutionary history, all the roles that the colour pattern plays in the ladybird's biology, the cost of changing colour, and features of the status of the species.

The evolutionary origin of a species is important because the ancestral pigment composition will be the template for future evolutionary changes. From a particular template, some changes in colour and pattern are more likely than others, simply because they are more likely to arise by mutation, or are less energetically costly.

The external colours and patterns may have a wide variety of roles. In ladybirds, defence is usually thought of as the major role of colour patterns. However, it is probable that colours and patterns also play a role in thermoregulation, species recognition and mate choice.

The cost of changing colour will not only involve the energetic cost of producing different pigments, it may also involve selective costs. To give a simple example, we may consider an aposematic species. Should a new colour pattern mutation arise, this mutation will only spread if it is advantageous. Here the advantage of the novel colour pattern has to be great enough to outweigh the reduction in fitness resulting from increased predation that results from looking different from the majority of the species. Similarly, if colour pattern is important in mate selection, individuals with novel rare colour patterns may be less successful in obtaining mates.

Finally, features of a species, such as degree of habitat or host plant specialisation, population size, or dispersal rate, will influence the likelihood of evolutionary change. For example, the colour pattern of a species that has a strong host plant preference will be constantly under selective pressures imposed by that particular host plant. Conversely, the colour pattern of a species that occurs in a wide variety of habitats, or on a diverse range of host plants, will not be exposed to selection imposed by one specific plant or habitat.

Population size is important in colour pattern evolution. Favourable mutations are more likely to arise in large populations simply because there will be more cases of the mutation. Perhaps more importantly, evolution occurs more rapidly in species in which population size fluctuates greatly than in species with more constant population size. This needs some explanation. The origin of all heritable variation is mutation. However, it is generally recognised that in sexually reproducing species, new variations are most often generated as a result of chromosomal recombination and fertilisation bringing new combinations of genes together in the same individual. Many heritable characteristics are affected by many genes (polygenes), each of which has a quantitative effect. Selection on these polygenic traits is most usually stabilising, which is to say that the average value of a trait confers the optimum fitness. Then, in a population of stable size, selection acting on each generation will eliminate a constant proportion of individuals at either end of the distribution of variability. Should the population start to increase in size, it implies that selection has been relaxed. It therefore follows that the cut-off lines indicating which individuals survive and which die are moved away from the average and optimum. More genetic variants therefore survive, and the degree of heritable variation in the population increases. The converse is also true. Should a population start to decrease in size, it implies that selection has become more intense, moving the cut-off lines closer to the average and optimum. Heritable variation thus declines while the population size is declining. It is important to recognise that it is only while population size is changing that the relaxation or the increased harshness of selection leads to a change in genetic variation. Once population size stabilises, even if at a new level, the selection cut-off lines will revert to their original positions and the level of genetic variation will stabilise again.

Population size may also affect evolution as a result of a process known as random genetic drift. This is the process by which the frequencies of alleles change as a result of chance because the gametes that are involved in fertilisation in each generation are only a sample of all the gametes produced. Chance deviations from expected frequencies have a greater effect in small populations than in large populations. Consequently, random genetic drift has a greater effect on evolution in species that frequently decline to very low levels, or in those in which new populations are founded by small numbers of individuals. Changes in the frequencies of alleles through random genetic drift in these two situations are known as genetic bottleneck effects and founder effects, respectively.

Taking all these factors into account, some general points may be made about the diversity of the colour patterns of ladybirds. Firstly, colour patterns that appear to us to be aposematic are most likely to occur in species that are liable to predation by organisms that possess colour vision and have the capacity to learn. The most obvious group of predators of this type is the birds. Small coccinellids are below the minimum profitable size of prey for many birds. Thus, it is likely that aposematic colour patterns are most likely to evolve in coccinellids that exceed a certain threshold size (Majerus, 1994a).

Secondly, due to the faster rate of evolution resulting from their greater population size fluctuations, aphidophagous species are likely to exhibit more diversity in their colour patterns than groups that feed on less ephemeral diets, such as scale insects or plant foliage (Majerus, 1994a).

Thirdly, host plant specialists will tend to have more complex colour patterns than host plant generalists because their colour patterns will be the result of selection imposed in more constant conditions. These general points may be illustrated with more specific examples.

The Colour Patterns of Smaller Coccinellids

The colours and patterns of many small species, for instance, those under about 3 mm in length, are perhaps the most difficult to explain. This is partly because the major selective influences on the colour patterns of these species have not been identified and partly because so little is known of the basic ecology of these species. Many of these species are rather nondescript, at least to the human eye. A small proportion of these species, for example *Hyperaspis pseudopustulatus* and *Brachiacantha felina*, do have obvious bright, contrasting colours and these are presumably aposematic. In other species showing two different colours, such as many of the scymnines, nephines and coccidulines, the efficacy of the colours as a warning is less tenable, as the colours are rather dull. One must question whether significant benefit would accrue to these small species from bright warning colours. It has long been known that, other things being equal, birds generally select the largest insect prey available, and that some insect prey are too small to be 'worth the effort'. Birds are one of the few groups in the animal kingdom that have colour vision in the same sort of visual spectrum as ourselves, although most birds see further into the ultraviolet spectrum than we do. If birds do not prey on these small coccinellids, because of their small size, irrespective of whether the coccinellids were edible or distasteful, perhaps the colours and patterns of these species, as birds and we see them, are not of great relevance. It may be that invertebrate predators and parasites, which see colour in different spectral ranges from ourselves, have been more influential in determining the colouration of these small coccinellids. It would certainly be of value to try to see these small ladybirds as other invertebrates might see them, by looking at them in ultraviolet light. That said, some of these small species are cryptic. The conifer specialist *S. suturalis* spends the winter tucked tightly into young pine buds. Here its bronzy colouration makes it very difficult to see. Similarly, the brown colouration of *Rhyzobius litura* (Fig. 4.7) serves as good camouflage in low herbage. Nonetheless, these speculations should serve to endorse rather than mask the fact that we know virtually nothing about the ecological significance of the colours and patterns of the smaller coccinellids.

Host Plant Generalists

The larger ladybirds can be split broadly into host plant generalists and specialists. For the purposes of considering colour patterns, the generalists include not only species such as *C. 7-punctata*, *Coccinella transversoguttata*, *Harmonia conformis*, and *H. convergens*, which are obvious habitat generalists, but also a variety of other species that occur commonly on a wide range of host plants, even if they are restricted to particular habitats. Thus, generalists include species such as *C. magnifica* and *Coccinella 5-punctata*, which, despite being restricted to very specific habitats, at least in parts of their range, may be found there on a wide range of host plants. Similarly, arctic and alpine species, such as *Hippodamia arctica* and *Coccinella alta*, may be found on any foodplant on which prey are available. Some species, which are predatory but also feed on pollen, such as *C. 14-pustulatus* and *Coccinula crotchi*, again may be found on a variety of different host plants. The same is true of some fully non-predatory ladybirds with many mycophagous species, such as *Psyllobora 22-punctata*, *Illeis galbula* and *Illeis koebelei*, feeding upon mildews on the leaves of a variety of host plants. For these species, there would be no advantage in evolving a colour pattern that was specifically adapted to one or a small number of host plants. A more generalised colour pattern that would serve on a range of host plants would be more beneficial. It is these species that show pure aposematic colouration most strongly, usually with just two strongly contrasting and sharply defined colours that have evolved primarily as warning defence mechanisms (Majerus, 1994a).

The majority of these generalist species are red, orange or yellow, in contrasting combination with black or white. However, other combinations, such as white and brown or orange and blue, are also employed by some species. Why all of these species are not red and black is not known. Indeed, we are so ignorant with regard to this conundrum that I hardly dare to speculate on the matter, except to say that phylogenetic history seems to have played some part. For example, many Chilocorini are black with reddish spots (Fig. 8.6), while many *Coccinula* species are black and orange. In the genus *Psyllobora*, yellow and black are predominant colours, although patterns vary. For example, *I. koebelei* (Fig. 8.7) has immaculate yellow elytra and jet black eyes; *I. galbula*

Fig. 8.6. *Chilocorus renipustulatus* mating pair.

Fig. 8.7. *Illeis koebelei.*

(Fig. 4.8) is yellow with a bold black zigzag bar across the elytra, while *P. 22-punctata* (Fig. 4.18) is yellow with black spots. Not only are all three members of the Psylloborini, but they are also all mycophagous, feeding on powdery white mildews. It may be that the yellow and black predominant colours reflect phylogenetic affinity. Alternatively, these colours may be defensively adaptive. It would be interesting to investigate the similarity of the pigments utilised by the different species. If the pigments are very closely similar, this may reflect common ancestry, making phylogenetic affinity a plausible explanation for the similarity. Conversely, if the pigments are different to any great degree, an adaptive explanation is more probable.

Host Plant Specialists

The colour patterns of the more host plant/habitat specialist species are in many cases easier to understand. Being restricted to one or a small number of usually similar species of host plants or trees allows them to evolve specific adaptations to those hosts. Of course, once specialised adaptations have evolved, a species will tend to be less fit if it moves to other hosts with different characteristics to which its adaptations are less appropriate. Thus, the specialisation is, in effect, self-reinforcing; specialisation to specific hosts restricts the species to these hosts and promotes further host-specific adaptations.

Some of the ladybirds that live among reed beds provide elegant examples. Many are extremely habitat specific, rarely being found far from reed-like plants. The colour patterns of many of these species are black on buff, with the markings generally being spots (e.g. *A. 19-punctata*), fused spots (e.g. *Anisosticta borealis*) or stripes (e.g. *Macronaemia episcopalis*) (Fig. 3.3).

The colour pattern of *A. 19-punctata* has been most closely scrutinised. In England, adults, after emerging from their pupae in late summer, are a pale fawn or buff colour, with 19 evenly spaced, usually discrete, small black spots. The ladybirds, which have just one generation per year, retain this colour through the late summer months, and into autumn, when they move to their winter refuges tucked between the leaves of the reeds. Here they remain, more or less inactive, for six or seven months. The buff and black colouration provides these overwintering beetles with extremely good camouflage on the dead, dried-out reed leaves, which rapidly become liberally spotted with the smuts of black mildews. *Anisosticta 19-punctata* retains this cryptic colouration throughout the winter. However, once they become active in the spring, they begin to change colour quite rapidly. The buff colour becomes flushed with red pigment and the ladybird takes on the red and black mantle of warning colouration (Fig. 8.8). The evolutionary rationale behind this change is that, during the spring the ladybirds must move from their winter retreats to seek aphids on the young green reeds emerging from the mud. Against this green backdrop, their cryptic colour pattern is no longer effective, so they employ the alternative stratagem of warning colouration (Majerus, 1994a).

Anisosticta 19-punctata, by changing colour, manages to be both cryptically and warningly coloured as appropriate at different times of its adult life. More remarkable are those species that manage to combine warning and crypsis in a single colour pattern.

It seems a contradiction to suggest that a warningly coloured species, which by definition is obvious to the eye, and memorable, can also be cryptic, but a number of ladybirds achieve this feat. The most striking example is the conifer specialist *H. 4-punctata* (Majerus, 1994a). The ground colour of this handsome

Fig. 8.8. *Anisosticta 19-punctata* summer colouration. (© Gilles San Martin.)

ladybird is somewhat variable, being some shade of orange, pink or reddish-brown, with diffuse longitudinal cream streaks and usually four or 16 black spots. When moving around on green pine needles, hunting for food, seeking a mate or laying eggs, it is easily seen (Fig. 8.9). However, when resting, it habitually takes up a position tucked away in the young conifer buds and shoots. Here it is superbly camouflaged (Fig. 8.10). *Harmonia 4-punctata* therefore gains benefit from both types of defence. When at rest, by taking up a position on a substrate that it matches in colour to a very high degree, its strategy is to avoid being detected at all. However, feeding and reproduction require movement. As moving objects are much easier to see than immobile ones, the ladybird is at risk when active, and then relies on its distasteful

Fig. 8.9. *Harmonia 4-punctata.* (© Gilles San Martin.)

Fig. 8.10. *Harmonia 4-punctata* camouflaged on conifer buds.

Fig. 8.11. *Myzia pullata.*

qualities and associated warning colours for defence. Other species that have evolved the same dual-purpose colour pattern defence include *Anatis ocellata*, *Anatis halonis*, *Myzia oblongoguttata* (Fig. 3.5), *Myzia pullata* (Fig. 8.11), *Myzia subvittata* and *Adalia conglomerata*. In *M. 18-guttata*, camouflage is achieved by keeping mainly to the crowns of mature, or semi-senile *Pinus sylvestris*, and resting on the little male cones.

It is probable that *T. 16-punctata* and *Tytthaspis gebleri* also manage to use the same pattern for both crypsis and warning (Majerus, 1994a). These species usually overwinter in large groups either among vegetation, such as dead grass stems, gorse foliage and dried reeds, or on flattish rather plain substrates, such as tree trunks, fencing posts (Fig. 6.1) or rocks. In the former set of situations, the black spots and stripes on the buff ground colour may act to disrupt the outline of the ladybird, making it difficult to see in the tangle of vegetation orientated at many different angles, and the confusion of shadows that such vegetation will throw. However, against a relatively plain substrate, these same black markings on a pale background will be easy to see, the more so as, in such situations, tightly packed groups of dozens, hundreds, or even thousands of ladybirds are the norm. Here the phrase 'safety in numbers' is very appropriate, for a predator attacking one individual, and learning of its unpalatability, will leave the rest of the group alone.

The colour patterns of some ladybirds are harder to explain from an evolutionary point of view. Several species of ladybird, such as *Halyzia 16-guttata*, *Calvia 10-guttata*, *Vibidia 12-guttata* and *Eocaria muiri* that live largely on the leaves of deciduous trees, are orange with white spots. That these species live in similar habitats, although they feed on different foods – *C. 10-guttata* and *E. muiri* are largely predatory, while *H. 16-guttata* and *V. 12-guttata* feed mainly

Fig. 8.12. *Halyzia 16-guttata.*

upon mildews – suggests that this colour combination is adaptive. It is possible that they also play the crypsis and warning colouration game. When on the underside of deciduous leaves, where these species tend to rest, they are all difficult to detect because, in the shade, their colours are not obvious and they become effectively just a black blob on the leaves, looking similar to the spots produced by certain viral blights. When active on the upper surface of leaves, the contrasting colours of these ladybirds are easy to see, and here these colours serve as a warning (Fig. 8.12).

Returning to the question of why all ladybirds are not alike, it is obvious that the selection pressures that have acted upon different species will be dependent upon the differences in the ecologies of these species. Moreover, as the theory of Müllerian mimicry is that species will converge on the colour pattern of the most common species, and this will vary from region to region, we may expect variety in the colour combinations involved in Müllerian complexes in different parts of the world. Add to that the differences in the ancestors of the different groups of ladybirds, and the possible use of colour and pattern in species recognition, and even if the reasons for the precise combinations of colours and patterns of the different groups are not known, we need not be surprised by the diversity of colour patterns.

There is, however, a second problem relating to colour pattern variation in ladybirds. Not only is there tremendous variation among species, but many ladybirds show extraordinary intraspecific colour pattern variation. If one believes in Darwinian evolution – the survival of the fittest – we must explore the ways in which this variation is maintained so that we can understand the solution to the question of how more than one colour pattern can be 'the fittest'.

Chapter 9: Variation and Evolution in Ladybirds

Heredity and the Environment

For my tenth birthday, my parents bought me a copy of the third edition of Professor E.B. Ford's book, *Butterflies*, originally published in 1945. Six months later, I had saved enough from my pocket money to buy Ford's companion volume, *Moths* (1955), from the same series. These two books, and particularly the sections on genetics and evolution, influenced my own approach to biological investigations. Two passages, one from each book introduced me to genetics. In the first paragraph of Chapter 9 of *Butterflies*, entitled 'Theoretical Genetics', Ford wrote that genetics 'provides opportunities for amateur scientific research of real value which, in my experience, a large number of collectors would be eager to undertake within the ordinary limits of their hobby if they could see more clearly how to conduct it. If one must find a reason for such work beyond the delight of satisfying one's curiosity on natural phenomena, that is easily done. For the principles of genetics are of universal application to living organisms, plant and animal. The breeder of butterflies who studies variation is obtaining evidence which may throw light upon problems arising in the culture of plants or the successful rearing of other animals. Moreover, certain aspects of such work have an especial bearing upon heredity in Man, whose blood groups are examples of polymorphism.'

He continues: 'The complaint is often made that genetics is a subject for specialists: not indeed too difficult for any ordinarily intelligent person, but requiring more intensive study than the amateur entomologist is prepared to accord it. There is no reason for this view. It is quite easy to master the essentials of genetics sufficiently to open up new and exciting possibilities to the collector of butterflies.'

In the introduction to Chapter 2, 'Genetics', in *Moths*, Ford goes further: 'There are few subjects in which the amateur can so easily and inexpensively make substantial contributions to scientific knowledge as in genetics... He can do so without disorganising his hobby or indeed very much extending it, even if his previous aim has merely been to amass a collection. He need only systematise his breeding of insects, keep his broods rigorously separate, and record his results accurately in the light of Mendel's laws and a few elementary extensions of them. Moreover, these are so simple that I have known a child of eleven master them in an afternoon and successfully apply them, without help, to the combination of different varieties of tame mice.'

Here was a direction for my hobby. It took me more than an afternoon, but, with the help of my mother and the public library, I soon grasped Mendel's laws, as applicable to ladybirds as to Lepidoptera. The wealth of colour and pattern variation found in many species of ladybird makes them model material for genetic studies, as does the relative ease with which several generations may be bred in a year, producing large numbers of offspring.

An understanding of the basic laws of inheritance is essential in considering variation within species. These laws are described and explained in many texts, including Ford's two New Naturalist books, and my books on *Ladybirds* (Majerus, 1994a), *Melanism* (Majerus, 1998b) and *Moths* (Majerus, 2002), therefore I shall not restate them. Rather, I shall describe the extent of colour pattern variation in ladybirds and its evolutionary role. Use of some related scientific terms is unavoidable. In addition, Table 9.1 defines the main types of genetic systems pertinent to variation in ladybirds, with examples of traits of each type.

Colour and Pattern Variation

Individual variation is a conspicuous phenomenon whenever organisms of the same species are carefully examined. We find little difficulty in recognising different people, through facial features, skin pigmentation, hair colour, shape and style, body configuration, height, weight and so on. We notice differences among ourselves more than variation in other organisms. Morphological variations are invariably detected when individuals of any species are carefully scrutinised. The most obvious features of ladybirds are their colours and patterns. Many species are extremely variable, having a great range of different forms. In others, colour and pattern are relatively stable.

Variation: Nature or Nurture?

Variation among individuals may be the outcome of differences in their genetic constitutions, or of differences in the environments to which they have been exposed, or of a combination of both. It is often difficult to discern whether a particular type of variation has a genetic or an environmental basis. This is the crux of the 'nature versus nurture' controversy with respect to human traits, such as intelligence or aggressive behaviour.

Table 9.1 Examples of Ladybird Traits Controlled by a Variety of Genetic Systems. The Control of the Trait is Given in Relation to the Typical or Wild-type Form, Unless Otherwise Indicated. Alleles are Autosomal Unless Otherwise Stated

Species and Trait	Mode of Inheritance	Expected Ratio in the Second Generation, Plus Notes
Extreme *annulata* form of *Adalia 2-punctata*	Single allele, with no dominance	1 extreme *annulata*:2 intermediate *annulata*:1 *typica*; intermediate *annulata* is somewhat variable
Harmonia axyridis f. *conspicua*	Single allele, with full dominance	3 *conspicua*:1 *typica*
Myzia oblongoguttata f. *lignicolor*	Single, fully recessive allele	3 *typica*:1 *lignicolor*
Vestigial-winged form of *Adalia 2-punctata*	Single, fully recessive allele; homozygote recessive has reduced viability	Variable, but at least 3 *typica* for every vestigial-winged individual
The *decempunctata*, *decempustulata* and *bimaculata* forms of *Adalia 10-punctata*	Alleles of a tri-allelic series, with a straight dominance hierarchy; the *decempunctata* allele being top dominant and the *bimaculata* allele the bottom recessive	Various. Note that no cross will give a 4:3:1 ratio of phenotypes
Pronotal and elytral patterns of *Adalia 2-punctata*	Controlled by genes close together on the same chromosome	Due to the linkage, three common elytral patterns, black, spotted and m-mark, are usually associated with melanic, *simulatrix* and *typica* elytral patterns, respectively
Pink form of *Aphidecta obliterata*	Single sex-linked recessive allele on the X chromosome	Frequency of pink males is higher than of pink females

Table 9.1 *(cont.)*

Species and Trait	Mode of Inheritance	Expected Ratio in the Second Generation, Plus Notes
Female mating preference for melanic males	Single dominant allele; expression is sex-limited	Although both males and females carry and transmit the allele, only females express it
Extent of black patterning in *Propylea 14-punctata*	Polygenic inheritance	Strong correlation between the mean level of melanism in parents and their progeny

We shall split variation into two classes, discontinuous and continuous. Variation is discontinuous if two or more discrete and obviously different forms occur, with all individuals being assigned to specific classes relatively easily. In *Myzia oblongoguttata*, the ground colour is either chestnut brown or dark chocolate brown, and intermediates between the two colour forms do not occur (Majerus, 1993). Discontinuous variation is usually under genetic control, the differences among forms being controlled by different alleles of one or a small number of genes. Performing a small series of specific crosses and scoring the progeny can demonstrate such control. Thus, the dark brown form of *M. oblongoguttata* (f. *lignicolor*) was shown in three generations of breeding to be controlled by a recessive allele of a single gene, with the more common chestnut coloured form being controlled by the alternative dominant allele (Majerus, 1993). Occasionally, discontinuous variation is produced by environmental factors, when one specific element of the environment passes a threshold condition at a crucial time in an organism's development, so triggering a particular developmental pathway. Such threshold effects are relatively rare, and can usually be uncovered by breeding under controlled conditions. This is the case with the alleles that control rate of pigment lay-down in *Adalia 10-punctata*, where a temperature threshold effects the expression of some alleles.

Variation is considered continuous when a range of forms exist, merging, one into another, with a full range of intermediates between the extremes. For example, measurement of the lengths of a large sample of *Coccinella 7-punctata* showed that the majority of individuals had lengths close to the average, with extremely short or extremely long individuals being rare, and all length classes between the shortest and longest being represented. The question is whether the variation in length is due to environmental factors, for example differences in temperature or available food during larval development, or to 'polygenes', where each of a large number of genes has a small effect on adult length, or to some combination of both. Here it is intuitively probable that both environment and genes play some part. Certainly, in humans, both affect height. Nevertheless, if both types of factor do affect *C. 7-punctata* lengths, can we say which has the greater influence? What we are really asking is how much of the observed variation is attributable to genetic and how much to environmental factors?

Heritability

The proportion of total variation under genetic control is termed the 'heritability' of the trait. Perhaps the simplest way of estimating heritability is to conduct what is called a mass selection experiment. The procedure involves measuring the lengths of a random sample of ladybirds to assess the total variation in length. From this sample, individuals of roughly the same length, from near one of the extremities of the distribution, are selected. These are allowed to breed among themselves and produce progeny. The lengths of all the progeny are then measured. Once we have the relevant data: L(total) – the average lengths of ladybirds in the whole original sample; L(parents) – the average length of ladybirds in the selected sample allowed to breed; and L(progeny) – the average lengths of their progeny, we can make an estimate of heritability by a simple calculation using these averages. We are concerned with two measures: the 'selection differential' (D) and the 'selection response' (R). D is the difference between the average lengths of the whole original population and that of the selected parent sample. R is the difference between the average lengths of the progeny produced by the selected parents and again that of the whole original population. If the average length of progeny is the same as that of the original sample, that is, R is zero, the variation must be environmental, for selecting parents of a particular length has had no effect on offspring length. Similarly, if the average length of progeny is equivalent to the average length of the selected parents, that is, D and R are the same, the variation must all be genetic. Such gains, either negligible, or conversely equivalent to D, will be rare. In most cases, some selection gain will be observed, but this value will be less than D. The heritability (h^2) is then obtained simply by dividing R by D. In terms of the average lengths, the calculation is:

$$h^2 = R/D = \{L(\text{Total}) - L(\text{parents})\}/\{L(\text{Total}) - L(\text{progeny})\}$$

Fig. 9.1. Variation in overwintering *Harmonia axyridis*. (© Nick Greatorex Davies.)

In the case of *C. 7-punctata* lengths, a mass selection experiment gave a heritability estimate of 0.61 (i.e. 61%) of the variation in length that is due to genetic factors.

At first sight, the most noticeable characteristics of a ladybird are its basic colour and its pattern. For example, *Harmonia conformis* is almost invariably orange with a set of 16 black spots. The main variation is in the size of the spots, but even here, the variation is not great. This species can be contrasted with *Harmonia axyridis*, in which the ground colour varies from yellow, through orange to red, and sports a huge variety of black patterns (Fig. 9.1). Essentially, there are five main pattern forms: spotted, banded, streaked, chequered and melanic, with many variations on each of these themes.

Harmonia axyridis is not unique in having a great variety of different forms. Many other coccinellids, such as *Adalia 2-punctata, A. 10-punctata, Propylea japonica, Cheilomenes 6-maculatus, Calvia 14-guttata* and *Harmonia testudinaria* (Fig. 9.2 and Fig. 9.3) show high levels of variability, which is almost exclusively under genetic control. However, not all variation is heritable and it is worth considering the mechanistic causes of non-heritable colour pattern variation before considering explanations of heritable variation.

Non-heritable Variation

One common type of non-heritable variation occurs most noticeably in species with a red ground colour. This is the occurrence of bronzy-brown or black patches on the elytra. These patches are often irregular in shape, asymmetrical between the two elytra and probably the result of disruption of

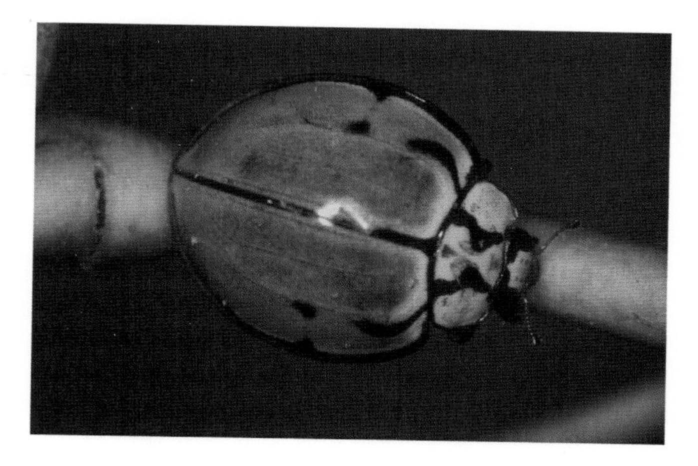

Fig. 9.2. *Harmonia testudinaria.*

Fig. 9.3. *Harmonia testudinaria.*

pigment production, resulting from injuries to larvae or pupae (Fig. 9.4). Often a line of black edges these injury marks. This is adaptive, for the black will be produced by the deposition of melanin pigments in the cuticle. Melanin is granular in form, and has a strengthening effect. In very rare cases, the whole of one elytron is abnormal, the other being normal. In reddish species, these individuals are easy to detect. 'Half-and-half' varieties of this kind (sometimes called bilateral mosaics) have been recorded in many species (Fig. 9.5). The abnormal side may be either the right or left elytron, and the pronotal pattern is unaffected. The possible inheritance of this type of condition was first investigated in *C. 7-punctata*. Two 'half-and-halves', one left-sided, the other right-sided abnormal, were allowed to mate. They produced over one hundred

Fig. 9.4. Disruption of pigment production resulting from injuries to *Coccinella 7-punctata*.

Fig. 9.5. *Harmonia axyridis* illustrating bilateral mosaic colouration. (© Jeff Winterbourne.)

progeny, all of which were completely normal. Twenty-three pairings between these progeny gave rise to over 1500 second-generation offspring, and again all were normal! The condition is therefore not inherited, a result that has been confirmed by breeding experiments with *A. 2-punctata* and *Coccinella magnifica*.

Quantifying Colour and Pattern Variation

In most ladybirds, patterns are more or less symmetrical, with pattern variation in spot number, spot size, spot fusions, and extent of melanism all occurring.

Some of the information on the extent of pattern variation has been obtained by examining museum specimens. However, colours change once the ladybird is dead. In particular, the lighter pigments may either lose brightness, or darken to a dirty yellow or orange colour. This has led to many descriptive errors in the literature, when species have been described simply on the basis of museum specimens. For example, *Anisosticta 19-punctata* and *Tytthaspis 16-punctata* are both described in taxonomic keys as yellow. While museum specimens often have a yellowish hue, this is not the case with live individuals. Similarly, some *Hippodamia* and *Coccinella* species described as orange are more usually red. In many coccinellid keys, the term testaceous is used for several species including, for example, *M. oblongoguttata* (Fig. 3.5), *Harmonia 4-punctata* (Fig. 8.9 and Fig. 8.10), *Halyzia 16-guttata* and *C. 14-guttata* (Fig. 9.6). Such a description can only derive from dead museum individuals. The description has the added drawback that many people do not know what colour testaceous is. Coleopterists should be wary of using colour traits given in taxonomic keys when identifying live coccinellids, unless the key is specifically based on the examination of live material.

The first step in studying colour pattern variation is to determine the normal extent of variation in one or more natural populations. This may be accomplished by first comparing the colours of wild individuals with a colour standard, such as the Ridgway *Standard Book of Color* (1912), and then drawing the patterning of large numbers of individuals. This process may be streamlined by having photocopied sheets of appropriate ladybird outlines (Fig. 9.7).

As mentioned previously, some species show very little variation. In some immaculate species, such as *Cycloneda polita*, *Mulsantina luteodorsa* and *Illeis koebelei*, colour pattern variation is almost confined to the markings on the

Fig. 9.6. Dark example of *Calvia 14-guttata*.

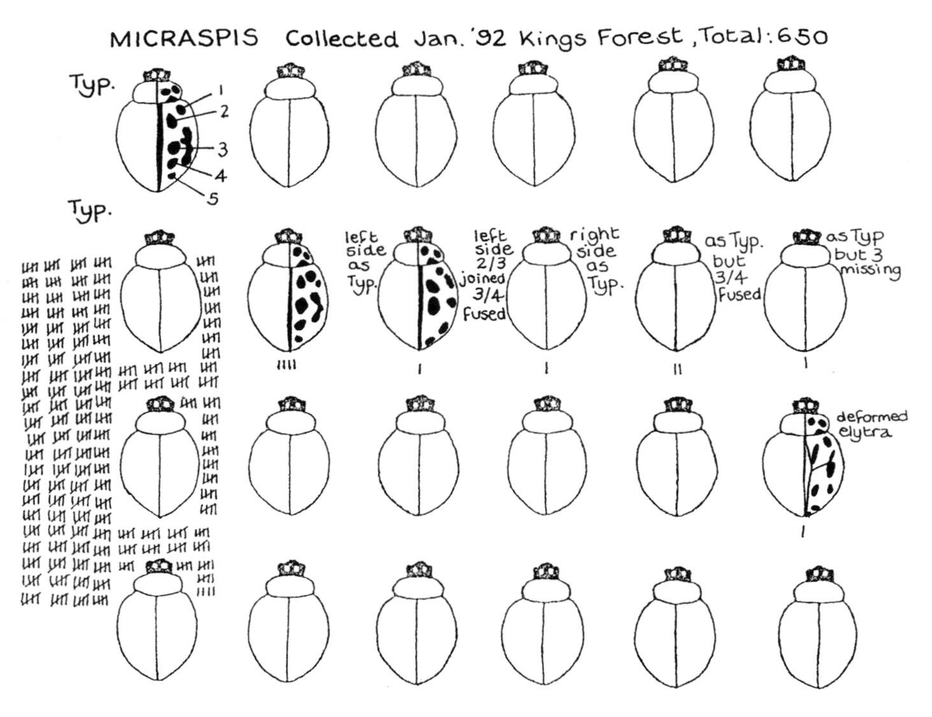

Fig. 9.7. Schematic of ladybird outlines for drawing the patterning.

pronotum. Most species of Chilocorini with black or dark ground colours and red or orange spots show little variation other than in the size of the pale spots. Thus, *Exochomus 4-pustulatus, Chilocorus cacti, Chilocorus orbus, Chilocorus stigma* and *Chilocorus 2-pustulatus* vary little. The same is true of the dark ground coloured *Hyperaspis 2-notata, Hyperaspis proba* and *Brachiacantha rotunda* being examples of consistent species. In some spotted species of Coccinellinae, such as *Psyllobora 22-punctata, Harmonia dimidiata, H. conformis*, and *T. 16-punctata*, again most colour pattern variation involves minute differences in spot size and shape. For example, in a sample of 650 *T. 16-punctata*, scored from an English population, and using the spot numbering system (Fig. 9.7), 640 were virtually identical.

In many other species, spot numbers are very variable. This can lead to a certain amount of confusion, when talking, for example, about an *A. 10-punctata* that has only two spots, particularly as ten-spotted examples of *A. 2-punctata* are also known. The situation is further complicated when spots are fused together. Counting then becomes difficult. Although spot numbers are often used in species names, the variation in spot number makes spot number counts poor indicators of species identity.

Geographic Spotting Variation

On a global scale, geographic trends occur in spot number and spot size. *Coccinella 5-punctata, C. 7-punctata* and *Hippodamia variegata* are widely distributed through Europe, Asia and North Africa. They exhibit rather similar geographic trends in the average amount of black on the elytra. The most red populations come from central Asia, with the average amount of black increasing in populations to the north, east and west. In *C. 7-punctata,* the increase in black is achieved primarily through an increase in the size of spots (Dobzhansky and Sivertzev-Dobzhansky, 1927), whereas in *C. 5-punctata* and *H. variegata,* it is achieved through an increase in the number of spots (Dobzhansky, 1933). Some workers have proposed that such consistent trends in geographic variations are correlated with climatic factors, such as temperature and humidity. In these three species, the variation appears to be correlated only with humidity, the least black populations occurring in the most arid areas. Correlation with temperature is poor, with, for example, populations of *A. variegata* having similar levels of pigmentation occurring in Ethiopia (mean annual temperature +30°C) and Yakutsk (mean annual temperature –10°C). Although the correlation between blackness and humidity suggests that the amount of black is adaptive, this has not been demonstrated experimentally.

The variation in spot number and size in these species, and many others, including many of the North American species of *Hippodamia,* is probably controlled by polygenic systems. In most, this has yet to be demonstrated. However, in *Propylea 14-punctata,* in which variation in the amount of black is largely a function of spot size, mass selection heritability studies have shown that the overall amount of black is polygenically controlled, with a heritability estimate of 0.78 (78%).

Not all spot number variants are controlled by polygenes. In Britain, *H. 4-punctata* normally has either 4 or 16 spots, this difference being controlled by a pair of alleles of a single gene. As we shall see later, many species of ladybird have a number of discrete forms, in which major genes (genes with a specific effect on phenotype) largely control spot number, although polygenes also play some part as modifiers.

Individuals in which some, or all, of the spots are fused together are known in many species. In some, such as *Brachiacantha felina, H. 16-guttata* and *Anatis rathvoni,* fusions are rare. In others, such as *P. 14-punctata, Naemia seriata, Hippodamia 15-maculata* and *Hippodamia sinuata* (Fig. 9.8 and Fig. 9.9), they are common, or indeed, the general rule. In *T. 16-punctata,* as virtually all individuals in many western European populations have the three lateral spots merging to produce a zigzag line (Fig. 4.1), this patterning is rarely thought of in terms of spot fusion. Similarly, the chequered pattern or 'bed socks'

Fig. 9.8. *Hippodamia sinuata.*

Fig. 9.9. *Hippodamia sinuata.*

chequered form of *A. 10-punctata* (Fig. 9.10) is, in essence, simply a 12-spotted typical individual with all the spots increased in size so that they fuse into the grid-like pattern.

Little is known about the inheritance of spot fusions. Fusion of spots is simply a consequence of an increase in the size of spots in *P. 14-punctata*, where spot size is controlled by a polygenic system (Fig. 9.11), and is independent of variation in the tone of the yellow ground colour (Fig. 9.12). Polygenic control of fusion resulting from increased spot size is also likely in species such as *H. 15-maculata*, *Brachiacantha ursina*, and *Brachiacantha bollii*. In other species in which fusions are common, the variation is less straightforward. Thus, in *Hippodamia 5-signata*, in which spotting varies from spotless to 11-spotted, there is no obvious correlation between spot number and spot

Fig. 9.10. *Adalia 10-punctata* f. *decempustulata*. (© Gilles San Martin.)

Fig. 9.11. Colour pattern variation in *Propylea 14-punctata*.

size. Here it is likely that the genetics of the variation involves both major genes and polygenes.

Geographic variation has been studied in a few ladybirds. For example, *Aphidecta obliterata* shows little variation in Britain, most individuals being brown and unspotted, or having a small oblique black mark near the back of the elytra. This is also true in North America, where *A. obliterata* has established in North Carolina from releases of German stock. Yet, in continental Europe, including Germany, a wide range of forms from unspotted to completely black occur (Fig. 9.13). Genetic studies on Luxembourg samples have shown that melanism in this species is inherited. Why this species is very much more variable in some parts of its range than in others has not been investigated.

Fig. 9.12. Two different tones of yellow in mating *Propylea 14-punctata*. (© Remy Poland.)

Fig. 9.13. *Aphidecta obliterata* variation in Germany.

Similarly, in Britain, *C. 14-guttata* is one of the least variable ladybirds. Of the many thousands of individuals of this species I have seen in Britain, only one has been significantly different from the typical maroon-brown ground colour with 14 discrete off-white spots (Fig. 9.14) (Majerus, 1994c). However, in North America, this ladybird is highly polymorphic. I well remember my

Fig. 9.14. Unusual colour form of *Calvia 14-guttata*.

first encounter with this species in Edmonton, Canada, in the company of John Acorn. The black-spotted pink, and white-spotted black beetles that he ascribed to *C. 14-guttata* were so far removed in colour and pattern from the *C. 14-guttata* with which I was familiar, that I found it difficult to believe that they were the same species. Yet subsequent breeding experiments between Canadian and European forms have shown that they interbreed freely producing viable and fertile offspring.

One final spotting phenomenon must be considered. This is the existence of compound spots made up of two colours, as occur in *Anatis mali*, *Anatis ocellata* and *Coccinella 3-fasciata*, in which black spots are usually surrounded by a ring of yellow. This type of spot occurs at high frequency in these species and as a rarity in some other red species with black spots, such as *A. 2-punctata* and *Coccinella 11-punctata*. In *A. 2-punctata*, a propensity to have yellow rings around the spots runs in families, and so is obviously inherited, although the control mechanism has not been analysed. In *A. ocellata* (Fig. 3.7), variation in the amount of black and yellow in the compound spots leads to ladybirds of very different appearances. Individuals with no yellow rings occur. Conversely, in some individuals, one or more of the spots have no black centres, leaving just creamy-yellow spots (Fig. 9.15). Work on this latter form has shown that this condition is controlled by the recessive allele of a single gene.

Black forms of ladybirds (and other organisms) are termed melanics, because their black pigments are called melanins. Chemically, these pigments comprise a variety of polymerised products of tyrosine. Not all these products are black. Some are yellow or brown. A biochemical pathway in which yellow melanin pigments constitute an intermediate step is probably responsible for

Fig. 9.15. *Anatis ocellata* lacking black spots.

the black pigments that make up the spots of many ladybirds. In *A. ocellata* and *A. mali*, yellow melanin is laid down in the spot positions first, and these pigments are converted to black from the centre of each spot outwards. In some, all the yellow is converted; in others, only some is converted, although in the majority of these, the reaction seems to continue throughout the ladybirds' lives, so that the width of the yellow ring gradually diminishes with age. This accounts for the fact that unringed individuals are less common in samples of recently emerged ladybirds than in post-winter samples. Those with 'blind spots', that is, spots without black centres, are presumably the product of the latter part of the biochemical pathway being locally inhibited.

Rare Varieties

Even in ladybirds that are relatively constant in colour and pattern, rare varieties occur as a result of genetic mutation. For example, *T. 16-punctata* has a very rare melanic form, f. *poweri*, in which the pronotum and elytra are completely black (Fig. 6.1). The rarity of this form may be gauged by the fact that I have recorded it only twice among many tens of thousands of individuals (Majerus, 1991d).

One rather painstaking way of detecting rare colour pattern varieties is simply to scrutinise large numbers of wild caught individuals. However, as most mutations are recessive when they arise, an alternative method of detecting rare varieties is to inbreed laboratory stocks. Lusis used this method to considerable effect, identifying a number of mutations in *A. 2-punctata*. These included f. *asiphonia* in which males lack the copulatory sipho, a form in which the wings are reduced to small stumps, a form with white eyes, and a form with albino larvae, a trait also seen in *A. 10-punctata*. In each case, Lusis found these traits to be controlled by single recessive genes (Zakharov, pers. comm.).

Many of the abnormal traits revealed by inbreeding do not occur in the wild, except as a rare recessive mutation. This is the case for a deep purple form of *A. 2-punctata* (f. *purpurea*) (Majerus et al., 1987) and for a wingless form of the same species (Marples et al., 1993; Ueno et al., 2004). Here, no reason for the existence of such forms need be sought. They arise simply as a result of chemical mistakes during the copying of the primary genetic material. Such mutations, if they have a significant effect on the appearance, behaviour or physiology of an individual, are likely to be detrimental and will be selected out of the population. However, that is not always the case. Occasionally, new mutations will be beneficial, and the mutant gene will begin to increase in the population, replacing its original allele. Consequently, at least for a time, both the new and old forms will be present in the population concurrently. As such, the population is said to be genetically polymorphic.

Colour Pattern Polymorphism in Ladybirds

Ford (1940) defined genetic polymorphism as: 'The occurrence together, in the same locality, at the same time, of two or more discontinuous forms of a species in such proportions that the rarest of them cannot be maintained merely by recurrent mutation.'

This definition excludes geographic races, subspecies, and seasonal variations. Additionally excluded are traits, such as size, in which variation is continuous and discrete forms cannot be picked out, and very rare forms that arise from time to time through mutation, but are selected out of the population.

Conspicuous genetic polymorphisms were problematic for the early evolutionary biologists. The difficulty was that, following Darwin's thesis of the 'survival of the fittest', it was not obvious how natural selection could promote two 'fittest' forms within an interbreeding population. Darwin (1859) considered polymorphisms in *The Origin of Species*, and specifically divorced natural selection from any involvement with their existence, writing: 'This preservation of favourable mutations, and rejection of injurious variations, I call natural selection. Variations neither useful nor injurious would not be affected by natural selection and would be left a fluctuating element as perhaps we see in the species called polymorphic.'

Darwin viewed conspicuous polymorphisms as selectively neutral, any changes in the frequencies of forms over time being a consequence of random genetic drift, and this was held as true by most evolutionary geneticists up to the 1940s. Sir Ronald Fisher and Professor E.B. Ford disagreed, believing that a balance of selective advantage and disadvantage maintained the different forms. The selection-drift controversy of the mid-part of the twentieth century (sometimes called the Fisher–Wright controversy after the two main protagonists, Fisher and Professor Sewall Wright) was enduring and often acrimonious (Provine, 1986). In the 1920s, Fisher showed mathematically that two or more forms could be maintained stably within a population if either a heterozygous genotype was fitter than either of its homozygotes (heterozygous advantage), or when the fitness of a form was inversely correlated to its frequency (negative frequency dependent selection) (Fisher, 1930). In the 1930s, Ford embarked on a research programme, which lasted into the 1960s, to collect evidence to assess whether the frequencies of forms of polymorphic species were influenced more by drift or selection (Ford, 1964). This programme included work on the scarlet tiger moth, *Callimorpha dominula*, the meadow brown butterfly, *Maniola jurtina*, the peppered moth, *Biston betularia*, the banded land snails *Cepaea nemoralis* and *Cepaea hortensis*, the African swallowtail, *Papilio dardanus* and the 2-spot ladybird, *A. 2-punctata*. In each case, selection rather than drift was shown to be primarily responsible for changes in the frequencies of forms. In most, the selective factors acting on and maintaining the forms were identified. However, in the case of *A. 2-punctata*, and the many other spectacularly polymorphic species of ladybird, although many selective factors have been implicated in the maintenance of the forms, there is still little consensus over the evolutionary reasons for the initiation and maintenance of polymorphisms.

The Problem of Polymorphism in Ladybirds

The colour pattern polymorphisms of ladybirds are puzzling because their existence is at odds with the theory of the evolution of aposematic colour patterns and Müllerian mimicry. Just as there is benefit in being one of many rather than one of a few in Müllerian mimetic complexes, it follows that, within a species, it should be advantageous for brightly coloured, chemically defended species to all look the same, so that predators only have to learn and remember one colour pattern to avoid.

Yet, the range of colour combinations found in ladybirds is tremendous. As described in Chapter 8, the ground colours cover almost the whole colour spectrum, including blues, lilacs and turquoise, and the colours of markings are almost as diverse. Among all this colour and pattern diversity within the family, one phenomenon is found in the vast majority of species that have been studied to any significant degree, namely the existence of melanic forms.

The word melanism derives from the Greek word *melos*, meaning black. However, the meaning in a scientific context has been widened; Kettlewell (1973) defined melanism as the occurrence in a population of a species of some individuals that are darker than the typical form due to an increase in the epidermis of certain polymerised products of tyrosine substances that produce the complex of pigments collectively known as the melanins.

In some ladybirds, darker or blacker forms occur as rarities; in others, the amount of black patterning varies with geographical location; while in still others, melanic and non-melanic forms occur together as genetic polymorphisms. I have already discussed some of the reasons why ladybirds are not all alike. Now I will focus on melanism to discuss the possible reasons for polymorphic variation in this aposematic group.

Types of Melanic Forms in Ladybirds

A ladybird may be considered melanic in one of two main ways. Most ladybirds have a two-colour pattern: essentially a ground colour with a pattern of markings of a contrasting colour. The markings may be darker than the ground colour as in *C. 7-punctata*, or lighter as in *M. oblongoguttata* (Fig. 3.5). Thus, it is possible to increase the overall darkness of an individual either by darkening the ground colour, or by altering the ratio of the two-colour elements in favour of the darker colour.

An example of melanism involving the ground colour is the *lignicolor* form of *M. oblongoguttata*. Similarly, *C. 14-guttata* has forms in which white spots pattern either maroon-brown or black elytra (Fig. 9.6). In some usually

Fig. 9.16. Melanic *Coccinella 7-punctata* mating with typical form.

red species, recessive mutants are known that cause the normal bright red colour to gradually darken through purple to almost black, as in *A. 2-punctata* f. *purpurea* and *C. 7-punctata* f. *anthrax* (Fig. 9.16) (Majerus et al., 1999).

Examples of melanic forms produced as a result of an increase in the extent of dark patterning are even more common. These include several species in which completely black forms occur only as extreme rarities, such as *Subcoccinella 24-punctata* f. *nigra*, *T. 16-punctata* f. *poweri* (Majerus, 1991d; Revels and Majerus, 1997), and *A. ocellata* f. *hebraea*.

The existence of melanic forms that occur only at very low frequency and are controlled by major genes is simply the result of the chance recurrence of such forms by mutation and their eventual elimination by selection because they are disadvantageous. Such rare variants need not concern us here.

In other species, the size of black spots and fusions between them increase to such an extent that they produce blotches or grid patterns. For example, Mader (1926–1937) depicted 26 pattern varieties of *Aiolocaria mirabilis*, from completely unmarked on the pronotum, apart from a black scutellary spot, to completely black, each variety shown having the black areas slightly enlarged compared to the previous one. The melanic variation here is continuous, and so cannot be considered as true genetic polymorphism. However, melanic polymorphism has been reported in many ladybird species: Fig. 9.17–19.19 and Figs. 9.20–9.22 showing two examples.

Three highly variable species, *A. 2-punctata*, *A. 10-punctata* and *H. axyridis*, have received most attention, both with respect to the genetics of melanism, and the underlying reasons for the evolution and maintenance of melanic polymorphism.

In *A. 2-punctata*, Mader (1926–1937) described and named over a hundred different forms from Europe and Asia, and other authors have illustrated many additional pattern variations subsequently (see Majerus, 1998b for review). In most populations, only a few forms reach frequencies above 1%. Briefly, European populations tend to be either predominantly *bipunctata*, or polymorphic for *bipunctata* and two melanic forms, f. *quadrimaculata* and f. *sexpustulata* (Fig. 9.23 and Fig. 9.24, respectively). Other forms may reach significant frequencies locally. In particular, a form named weak *annulata* (Majerus, 1994a), a genetically heterogeneous form comprising heterozygotes between the *bipunctata* allele and at least three comparatively rare alleles (*simulatrix*, *duodecempustulata* and extreme *annulata*), has been recorded at over 5% in many parts of Europe, and f. *frigida* is common in the north of Scandinavia. This latter form is similar to f. *bipunctata*, but with the black spot being flanked by additional spots or extended into a bar (Fig. 9.25). In central and eastern Asia, the *bipunctata* form predominates in most populations, sometimes together with a different melanic, f. *sublunata*, while *quadrimaculata* and

Figs. 9.17–9.19. Colour form variation in *Coelophora inaequalis*.

sexpustulata are rare or absent. In some Asian populations, other forms, such as *simulatrix* and *duodecempustulata*, reach significant frequencies. In American populations, five forms attain frequencies in excess of 1%. These range from

Figs. 9.20–9.22. Colour form variation in *Menochilus 6-maculatus*.

f. *melanopleura*, which has unpatterned red elytra, through *bipunctata*, *coloradensis* and *annectans*, which are all red with varying numbers of black spots, to f. *humeralis*, a melanic form that appears indistinguishable from

Fig. 9.23. *Adalia 2-punctata* f. *quadrimaculata*.

quadrimaculata (Fig. 9.26). Despite the apparent similarity between *humeralis* and *quadripunctata* on either side of the Atlantic, their genetic control is different, *humeralis* being genetically recessive to *bipunctata*, while *quadripunctata* is dominant.

In Eurasia, *A. 10-punctata* exhibits even more variation than *A. 2-punctata*, largely because it has more variation in colours involved in the patterns. Essentially, *A. 10-punctata* has three basic morphs. Form *decempunctata* is orange with zero to 15 dark dots (Fig. 9.27), *decempustulata* has a dark grid like pattern on a paler ground colour (Fig. 9.10 and Fig. 9.28, male), and *bimaculata* is dark all over the elytra, with the exception of a pale flash at the anterior outer margin of each elytron (Fig. 9.28, female). Each of these basic forms has many variations on a theme. The pale colour may vary from cream or pale beige through dull yellow, orange, light tans and browns to bright red, while

Fig. 9.24. *Adalia 2-punctata* f. *sexpustulata*.

Fig. 9.25. *Adalia 2-punctata* f. *frigida*.

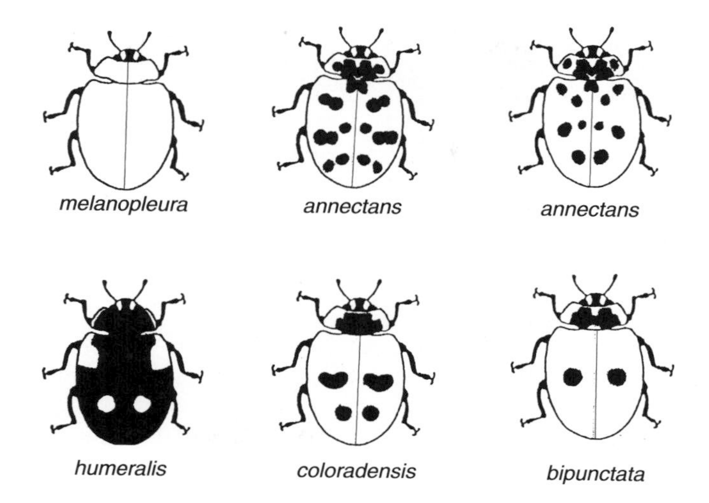

Fig. 9.26. *Adalia 2-punctata* f. *melanopleura*, f. *annectans* (form 1), f. *annectans* (form 2), f. *humeralis*, f. *coloradensis* and f. *bipunctata*.

Fig. 9.27. *Adalia 10-punctata* f. *decempunctata* mating pair (bottom: female; top: male).

the dark colour varies from mid-tans, maroon, purple, darker browns, to black. The strength of the markings also varies considerably. Furthermore, the rate of pigment lay-down can be very fast or extremely slow, even when temperatures are controlled. A single gene has been shown to play a major role in the rate of pigment deposition (Majerus, 1994a).

If just the three basic forms of *A. 10-punctata* are considered, the pattern of temporal and spatial variation is very different to that in *A. 2-punctata*, for all populations appear to support all three forms at significant frequencies, *decempunctata* usually being most common and *bimaculata* most scarce. Furthermore, the frequencies of these forms do not vary substantially, either

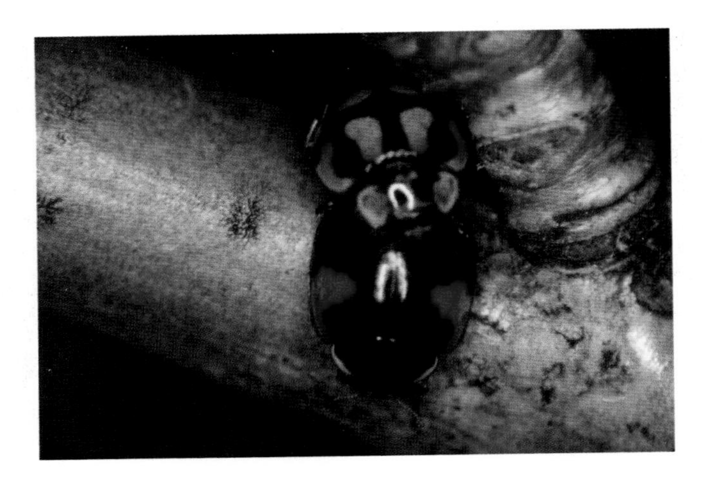

Fig. 9.28. *Adalia 10-punctata* f. *bimaculata* and f. *decempustulata* mating pair.

temporally or spatially, and changes that do occur take place slowly (Brakefield and Lees, 1987; Majerus, 1994a).

The harlequin ladybird, *H. axyridis*, is nearly as variable as the two *Adalia* species. The patterns on the elytra of this species are bold combinations of black and orange or black and red. The form that might be considered typical is orange with 18 black spots. However, some or all of these spots may be absent, or they may be extended so that some are fused together. In western populations, such as those from the Altaysk Republic and around Lake Baikal, a chequered patterned morph, f. *axyridis*, is the most common, although this form declines further east and is scarce in Japan. A range of melanic forms occurs. These are characterised by having most, or all, of the lateral margins of the elytra black, with the interior of the elytra sporting various bold orange or red markings. In f. *conspicua*, a bold red (or more rarely orange) spot is present towards the front of each elytron. Form *spectabilis* has two red spots on each elytron, and f. *aulica* has a large red patch that covers the central section of the elytra.

Dobzhansky surveyed many populations of *H. axyridis* in the 1920s, demonstrating that this species shows great geographic variation in the frequencies of the various forms (Dobzhansky, 1933; Majerus, 1998b). Work in Asia but outside the former Soviet Union has also shown considerable geographic variation in form frequencies. For example, Komai (1956), noting that the frequency of melanics is greater in southern locations, suggests *H. axyridis* exhibits climatic adaptation. Variation in this species has the added component that melanic frequencies alter seasonally (Tan, 1949). Interestingly, in North America, where *H. axyridis* became established and has spread dramatically over the last two decades, the nominate form, f. *axyridis*, is absent, and melanic forms are rare. Orange ladybirds, with from zero to 21 black spots, dominate American populations, with females generally being more strongly marked than males. Similarly, in South Africa f. *succinea* dominates. The lack of f. *axyridis* and the rarity of melanics are presumably a result of the gene pool of introduced individuals being dominated by alleles that produce orange spotted or unspotted beetles. In Britain, where *H. axyridis* arrived in 2004 (Majerus, 2006), the f. *succinea*, f. *conspicua* and f. *spectabilis* are all present, but f. *axyridis* is absent. Survey work has shown that the frequencies of the two melanic forms have declined significantly since establishment. Interestingly, the temperature at which individuals of f. *succinea* pupate can influence the spot size (Michie et al., 2010) and, at very cool temperatures, individuals of this predominantly orange colour form can appear almost melanic.

The Genetics of Melanic Polymorphism in Ladybirds

The genetics of many of the varieties of these species has been described (see Majerus, 1994a, 1998b for reviews). In each case, a single locus with multiple alleles appears to control most of the common forms. In *A. 2-punctata*, Lus (1928, 1932) identified 12 alleles of a single gene in European and Asian material. The dominance relationships between these alleles were relatively straightforward, at least where alleles occurred in the same population. The order of dominance was s^{l} (the top dominant), s^{lu}, s^{m}, s^{p}, s^{i}, s^{a}, s^{il}, s^{t}, s^{o}, s^{s1}, s^{s2}, s^{d} (the bottom recessive) (for allelic notation see Fig. 9.29A, B). Throughout the series, there is a tendency for the degree of melanism to be correlated to the genetic dominance of the allele responsible, such that melanics are dominant to the non-melanics. However, the correspondence is not exact. The order of decreasing melanism in respect of forms homozygous for the alleles described by Lus is: *sublunata* (s^{l}), *lunigera* (s^{lu}), *quadrimaculata* (s^{m}), *sexpustulata* (s^{p}), *duodecempustulata* (s^{d}), *annulata* (s^{a}), *ocellata* (s^{o}), *simulatrix 1* (s^{s1}), *simulatrix 2* (s^{s2}), *bipunctata* (s^{t}), *impunctata 2* (s^{i}), *impunctata 1* (s^{il}). In *H. axyridis*, a similar dominance hierarchy of the multiple alleles exists, again with melanic forms usually being dominant over non-melanics. Conversely, in *A. 10-punctata*, the top dominant allele is the non-melanic f. *decempunctata*, followed by f. *decempustulata*, with the melanic f. *bimaculata* being the bottom recessive (Lus, 1928; Majerus, 1994a).

Analysis of European populations has shown that the main colour pattern locus of *A. 2-punctata* identified by Lus is not, in fact, a single gene, but a group of genes all located close together on the same chromosome, that is, a super-gene. The close proximity of the genes that make up the super-gene means that alleles of the different genes are genetically linked and are usually inherited together, only not being inherited as a unit following an exchange of genetic material between chromosomes when a cross-over event occurs during meiosis. The super-gene in *A. 2-punctata* contains at least four and

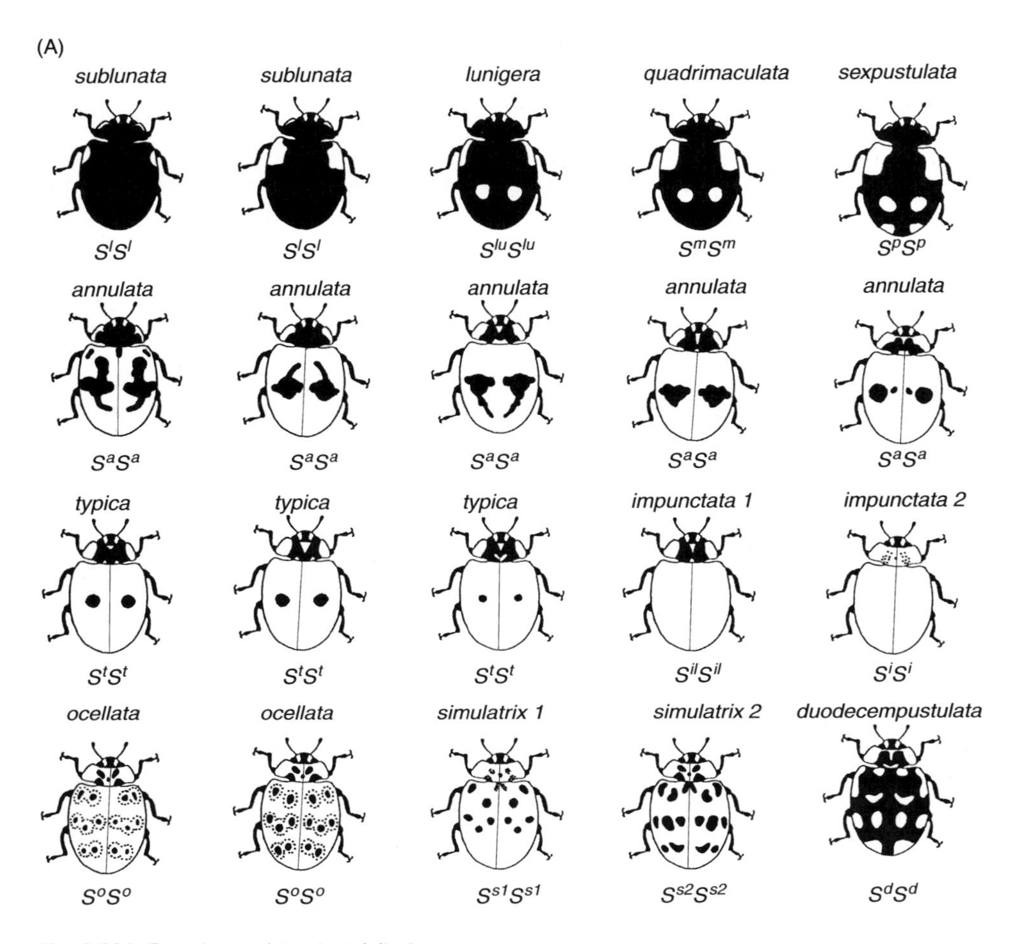

Fig. 9.29A, B. Polymorphism in *Adalia 2-punctata*.

possibly more very tightly linked genes (Majerus, 1994a). At least three genes affect elytral colour patterns and at least one affects the pronotal colour pattern. Why four genes affecting similar morphological traits should be located very close together on the same chromosome is not known, but is unlikely to be coincidental. It is possible that selection has acted to draw relevant genes together from different parts of the genome. However, the fact that three of the four genes within the super-gene affect elytral colour pattern raises the possibility that these genes have arisen by gene duplication. Unequal crossing-over during meiosis, whereby homologous chromosomes do not align absolutely precisely, may lead to two or more copies of the same gene occurring next to each other on a chromosome. Once they have arisen, these duplicate genes may evolve independently to one another, with new alleles arising by mutation at each locus within the super-gene.

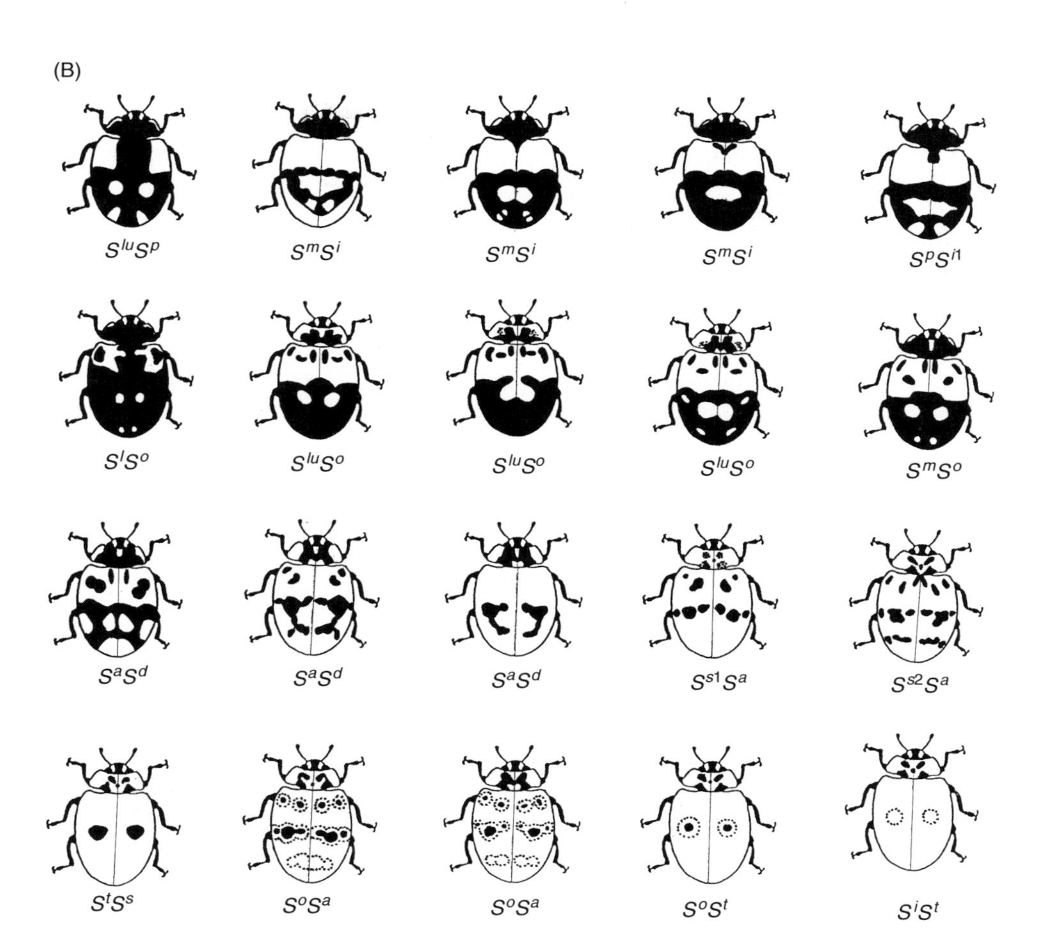

Fig. 9.29A, B. (cont.)

There is some evidence to suggest that the main colour pattern gene in *A. 10-punctata* and of some of the major genes involved in the mosaic inheritance in *H. axyridis* (Tan and Li, 1934; Hosino, 1936) are also super-genes, with rare phenotypes that may be recombinants occasionally appearing in laboratory crosses (Majerus, 1998b).

While a super-gene is responsible for most forms in these species, some other unlinked loci cause some of the varieties. In English *A. 2-punctata*, the recessive allele of an unlinked gene increases the area of melanic markings of some forms, and a recessive allele at another locus further extends black patterning, producing an all-black form (Majerus and Kearns, 1989).

There is also evidence that modifier genes are involved in the dominance of several of the alleles in the super-gene. Hence, for example, while *quadrimaculata* is dominant over *bipunctata* in Europe (Hawkes, 1920), the phenotypically indistinguishable *humeralis* in America is recessive to *bipunctata* there

(Palmer, 1911, 1917). Furthermore, breeding between Scottish and southern English or Parisian *A. 2-punctata* produces breakdown of dominance, suggesting that different sets of dominance modifiers have become fixed in these different populations (Majerus and Kearns, 1989). One intriguing finding with respect to the dominance relationships of melanic and non-melanic forms in *A. 2-punctata* arose from studies of New Zealand samples. *Adalia 2-punctata* was introduced into New Zealand in the early days of colonialism, probably from Europe. Melanic forms occur and are of two basic types. Some are indistinguishable from British *quadrimaculata* and *sexpustulata*, while others were similar to melanic *annulata*. Breeding experiments revealed that all types of melanics were controlled by a single allele of the main colour pattern locus, with the melanic allele recessive to *typica* (Majerus, unpubl. data). Assuming that the original introductions were from Britain or Europe, these findings present two possibilities. Firstly, that the melanics in New Zealand may have arisen through selection acting on recessive *annulata*-type alleles of the supergene and their modifiers, in the absence of the *quadrimaculata* allele, to create a stable melanic form. Alternatively, selection may have acted against the expression of melanic alleles in heterozygotes, gradually reversing the dominance of the alleles involved. This latter possibility seems unlikely, because, were the melanic form selected against, being originally dominant, it would have simply been selected out of the population.

In *A. 10-punctata*, the number of dark dots on the *typica* form is controlled partly by the major colour pattern locus, and partly by a polygenic system (Majerus, 1994a). In addition, the colours of both the lighter and darker elements of the pattern in this species vary greatly. The colours of most ladybirds darken as they age, because pigments continue to be synthesised and laid down throughout the ladybirds' adult lives. In *A. 10-punctata*, the lay-down rate is highly variable for both dark and light pigments, and is influenced by both genetic and environmental factors. Analysis of this variation has shown that the lay-down rate of light colours (mainly carotenoids) is independent of that of the dark colours (melanins). Each is controlled by two alleles of a single gene, one for carotenoids, and the other for melanins. In ladybirds reared at 24°C, the slow lay-down alleles are dominant to the fast lay-down alleles for both genes (Jones and Majerus, unpubl. data). However, if pupae are kept at 12°C, the melanin pigments in resulting adults develop rapidly, irrespective of their genetic composition (Jones and Majerus, unpubl. data). The carotenoid and melanin genes are closely linked on the same chromosome, either both fast alleles, or both slow alleles, usually being found together. They are not linked to the main colour pattern gene. The result is four combinations: carotenoids fast/melanins fast; carotenoids fast/melanins slow; carotenoids slow/melanins fast; and carotenoids slow/melanins slow, with the first and last classes being most frequent due to the genetic linkage between the pigment lay-down rate loci (Majerus, 1994a).

The Maintenance of Melanic Polymorphism in Ladybirds

Many researchers have tried to fathom the factors that have influenced the evolution and maintenance of melanic polymorphism in ladybirds, with most effort being devoted to *A. 2-punctata*. As yet, this work has not provided a definitive explanation of the variation in the frequencies of forms across time or space. Rather, the data from both field and laboratory studies leads to contradictory interpretations. Factors proposed in explanation include mimicry, thermal melanism, balancing selection acting upon co-adapted gene complexes and reproductive factors deriving from genetically controlled mating preferences and female-biased population sex ratios resulting from the presence of male-killers. These factors are not mutually exclusive, but it is easiest to consider each individually before discussing ways they may interact.

Before doing so, what constitutes a melanic form in *A. 2-punctata* must be defined. If we were to follow Kettlewell's (1973) definition, almost all forms of *A. 2-punctata* would be melanic, non-melanics being forms such as *bipunctata* itself, the American *melanopleura*, the Asian *impunctata* and the rare British varieties 'mini-spot' and 'halo' (Majerus, 1994a). Here, however, I shall arbitrarily declare that those forms in which more than 50% of the elytra are black are melanic, while those in which red covers more than half are non-melanic. With this designation in mind, two categories of melanics can be defined on the basis of the frequencies that they achieve. These are the common melanics that comprise more than 1% of some populations and include *quadrimaculata* and *sexpustulata* in Europe and western Asia, *bimaculata* in central Asia and *humeralis* in America. The other class, the rare melanics, include forms such as *duodecempustulata*, strong spotty, extreme *annulata*, and melanic *annulata*, among others. None of these have been recorded as comprising more than 1% of any large population.

Mimicry and Polymorphism

What factors are responsible for maintaining melanic polymorphism in *A. 2-punctata*? As explained previously, if *A. 2-punctata* is aposematic, it is presumably also part of the 'black spots on a red background' Müllerian mimicry complex, and should, according to the theory of Müllerian mimicry, be monomorphic. However, what if *A. 2-punctata* is not protected from predation by a combination of defensive chemicals and memorable colours? In that case, it may be an edible Batesian mimic rather than an unpalatable Müllerian mimic, and in special circumstances, Batesian mimicry may lead to polymorphism. If a non-mimetic species evolves a mimetic form, polymorphism may result if the initial advantage that the mimic gains from its resemblance to the model disappears before the mimic completely replaces the non-mimetic form. This may happen because the selective advantage of a Batesian mimic declines as it becomes more common. The advantage of resembling an unpalatable model depends on predators having learnt to avoid the model. As the mimic becomes more common relative to the model, the probability that a naive predator will come upon and try to eat a mimic before it has encountered and learnt to avoid the model will increase. This leads to the negative frequency dependent selection that may maintain polymorphism in Batesian mimic species. Furthermore, polymorphism will not be restricted to just an ancestral non-mimetic form and a single mimetic form. If several model species are available, each may become the object of a mimetic form.

Brakefield (1985a) proposed that *A. 2-punctata* might be an example of a polymorphic Batesian mimic. He argued that *A. 2-punctata* might not be toxic to birds. This proposition was confirmed by comparative tests, in which *A. 2-punctata* and *C. 7-punctata* were forcibly fed to *Parus caerulus* chicks (Marples et al., 1989). While *C. 7-punctata* was lethal to the chicks, *A. 2-punctata* was not. Brakefield (1985a) then proposed that the non-melanic forms of *A. 2-punctata* may be Batesian mimics of black on red species, such as *C. 7-punctata*, while the melanic forms are Batesian mimics of red on black species such as *Exochomus 4-pustulatus* and *Chilocorus renipustulatus*. However, while *A. 2-punctata* is not poisonous to birds (Marples et al., 1989), they also showed that it is unpalatable. Recognising this possibility, Brakefield (1985a) also speculated that *A. 2-punctata* may be a Müllerian mimic, being involved in two Müllerian mimicry complexes. This latter explanation seems less realistic, because Müllerian mimicry is positively, not negatively frequency dependent. In each population, the expectation is that *A. 2-punctata* should mimic the most common colour pattern combination, either black on red or red on black, rather than being polymorphic for both. Polymorphism will result only in areas where monomorphic populations for the alternative colour forms meet and cross breed, as in the polymorphic hybrid zones of the various geographic forms of the Müllerian mimetic butterflies *Heliconius erato* and *Heliconius melpomene* in tropical South and Central America.

If it seems unlikely that Müllerian mimicry is involved in the maintenance of melanic polymorphism in *A. 2-punctata*, the question of whether *A. 2-punctata* is aposematic becomes critical. Some birds can and do prey upon *A. 2-punctata*, but most only do so very rarely, under conditions of hardship. The selection pressure imposed by this normally low level of predation will be slight, and is unlikely to be sufficient to lead to the evolution of true polymorphic Batesian mimicry (Majerus, 1998b; Majerus and Majerus, 1998). Therefore, although it is impossible to be sure that advantageous resemblance to well-protected aposematic species was not influential in the past evolution of polymorphism in *A. 2-punctata*, it is very unlikely that Batesian mimicry significantly influences the frequencies of forms now. Müllerian mimicry may influence form frequencies, but it will be acting against the maintenance of rare forms, and will not play any part in maintaining polymorphism. Taking all these threads together, my view is that the case for mimicry being influential in the evolution of melanic polymorphism in ladybirds has not been made.

Thermal Melanism

Timofeeff-Ressovsky (1940), working in Berlin, recorded the numbers of melanic and non-melanic *A. 2-punctata*, in spring and autumn samples, from 1929 to 1940 (Fig. 9.30). The data show that melanics decline over the winter, and increase in summer. The explanation of this cyclical change in frequencies is based on the thermal properties of dark compared with light surfaces. Brakefield and Willmer (1985), using minute thermocouples placed on and under the elytra of *A. 2-punctata*, have shown that black-patterned areas heat and cool faster than red areas. Timofeeff-Ressovsky (1940) proposed that melanics warm and become active earlier on spring and summer mornings, giving them a competitive advantage over non-melanics, when replenishing exhausted food and fluid reserves, finding mates, and laying eggs (Benham et al., 1974). As *A. 2-punctata* produces three generations a year in Berlin (Timofeeff-Ressovsky, 1940), melanics may gain a sufficient advantage through the summer, explaining the observed summer increase in melanic frequency. Furthermore, the thermal properties of black surfaces also mean that, in the winter, melanics are subject to greater fluctuations in temperature than are non-melanics. Overwintering ladybirds are particularly susceptible to rapid changes of temperature (Hodek, 1973), so these greater fluctuations in temperature may cause higher winter mortality in melanics than in non-melanics, again as observed. Thus, this is a case of thermal melanism.

However, other workers have not found such seasonal changes in the frequencies of the forms (see Majerus, 1998b for reviews). Brakefield's (1985b) work in Holland illustrates the difficulties well, for while melanics mated earlier in the spring, supporting the thermal melanism hypothesis, he also showed that non-melanics suffered significantly higher winter mortality than melanics. This latter finding is directly contradictory to Timofeeff-Ressovsky's hypothesis. Furthermore, in Britain, protracted studies of the mortality rate of

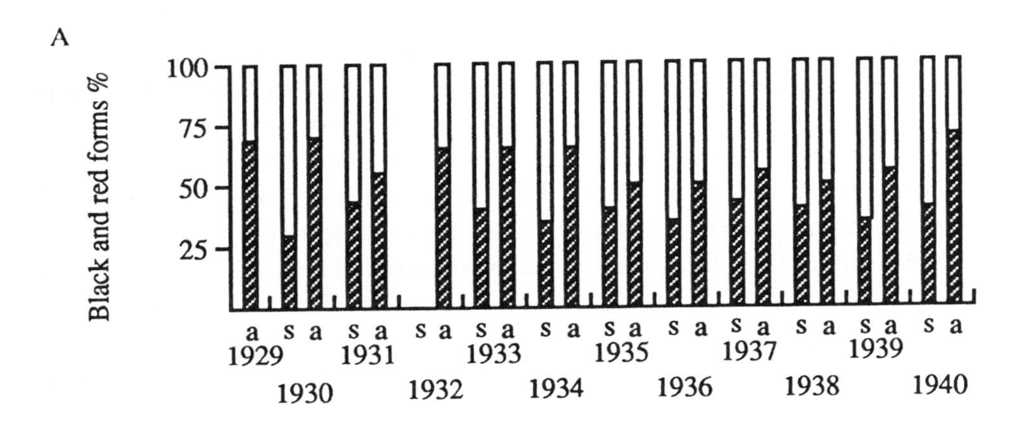

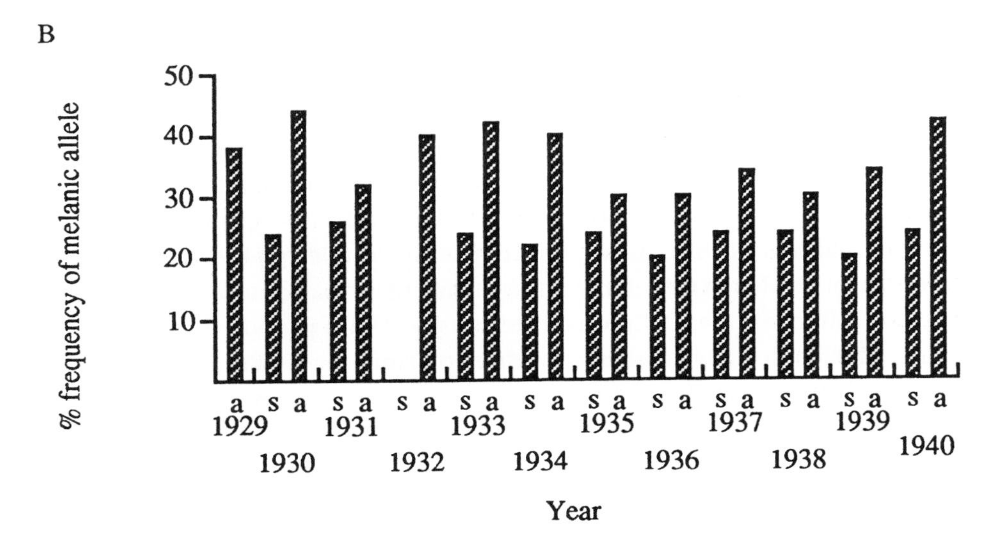

Fig. 9.30. A: Percentages of black (shaded bar) and red (unshaded) *Adalia 2-punctata* recorded in Berlin, in spring and autumn samples, from 1929 to 1940. B: Percentages of Melanic alleles in the population. Adapted from Timofeeff-Ressovsky (1940).

f. *bipunctata* and *quadrimaculata*, at natural overwintering sites, showed no morph-specific mortality differences (Majerus, 1994a; Majerus and Zakharov, 2000).

This lack of concordance between the different studies undermines the thermal melanism hypothesis. Moreover, in north-central Italy, Scali and Creed (1975) found that melanic frequency decreases with increasing altitude, concluding that melanic frequency was positively correlated with temperature, a finding counter to the thermal hypothesis. Furthermore, Lusis (1973) recorded the melanic f. *sublunata* to comprise up to 96% of some populations in extremely hot summer, cold winter regions of central Asia.

Creed (1966, 1971a) showed that melanic frequency was correlated with smoke pollution levels. He found that in the 1960s, the frequencies of melanics had declined by about 15%, in the industrial Midlands of England near Birmingham, shortly after the clean-air legislation of 1956, and the consequent introduction of smokeless zones (Creed, 1971b). This decline continued into the 1980s (Brakefield and Lees, 1987), and similar declines were recorded for eight sites in industrial north-west England (Bishop et al., 1978).

Explanations for this decline have varied. Creed (1971a, b) believed that atmospheric soot pollution was directly detrimental to the fitness of non-melanics compared with melanics. However, Muggleton et al. (1975) and Brakefield and Willmer (1985) proposed that the correlation between smoke pollution and melanic frequency was the result of soot particles having a shading effect, thereby reducing sunshine strength at ground level, with melanics gaining a thermal advantage. Here they are echoing Timofeeff-Ressovsky (1940) and Lusis (1961), who found melanic frequencies to be highest in industrial areas and regions with a humid maritime climate, and therefore fewer sunshine hours.

Irrespective of whether melanic frequency is positively correlated with smoke levels, or negatively correlated with sunshine levels, such correlations only provide circumstantial support for the thermal melanism hypothesis. They do not provide proof of either cause or effect, and should not be deemed to do so until contradictory data have been explained.

Extensive surveying during the Cambridge Ladybird Survey, between 1984 and 1994, showed that the frequencies of melanics, while still declining in the Midlands and northern England, were increasing in southern England and East Anglia (Majerus, 1995). It is difficult to envisage why the frequencies of melanic forms should be changing in opposite directions, were only thermal factors affecting the morph frequencies.

In a recent study, thermal melanism was evoked as a hypothesis to explain the spread of the invasive ladybird *H. axyridis* across Great Britain (Purse et al., 2014). Using the large-scale and long-term datasets accrued through volunteer recording of ladybirds, it was apparent that the different colour forms of *H. axyridis* spread to similar extents, although increased sunshine significantly enhanced the spread of the non-melanic form, f. *succinea*; perhaps a tantalising hint of the relevance of thermal melanism.

Female Mating Preferences and Negative Frequency Dependent Selection

Melanic polymorphism is widespread in *A. 2-punctata* and many other coccinellids. There is evidence that this polymorphism is long lived, melanic and non-melanic forms both being present in the oldest museum collection of beetles in existence (Pope, pers. comm.). This being the case, factors that will maintain stable polymorphism must be sought. Despite considerable breeding data, there is no evidence of heterozygote advantage maintaining the polymorphism. Therefore, some mechanism that would produce negative frequency dependent selection seems implicated. Advantage or disadvantage through thermal melanism will not be dependent on form frequency. Therefore, while thermal properties of melanic and non-melanic forms may affect the frequencies of the forms, they cannot maintain the polymorphism. The same is true of Müllerian mimicry.

Negative frequency dependence may be the outcome of a variety of ecological or behavioural systems. If predators form search images for prey types that they encounter often, it may be beneficial to become polymorphic, spreading a prey species across a variety of different images. Rarer types then gain an advantage because predators are less likely to form a searching image for a rare prey type. As *A. 2-punctata* is at least partly aposematic, this source of negative frequency dependent selection seems unlikely to play a significant role here. Similarly, the negative frequency dependence inherent in Batesian mimicry may be discarded.

A third type of negative frequency dependence derives from the coevolutionary arms races between hosts and parasites. Hosts will evolve immunity or other defence mechanisms against prevalent parasites, but less so against rare parasites. As a common form of parasite becomes more rare because of host defences against it, more rare forms, against which the host has not evolved defences, will become more common. The host will then evolve defences against the previously rare, now common parasite, and if there is a cost to defence, may lose its defences against the previously common, now rare parasite. The result is a cycle of increase and decrease in both the host defences and the prevalences of the different forms of parasite. Crucially, parasites must attack common forms of host more than rare forms, with negative frequency dependent selection resulting. The parasites of *A. 2-punctata* include *Coccipolipus hippodamiae*, *Laboulbeniales* fungi and a suite of male-killing bacteria. It is possible that the costs imposed by these parasites vary in the different morphs of *A. 2-punctata*, and impose negative frequency dependent selection on the forms. As yet, no significant work has been conducted on interactions between these parasites and different colour pattern forms of their hosts. Such work is urgently needed, for Majerus (2003) proposed that there is a connection between one of these parasites and colour pattern polymorphism in aposematic insects.

Negative frequency dependent selection may also derive from many types of non-random mating. Lusis (1961) provided the first evidence that mating in

A. 2-punctata is not random with respect to colour form. Muggleton (1979) showed that the biases in Lusis' data and further data of his own were negatively frequency dependent, rare forms gaining a mating advantage. This type of non-random mating is sufficient to maintain the polymorphism (O'Donald and Muggleton, 1979). Lees (1981), noting this, suggested that the principal factor that maintains melanic polymorphism in *A. 2-punctata* may be non-random mating, or some other factor unrelated to any environmental factor, and that environmental factors, such as climate, or the abundance of other coccinellid species present, simply affect the frequency at which a stable equilibrium is maintained.

Subsequently, *A. 2-punctata* has been shown to exhibit a variety of different patterns of non-random mating: assortative mating (O'Donald et al., 1984), female preference to mate with *quadrimaculata* and *sexpustulata* males, irrespective of their own genotype (Majerus et al., 1982a, b; O'Donald and Majerus, 1988, 1992), and variation in the responses of females to males of different sizes (Kearns et al., 1992; Haddrill, 2001). The single gene female mating preference for dominant melanic genotypes is pertinent here, because the benefit that accrues to preferred males will necessarily be negatively frequency dependent (O'Donald, 1980), and thus will maintain polymorphism. This female mating preference for melanic males is itself a genetic polymorphism; some females have the preference, others lack it. Furthermore, this mate preference is also temporally and spatially variable.

Variable mating preferences are not confined to *A. 2-punctata*. Osawa and Nishida (1992) showed that in *H. axyridis*, both sexes have mate preferences, female mating preferences generally being stronger than those of males. Moreover, the preferences varied from season to season. In the spring, females preferred non-melanic males, while in the summer melanics were favoured. By manipulation experiments, Osawa and Nishida showed that female *H. axyridis* use male colour pattern in making their choice, although other features were also implicated. Here again, the negative frequency dependence resulting from mate preference will help maintain balanced polymorphism. In addition, the seasonal variation in mating preferences will lead to cyclical changes in the frequencies of melanics and non-melanics. It is thus feasible that the oscillations in melanic frequency in *A. 2-punctata* reported by Timofeeff-Ressovsky (1940) could be a consequence of changing mating preferences.

The Recurrence of Rare Phenotypes

The melanic polymorphism in *A. 2-punctata* is a highly fluid system, probably acted upon by multiple selective factors. The import and intensity of the various selective factors undoubtedly vary in both time and space. Imposing a highly dynamic selective regimen upon the multiple allelic super-gene that controls most of the colour pattern variation in *A. 2-punctata* may allow this species to respond rapidly to changing environmental circumstances. One manifestation of this genetic control system is that occasional crossing-over within the super-gene generates novel genotypes (Majerus, 1994a). Some of these recombinants are melanic; others are non-melanic. Furthermore, although in many populations these recombinants are rare, in others they reach frequencies of several per cent. In a few regions of Asia, much of North America, and New Zealand, the common forms other than *bipunctata* that occur at significant frequencies are forms that in Europe would be considered recombinants.

In most studies of colour pattern polymorphism in *A. 2-punctata*, all forms that are more than half dark are lumped together as melanics, with the other forms being assigned to a single non-melanic class. This may be misleading. In many populations, the less common forms, if added together, comprise in excess of 10% of the total. Yet, these genotypes are usually neglected, and thereby the underlying fluidity of the colour pattern super-gene is ignored.

Little experimental attention has been paid to those forms that occur at low frequencies in Europe, even if these forms are common elsewhere. Sergievsky and Zakharov (1989) suggested that the maintenance of a wide range of genetic variants might be crucial to the survival of *A. 2-punctata* populations during periods of environmental stress. They proposed that the initial response to a variety of types of environmental stress is an increase in melanic frequencies. This increase results from the destabilisation of a previously stable co-adapted gene complex by novel selective factors. In *A. 2-punctata*, this co-adapted gene complex is, in essence, the colour pattern super-gene. Subsequently, melanic frequencies decline again, as the genome as a whole, and the super-gene in particular, re-establishes stability, finely adapted to the new circumstances. This hypothesis has some very interesting features, although it is not obvious why the initial increase should be specifically an increase in melanism. Firstly, the hypothesis creates a role for the maintenance of a fluid system controlling colour pattern variation. This fluidity is not exclusive to *A. 2-punctata*, for in other coccinellids showing high degrees of colour pattern polymorphism, such as *A. 10-punctata, H. axyridis, C. 6-maculatus, P. japonica* and *Coelophora inaequalis*, there is evidence that the colour pattern variation is in large part also controlled by multiple allelic super-genes.

There is one problem with Sergievsky and Zakharov's hypothesis. It seems to suggest that maladaptive alleles are maintained in a population because they may become beneficial in the future, should the environment change. However, selection is generally thought to be non-predictive. Selection acts now, and is imposed by current pressures, not those that may or may not come to pass in the future.

Two, not mutually exclusive, explanations of the occurrence of rare forms have been proposed (Majerus, 1994a). The first depends on the cannibalistic tendencies of neonate larvae. It was reported by Lusis (1947a) that *A. 2-punctata* carries an exceptionally high load of lethal recessive alleles, manifest by very high inbreeding depression seen in laboratory cultures. Similarly high levels of inbreeding depression have been observed in many of the aphidophagous coccinellids that we have cultured over the last 20 years, including *H. axyridis*, *A. 10-punctata*, *C. 6-maculatus* and *P. japonica*. Normal expectation would be that these deleterious recessive alleles that lead to inbreeding depression should be purged from populations by natural selection. However, due to the cannibalistic behaviour of neonate larvae towards unhatched eggs, and the resource advantage that the cannibalisers gain, there is a rational mechanism to explain the occurrence of this high frequency of recessive lethals. Here, individuals heterozygous for a recessive lethal may gain an advantage if recessive homozygotes, which are thus lethal, are present in the clutch, for they gain a resource advantage in much the same way that female larvae infected with a male-killer gain from eating their dead brothers. This then is a very specialised form of heterozygous advantage. The lethal recessive is maintained in the population at low frequency because the heterozygote is fitter than the normal homozygote. In some ways, this is similar to the maintenance of the sickle cell anaemia allele in parts of the world with a high incidence of malaria. There, the sickle cell allele, which is usually lethal when homozygous, is maintained in the population because those heterozygous for the sickle cell allele and the normal allele are resistant to malaria, whereas normal homozygotes are not.

In the context of the more rare colour pattern phenotypes, let us assume that ladybirds carrying and expressing rare recessive colour pattern alleles have reduced fitness compared to the predominant phenotypes. If this reduction in fitness takes the form of slower embryonic development, then the tendency for newly hatched larvae to cannibalise unhatched eggs of their own clutch will produce the same advantage to heterozygotes as just described for recessive lethals. This explanation only holds for rare recessive alleles. Dominant alleles with lower fitness will be expressed and therefore they will be rapidly eradicated from the population.

Most of the rare colour pattern alleles of *A. 2-punctata* are recessive, or recessive to the *bipunctata* and common melanic alleles. Thus, sibling cannibalism may act here in this way if the disruption of the super-gene results in slower embryonic development. The observation that pigment lay down in most of these forms is slower than in the common *bipunctata* and melanic forms supports this idea.

The second explanation is genetic. As the main colour pattern gene is a super-gene, recurrent recombination within the super-gene will continually give rise to new alleles. The frequencies at which recombinants are produced will be a function of the distances between the various genes within the super-gene.

Male-killers and Host Evolution

One strange feature of the natural history of many aphidophagous coccinellids is the male-killing bacteria that they harbour. It is feasible that these male-killers, and the biased sex ratios that they cause in their hosts, play a role in the evolution and/or maintenance of colour pattern polymorphism in ladybirds. Certainly, there seems to be some form of association between them, for in each of the ladybirds showing high degrees of colour pattern polymorphism that have been assayed for male-killers, they have been found. Therefore, before summarising the factors that may influence colour pattern polymorphism in ladybirds, I turn to the evolutionary implications of male-killers.

In most sexually reproducing animals, males have the potential to reproduce faster than females, so females are the limiting sex. Consequently, males compete with one another for access to females while females choose between males. However, in populations in which the sex ratio of adults is female biased due to the presence of male-killers, the strength of selection for males to compete and for females to choose is reduced. The extent of this reduction will depend on the prevalence of the male-killer and its vertical transmission efficiency. In general, the greater the female bias, the greater this reduction will be, although the mating system, and in particular the potential level of promiscuity of both males and females, will mediate the reduction (Majerus, 2003). In highly female-biased populations, if males rather than females limit the reproductive rate of the population, full sex role reversal may result. Here females compete with one another for access to males and males choose between females (Trivers, 1972).

The biases in population sex ratios caused by male-killers provide perfect stages on which to assess theories of the evolution of a variety of aspects of mating behaviour and reproductive strategies. The taxonomic diversity of male-killer hosts, variation in vertical transmission efficiencies, and heterogeneity in environments, mean that a range of female-biased sex ratios, both within and among species, are available as testing grounds. Indeed, recent studies on some of these testing grounds have shown that male-killers do affect host reproduction in a variety of ways.

The Limiting Sex and Sex Role Reversal

Sex role reversal has been reported from a small, but taxonomically diverse array of organisms, including birds such as jacanas, the dance fly, *Empis borealis* (Sivinski and Petersen, 1997), and the African butterfly *Acraea encedon* (Jiggins et al., 1999). In the latter case, the female bias in the population sex ratio is due to a high prevalence of male-killing *Wolbachia* (Jiggins et al., 1998). In some colonies, over 95% of the population are female, and over 80% of females die virgin because of the rarity of males. Here, females, rather than males, congregate on lekking grounds and males visit these female leks to find a mate.

This degree of impact on mating has not been reported in ladybirds. This may be because female biases in ladybirds are not as extreme as in *A. encedon*. Although the populations of ladybirds harbouring male-killers are female biased, the maximum proportion of females reported is only 81%. Given the potential promiscuity of male ladybirds (highest number of reported matings for a male *A. 2-punctata* in captivity is 61), even in the most female-biased populations, all females should be able to find a mate.

However, although full sex role reversal has not as yet been recorded in a ladybird, some effects of sex ratio biases have been. For example, in the two species of ladybird bearing male-killers that have been most studied, *A. bipunctata* and *H. axyridis*, a variety of mate choice patterns, including weak male choice among females, have been recorded (Osawa and Nishida, 1992; Majerus, 1998b).

Sex Ratio Distortion and Male Investment in Copulation

Even if strong evidence of males choosing among females is lacking, some effects of sex ratio distortion on males are supported by empirical evidence. As already noted, male *A. 2-punctata* often complete more than one mating cycle while copulating, passing two or three spermatophores into the same female. This apparently wasteful behaviour may be a form of sperm competition, with those depositing more spermatophores into a female gaining advantage through higher paternity simply by force of numbers. If this is the case, in female-biased populations the selection pressure for males to impregnate females with an excess of sperm will be reduced because of the reduced level of male competition. This decrease in the strength of selection may result in male investment per copulation being inversely correlated to the proportion of males in the population.

This conjecture has been tested via analysis of the copulatory behaviour of males from three populations with differing population sex ratios. Males from two populations, over 8000 kilometres apart, both harbouring male-killers at low prevalence and with population sex ratios close to 1:1, pass on average just under two spermatophores per copulation. Conversely, males from a population between the two, in which half the females are infected with male-killers, and in which the population sex ratio is two females per male, pass on average only just over one spermatophore per copulation. Interestingly, the average number of sperm per spermatophore did not vary significantly between these populations. Thus, of the two mechanisms by which male *A. 2-punctata* might reduce their investment per copulation – reduced number of spermatophores per copulation, or reduced number of spermatozoa per spermatophore – only the former was observed (Majerus, 2003).

It would be interesting to investigate male investment in *C. 6-maculatus* populations that have different sex ratios due to male-killing because the males pass sperm freely, without manufacturing a spermatophore.

Male-killers and Sexually Transmitted Diseases

Population sex ratios may have a significant impact on the epidemiology of sexually transmitted diseases. *Adalia 2-punctata* hosts both a suite of male-killing bacteria and two sexually transmitted disease agents: the mite *C. hippodamiae* and fungi of the genus *Laboulbeniales*. In some populations in which *C. hippodamiae* occurs, population sex ratios of *A. 2-punctata* vary from equality to highly female biased, with the proportion of males in some being less than 30%. In very female-biased populations, the prevalence of *C. hippodamiae* is likely to increase more rapidly than in 1:1 sex ratio populations. This is because males, due to their relative scarcity here, will have more mating partners and therefore more chance to pass on the mite than those in normal populations.

Intragenomic Conflict and Population Sex Ratios

The existence of inherited material that is passed to the next generation through one sex only, as is the case with male-killers, sets up a conflict between such genes and those that are inherited equally from both parents. There will thus be conflict between cytoplasmic male-killers and autosomal nuclear genes. This conflict will impose selection in favour of nuclear genes that suppress or reduce the pathogenicity of male-killers. This selection would act primarily through males. In addition, sex ratio biases in populations should impose selection in favour of the more rare sex, thereby restoring the sex ratio towards equality (Fisher, 1930; Hamilton 1967). Although nuclear genes that act against male-killers would help to do this, selection could also favour the production of the more rare sex, in this situation males, by females that lack the male-killer. Uninfected females that produced more sons than daughters in female-biased populations would, by dint of the skew in the number of matings available to the more rare sex, produce more grandchildren than females that produced equal numbers of males and females.

Ladybirds may show both of these responses to the genetic conflicts induced by male-killers. The evidence supporting the conjecture that some females may produce more males than females is currently tenuous, simply comprising reports of significant male biases in some families of *H. axyridis* and *Coccinula sinensis* (Majerus, 2003). The evidence for nuclear genes that act against male-killers is stronger.

Cheilomenes 6-maculatus harbours a male-killing γ-proteobacterium. In crosses designed to investigate the inheritance of the male-killing trait, it was found that the origin of the male mated to females from male-killing lines influenced progenic sex ratios. Experiments that involved swapping males between females showed that some males caused infected females to produce both sons and daughters, while the same females with other males produced only or predominantly daughters (Fig. 9.31). Experiments further showed the existence of a dominant allele of an autosomal gene that rescues males from the killing effect of the bacterium. This allele is inherited in a normal Mendelian fashion, through both male and female lines, and acts as a rescue gene whether inherited from a male progeny's father or mother. This rescue

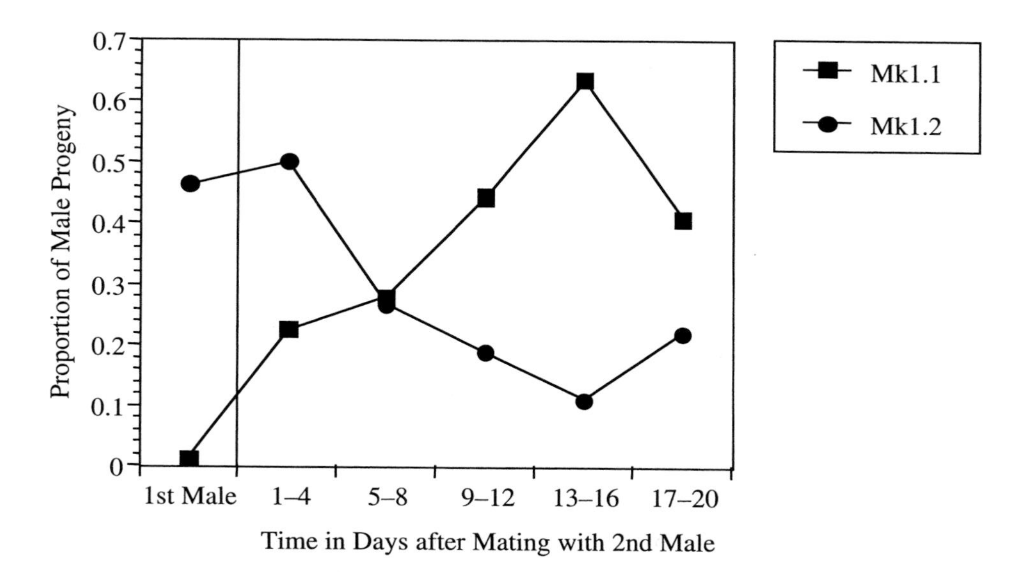

Fig. 9.31. Experiments that involved swapping males between females showed that some males caused infected females to produce both sons and daughters, while the same females with other males produced only or predominantly daughters.

gene does not kill the bacterium. Females from male-killer-rescued lines, when mated to males lacking the rescue gene, produce female-biased families. Furthermore, molecular genetic analysis has shown that rescued males still harbour the bacterium, but do not transmit it (Majerus, 1999, 2003).

It is probably not coincidental that the male rescue gene in *C. 6-maculatus* is a single gene. Hurst et al. (1997b), discussing selection against male-killer infection, argue that polygenic modifiers acting to delay male death in infected individuals, would not evolve. The reasoning is that early death of males is beneficial to female siblings through resource reallocation. Modifier genes that retarded death of infected individuals, such that some infected males die before reproducing, while others survived to reproduce, would confer both a fitness increase through the survival of some males, and a fitness loss as a result of the waste of nutrients used by infected males that develop to some extent but die before reproducing. The lack of resource reallocation from these dying males to female kin would prevent the spread of the modifiers.

Male-killers and Mitochondrial DNA

The evolution of suppressor genes or rescue genes, and the possibility of uninfecteds producing an excess of males to restore the 1:1 sex ratio, are not the only effects that male-killers have on their hosts' genomes. Invasion by a male-killer will have a considerable impact on the variability of mitochondrial DNA (mtDNA). Here the impact of a male-killer is simply a consequence of its mode of inheritance. Male-killers are maternally transmitted, which is also the case for mitochondria. Therefore, if a male-killer invades a new host and begins to spread, as it spreads, it will pull the mitochondria of the first female host that it invades along with it. As both male-killers and mitochondria are inherited in the cytoplasm of eggs, they will be genetically linked. This means that the mitochondrial type in the first host of the invading male-killer will hitchhike on the success of the spreading male-killer. Moreover, as the vertical transmission efficiency of male-killers is never perfect, some uninfected hosts will also bear this mitotype. Eventually, because of the advantage to male-killing and their imperfect vertical transmission, all mitochondria in the host population should be descendants of the mitotype of the host originally invaded (Johnstone and Hurst 1996). It follows that if several hosts were invaded by male-killers, there would be several successful hitchhiking mitotypes. The result of these selective sweeps will be that the level of mtDNA variability will be severely reduced. The level of variability in such a population will be a consequence of the variation in the mitotypes of uninfected matrilines invaded by male-killers and mutations in mtDNA that have occurred since male-killer invasion.

This effect on mtDNA variability has consequences for the growing number of evolutionary geneticists who use mtDNA variation to study changes within and between populations. For example, many studies of population structure, gene flow and taxonomy use levels of mtDNA sequence variability because mtDNA has been thought to be selectively neutral and because it is inherited from just one sex. However, the presence of male-killers, now or in the past, in any species investigated in this way, will profoundly affect the mtDNA data obtained, and should greatly influence its interpretation (Majerus, 2003).

Male-killers may influence mtDNA in another way. As male-killers are detrimental to their hosts, selection on the host should act to reduce their vertical transmission. This might be achieved by increasing the rate of cell division in host germ-line cells, thereby reducing the time that the bacterium has to replicate to a degree sufficient to ensure that each host daughter cell receives the male-killer. When cells divide, mitochondria also have to be replicated. Faster cell division may impose selection for a reduction in the size of the mitochondrial genome to allow more rapid replication. Thus, selection acting to reduce the vertical transmission of male-killers may favour loss of those parts of the mitochondrial genome that have no important function. This may have consequences for the use of mtDNA in phylogenetic studies using molecular clocks, because mtDNA evolution may be accelerated in male-killer infected lineages.

Aposematic Polymorphism and Male-killing

I do not believe much in biological coincidence. Thus, when I see an unusual association that is repeated several times, it causes me to pause and wonder. When the association involves some feature that presents a biological conundrum, my interest is increased. Such is the apparent association between the unexpected occurrence of colour pattern polymorphisms in aposematic insects and the presence of male-killers.

The rationale for aposematism and indeed Mûllerian mimicry is that selection should favour uniformity. Therefore, why are some aposematic ladybirds polymorphic? As said earlier in this chapter, there are a variety of selective mechanisms that may lead to polymorphism, and as long as the selective advantages incurred from these is greater than the disadvantage of an aposematic species not all looking alike, these factors may maintain polymorphisms in these species. Yet, as has also been said, in no species is there empirical evidence providing an unequivocal explanation of the evolution and maintenance of colour pattern polymorphism. Certainly, after over 20 years of study of *A. 2-punctata*, I do not have the feeling that any of the various factors suggested to have an effect on the polymorphism in this species – female

Fig. 9.32. Three colour forms of *Propylea japonica*.

choice, thermal melanism, co-adapted gene complexes, heterozygote advantage through sibling consumption of embryos homozygous for deleterious lethals – can, on their own, account for the patterns of polymorphism seen.

Every one of the polymorphic ladybirds that has been assayed for male-killers has been found to have them. The list includes *A. 2-punctata*, *A. 10-punctata*, *H. axyridis*, *H. 4-punctata*, *P. japonica* (Fig. 9.32), *C. 6-maculatus*, *C. inaequalis* and *C. sinensis*. Interestingly, three species of polymorphic aposematic butterfly have been assayed for the presence of male-killing bacteria, and in each of the three, male-killers have been found. In both *Acraea encedon* and *Acraea encedana*, male-killing *Wolbachia* have been identified (Hurst et al., 1999b; Jiggins et al., 2000a), while in *Danaus chrysippus*, a male-killing *Spiroplasma* has been detected (Jiggins et al., 2000b). The last of these species is particularly interesting because, over much of its range, this widespread tropical species is monomorphic. Only in East Africa is it polymorphic, and only here does it harbour a male-killer (Majerus, 2003). It would certainly be worth assaying other polymorphic ladybirds, such as *Olla v-nigrum*, for the presence of male-killers.

The presence of male-killers, by their 'ultra-selfish' behaviour and by distorting population sex ratios, may have considerable influences on the evolutionary genetics of their hosts. Studies of these effects are as yet in their infancy. Nonetheless, already we have seen that male-killers can have a bearing on male investment, levels of male testing by females, male choice and female competition. It has also been speculated that assortative mating choices based on colour pattern, as have been recorded in *A. 2-punctata* (O'Donald et al., 1984), may provide males with a means of selecting uninfected female partners (Majerus, 2003). No doubt, other effects will be identified in the near future.

Before the link between male-killing and colour pattern polymorphism was first made, I had already expressed my opinion that the selective factors causing the evolution and maintenance of colour pattern polymorphism in ladybirds would be complex and fluid (Majerus, 1998b). I then envisaged that the morphs of a species might at any instant be maintained in a delicate state of equilibrium by a tenuous balance of a complex set of selective factors, or they may be changing in frequency as such balances moved slightly out of kilter, as first suggested by Sergievsky and Zakharov (1989). The situation was likened to trying to balance a ball bearing on a warped and undulating piece of cardboard. While it is possible to hold the card still enough to keep the ball-bearing in one of the shallow depressions for some time, even small movements of the card may shift the ball out of an adaptive depression and set it moving towards some other delicate adaptive equilibrium. Here the situation approaches the concept of Wright's (1982) 'shifting balance' theory, but with the adaptive topology being only slowly undulating, without sharp, well-defined adaptive peaks. The range of colour pattern forms that may be produced by the multiple allelic super-genes that control most colour pattern variants in the more spectacularly polymorphic species will ensure that an array of phenotypes are generated each generation for selection to act upon.

Some of the unusual features of the biology of ladybirds suggest that display of a considerable degree of mental flexibility and lateral thinking may be necessary if we are to formulate a convincing explanation of colour pattern polymorphism in these aposematic ladybirds. The variety of potential selective factors that may be involved will require painstaking monitoring and experiments if the role of each is to be assessed. More difficult still may be addressing the ways in which different selective factors interact. That the different factors will interact is not in doubt. To take one example, we may consider thermal factors and male-killers.

Male-killing bacteria are temperature sensitive; infected female ladybirds exposed to temperatures around 35°C for a period of several days lose the male-killer (Hurst et al., 1992). If melanic *A. 2-punctata* heat up more than non-melanics, the bacteria may have lower prevalence in melanics than non-melanics. As male-killing bacteria impose a cost on their female hosts, in terms of lower fecundity and shorter lifespan (Majerus, 2003), the increase in melanics recorded by Timofeeff-Ressovsky (1940) over the summer in Berlin might be explained due to this lower male-killer prevalence. Furthermore, Tinsley (2003) has shown that winter mortality of male-killer infected female *A. 2-punctata* is greater than that of uninfected females. As melanics may be more prone to winter mortality than non-melanics in some climates, then a winter fitness hierarchy might be: uninfected non-melanic > uninfected melanic = infected non-melanic > infected melanic. This would lead to an overall decline in the frequency of melanics over the winter, as observed in Berlin. Here, crucially, the prevalence of male-killers, their levels of temperature sensitivity, the thermal dynamics of the melanic and non-melanic forms, and the climate will all have an impact on the frequency of the forms.

The association between male-killing and these polymorphisms, and the theoretical possibility that male-killers may interact with other factors affecting colour pattern polymorphism in ladybirds, adds just one more complicating factor to an already complicated phenomenon. I very much doubt that even with this latest addition, the list of factors affecting colour pattern polymorphism in ladybirds is exhaustive.

Chapter 10: Ladybirds and People

Of all insects, ladybirds have a special place in human affection. Yet our relationship with them is not simply one of an observer looking down at a small, rather attractive and sometimes slightly comical creature. We also use and, sadly, sometimes misuse ladybirds. One of the reasons for the popularity of ladybirds is that, as a result of their dietary preferences, they are friends of farmers and gardeners, the majority of species feeding upon insects or mites that damage plants. They are thus considered beneficial insects, providing an essential ecosystem service, and have been harnessed as such. It should, therefore, be in our interests to care for and conserve ladybirds.

In this final chapter I will explore some of the interactions between humans and ladybirds. I will start with their role as biological pest control agents, before examining the threats we pose to them, through the introduction of competitors into their habitats, habitat destruction, pollution and the use of chemical pesticides. Finally, I will consider whether there are ways that we can conserve these helpful insects.

Biological Control

The use of ladybirds as biological control agents has been extensively and recently reviewed (see Michaud, 2012). The account that follows is far from exhaustive but provides reflections, particularly from a conservation perspective, highlighting gaps in understanding and future research priorities.

The Ladybird that Began Biological Control

Coccinellids have been widely used as biological pest control agents, as beneficial effects of ladybirds have long been recognised. Indeed, it was a ladybird that laid the foundations of modern biocontrol, over 30 years before Smith (1919) first coined the term biological control. In 1887, the citrus industry of California was under serious threat from damage caused by a coccid, the cottony cushion scale, *Icerya purchasi*, an accidental import into the United States from Australia. Looking to find ways to reduce the ravages of this coccid, the federal entomologist C.V. Riley employed Albert Koebele to assess whether any of this scale insect's natural enemies, in its country of origin, might be used against it in California. Among the candidate insects that might impose a degree of control was the vedalia ladybird, *Rodolia cardinalis*. Between late 1888 and mid-1889, a little over ten thousand of these red and black ladybirds were brought from Australia and released in California.

The introduced ladybirds survived and reproduced well, establishing populations throughout the citrus groves. The decrease in coccid populations was immediate and dramatic, with the result that 1890 saw a three-fold increase in the orange crop over that of 1889. The cost of the introduction project was just $1500, with a return, in just over a year, running into millions of dollars. *Rodolia cardinalis* and *I. purchasi* still both live in the citrus groves of California, the ladybird keeping the scale insect at densities too low to be of economic importance. Here an ecological balance between pest and predator now exists, so that the pest is no longer a problem. It was the startling success of *R. cardinalis* that began the biocontrol 'bandwagon' that ran on through the first half of the twentieth century, until it was slowed by the development of cheap and effective synthetic insecticides.

Types of Biological Control

The definition of biological control includes any instance whereby an organism's population density is reduced by its predators or parasites to a lower average level than it would otherwise attain (de Bach, 1964). Importantly, this definition does not require human involvement.

Luck (1990) summarised three main methods of biological control:

(i) Classical biological control, in which non-native natural enemies are introduced to cause a permanent reduction in the numbers of a pest, usually of foreign origin.

(ii) Augmentative biological control, in which natural enemies, mass reared in captivity, are periodically released to supplement native populations, or to found new enemy populations for a limited period, for example, a growing season.

(iii) Conservation of native predator or parasitoid populations, in which naturally occurring enemies are identified, and husbanded in the field by means of appropriate management practices.

Classical Biological Control

Rodolia cardinalis is not the first documented case of biological control involving a ladybird. That honour goes to *Coccinella 11-punctata* (Fig. 6.4), which was transported from Britain to New Zealand in 1874 for aphid control. Indeed, Kirby and Spence had pointed out the potential for using ladybirds to control aphids as early as 1815. The results of introducing *C. 11-punctata* into New Zealand were not as dramatic as obtained with *R. cardinalis*, but *C. 11-punctata* is now generally distributed, and often common, in New Zealand. Its impact on pest aphids might have been greater if the parasitoid wasp *Dinocampus coccinellae* had not been inadvertently introduced with it.

Since its success in the United States of America, *R. cardinalis* has been successfully introduced into over 40 other countries to control cottony cushion scale and other species of scale insect. On each occasion, substantial, if not complete control has been achieved. Despite this, other introductions have been less effective. Following the importation of *R. cardinalis*, Koebele sent 46 more species of coccinellid from Australia to America during 1891 and 1892, but very few of these established successfully (Hagen, 1974).

The majority of examples of successful classical biological control using ladybirds have involved the control of scale insects. De Bach (1964) lists a number of *Icerya* species and 11 other coccids that have been successfully controlled, the controlling species being *Chilocorus cacti*, *Chilocorus distigma*, *Chilocorus nigritus*, *Chilocorus politus*, *Cryptognatha nodiceps*, *Cryptolaemus montrouzieri*, *Rhyzobius ventralis*, *Rodolia pumilla* and *Telsimia nitida*. Some of these species have been used against scale insects with great success. Thus, for example, *C. nigritus* (Fig. 10.1) has been introduced to 21 different countries, and has established in 19 of them (Omkar and Pervez, 2003).

Fig. 10.1. *Chilocorus nigritus.*

Other examples of successful biological control of coccids include the use of *Rodolia limbata* to reduce densities of *I. purchasi* attacking breadfruit trees on various tropical islands in the Pacific (Dixon, 2000). *Hyperaspis raynevali* was transported from Guyana to Zaire to act against the cassava mealybug *Phenacoccus manihoti* (Fabres et al., 1991), while the Australian ladybird *Diomus flavifrons* was used to control the citrus mealybug *Planococcus citri* in the United States (Meyerdirk, 1983). Additionally, from Australia, *Halmus chalybeus* has been used to control the olive scale *Saissetia oleae* on citrus; *R. ventralis*, used against the gum tree scale *Eriococcus coriaceus*, which was destroying large areas of eucalyptus plantations; and *C. montrouzieri*, imported to control used mealybugs (Majerus, 1994a). In Europe, a range of coccinellids has been introduced to control pests in citrus groves. These include *Rhyzobius forestieri* to control *Coccus pseudmagnoliarum* in Greece (Katsoyannos, 1984), *C. montrouzieri* to control *P. citri* in France, Spain, Greece and Italy (Panis, 1981; Ragusa and Russo, 1988), and *Serangium parcesetosum* to control the whitefly *Dialeurodes citri* (Malausa et al., 1988).

In the 1980s, a ladybird was used as a conservation tool on the island of St. Helena, which had been invaded by the scale insect *Orthezia insignis*. This coccid attacked and almost wiped out the island's national tree, an endemic gumwood *Commidendrum robustum*. The ladybird *Hyperaspis pantherina* was introduced into remaining stands of the tree, and successfully controlled the coccid, saving the host tree from extinction (Fowler, 2004).

There are rather fewer examples of ladybirds being successfully used against other pest species, such as aphids, adelgids, mites or herbivorous beetles in the field, although many attempts have been made. *Harmonia axyridis* has been used in various parts of the world to control aphids on a variety of crops (Roy et al., 2016). Successful control has been achieved on a number of occasions on pecan, maize, alfalfa, tobacco, winter wheat, soybean, strawberries, cotton, and in both apple and citrus orchards.

A further interesting case is that of the release of *Brumus 8-signatus* and *Semiadalia 11-notata* against the weevil *Phytonotus posticus*, a pest on lucerne in central Asia. This case was significant because the predators were used against alternative rather than primary prey. Although complete control was not achieved, crop yield did increase significantly (Yakhontov, 1960b).

Although ladybirds have long been recognised as beneficial, the success of aphidophagous species as biocontrol agents has not been great. In most cases the ladybirds have failed to establish, and of those that did establish, many have failed to keep aphid populations below economic damage thresholds (Bartlett, 1978; Greathead, 1989; Dixon, 2000). For example, Gordon (1985) reported that of 179 species intentionally introduced into North America, only 16 had become established. Moreover, these have included species that are not primarily aphidophagous, such as *Chilocorus 2-pustulatus*, used against the olive scale, *Chilocorus kuwanae* and *Exochomus 4-pustulatus*, used to control various coccids, and *Aphidecta obliterata*, used to control balsam woolly adelgid. Indeed, many releases of *H. axyridis* failed. In North America, *H. axyridis* was first released as a biocontrol agent in the United States in 1916 (Gordon, 1985), but despite numerous further releases there and in Canada and Mexico over the next 70 years (Koch, 2003), it was not until 1988 that it was reported as established in America (Chapin and Brou, 1991).

In summary, although ladybirds have been used as classical biological control agents for well over a century, they have most commonly been effective against coccids (Dixon, 2000).

What Factors Make for a Successful Biological Control Agent?

In assessing the benefits of classical biological control programmes, we need to consider two questions. Firstly, does the control species establish? Secondly, if the control species does establish, does it result in economically beneficial reductions in damage caused by the target pest? It is the second of these questions that has received most attention.

A range of factors affects the likelihood that an introduced coccinellid will establish. Perhaps paramount among these factors is the climate, as it pertains to reproductive rate, growth rate and survival during unfavourable periods. Other factors that may also be influential are host availability, presence and abundance of natural enemies, presence of dormancy sites, dispersal characteristics, the size of the release population and the use of insecticides in the release area. To assess the importance of these factors, successful and unsuccessful introductions of ladybirds may be compared. It is here that the differential success of coccidophagous and aphidophagous coccinellids is most obvious.

The characteristics of a successful biocontrol agent were outlined by Luck (1990) as: a predilection for the target species, temporal and spatial synchrony with the target species, ability to search effectively for the target species, ability to successfully capture the target species, a propensity to aggregate in areas with high pest prey density, appropriate dispersal characteristics and rapid reproduction relative to the target species. In addition, there should be ecological and climatic similarity between the place of origin of the control agent and the region into which the agent is released.

Hopefully, we may assume that biocontrol programme managers will have ensured that the ladybird chosen will eat the target species, and that stocks are derived from regions that are similar enough to the target region to give imported ladybirds some likelihood of survival. We must then examine the other conditions. Both adult and larval ladybirds of most species certainly exhibit prey-searching behaviour. Although they will search plant surfaces more or less randomly at first, once they have found prey they restrict their searching area. This behaviour usually results in concentrations of ladybirds in areas where prey are most dense. Furthermore, oviposition is at least partly correlated to prey densities (Hemptinne et al., 1993). Ladybirds also have appropriate dispersal characteristics. Adults can fly, and larvae of aphidophagous species are relatively mobile. These features thus all seem in favour of the success of aphidophagous coccinellids in control programmes.

Failures are therefore attributed to the remaining factors. Firstly, in some cases there is a lack of synchronisation between control agent populations and initial aphid outbreaks early in the season (Coderre, 1988). Densities of aphids early in the season are too low to promote maturation of ovaries and oviposition by many aphidophagous coccinellids (Hemptinne and Dixon, 1991). Thereafter, the reproductive rate of ladybirds is too slow, compared with that of the aphids, to affect full control. Although a female ladybird can produce many hundreds of eggs, development from egg-laying to egg-laying in continuously breeding aphidophagous species usually takes between two weeks and two months, depending on climate. This is too long, as many aphid colonies establish, peak, and diminish in response to changes in host plant quality within this sort of time-span, particularly on annual crops.

Dixon (2000) compared the relative reproductive rates of a number of pest types and specific ladybird predators at 20°C. He found that coccid predators developed faster than their prey, aleyrodids and psyllids developed at a similar rate to their predators, while aphids, adelgids and acarids developed faster than their predators. The difference was greatest for coccid- and aphid-feeding ladybirds, the relative developmental rate of the latter compared to its prey being almost five times slower than the former.

Another reason for the lack of success of ladybirds in controlling aphids is that few aphidophagous species are prey-specific. Many ladybirds released in classical biocontrol programmes simply switch to feeding on aphids other than the target species. Again, this is in contrast to coccidophagous species, which tend to have a much narrower prey range (Dixon, 2000).

In most classical biocontrol programmes involving coccinellids, releases are aimed at a specific target pest. Because aphidophagous ladybirds will switch from one prey to another as the populations of different aphids wax and wane, predicting whether released populations will establish is difficult. This makes advanced assessment of the overall benefits and costs of introduction programmes of aphidophagous species uncertain (Dixon, 2000).

The wide prey range of many aphidophagous coccinellids also means that released aphidophagous ladybirds may attack non-target organisms. Moreover, released coccinellids, if they establish successfully, are likely to come into competitive conflict with indigenous aphid predators and parasites. This conflict raises a conservation issue, which I will return to presently.

Many of the best known and most abundant generalist aphidophagous ladybirds, such as *Adalia 2-punctata, Coccinella 7-punctata, H. axyridis* and *Propylea 14-punctata* have been released in many parts of the world. The success of these releases has been variable, the majority failing completely. However, these species are all now found outside of their natural ranges. For example, *A. 2-punctata*, which is a native of the northern hemisphere, is now found in Japan, New Zealand, South Africa, Argentina and Chile. Introduced aphidophagous species that will feed on aphids on a wide range of host plants are undoubtedly of benefit against aphid pests, even when the specific target they were introduced against has not been effectively controlled. However, the level of benefit is very difficult to assess economically. Both naturally occurring and introduced aphidophagous coccinellids are attracted to high concentrations of aphids, and by feeding on these they certainly reduce the damage done by aphids on a very wide range of plants. For example, Hurej (1988) studied *Aphis fabae* on sugar beet in Poland in the 1980s. In three of four study years, natural populations of ladybirds effected sufficient control to prevent any economically significant loss in yield. Only in one year, when there was an exceptional abundance of spring aphid colonisers, was control not achieved. Similarly, Ofuya (1991) has shown that in Nigeria, the ladybirds *Cheilomenes lunata* and *Cheilomenes vicina* naturally control the cowpea aphid *Aphis craccivora*, a pest of the cowpea, and *Vigna unguiculata* on the groundnut *Arachis lypogaea* and other grain legumes in the tropics and subtropics. Thus, generalist aphidophagous species, including such species as *C. 7-punctata* and *H. axyridis*, have importance as aphid controllers both in their native range and in new ranges where they have established following intentional or accidental introduction by humans. That an economic value cannot easily be put on this benefit is unfortunate, because as we shall see, there may also be costs that result from introductions of these generalist species.

Augmentative Biological Control

The ladybirds released as biocontrol agents are usually non-native species, often mass reared in captivity. However, in the United States, *Hippodamia convergens* provides a unique example of another type of biocontrol. The unusual dormancy behaviour patterns of *H. convergens*, and particularly the aggregative winter behaviour of this ladybird, have been exploited for over a century. The winter congregations of this species may be massive, comprising many millions of individuals which remain inactive in their upland refuges until the spring, when they migrate back down to the lowlands, there to feed and reproduce. Many overwintering sites support aggregations year after year, and are visited by commercial companies, which collect the ladybirds for sale as aphid controllers. This practice began early in the twentieth century, and continues to this day. Early harvesters used brushes and scoops to collect the ladybirds. It was said that between 50 and 100 pounds of ladybirds could be collected by two people in one day. At 1500 ladybirds to the ounce, a day's crop could exceed two million ladybirds. Nowadays vacuum cleaners are used to harvest the beetles. The ladybirds are bottled and placed in cold storage at around freezing point until the spring, when they are sold. Amateur gardeners, market gardeners and organic farmers are the main buyers. The ladybirds may be obtained by the pint or gallon, either by mail order or through garden centres. For commercial use, recommended applications are: cotton – six gallons per acre, fruit trees – two gallons per acre, and market gardens – one gallon per five acres. A gallon is about 30 000 ladybirds. They should not be hand-cast, as seed is sown, but carefully placed in a handful towards the base of the plants. They should be released in the early morning or late evening, on coolish days, as the ladybirds are less likely to fly from the area where they are required.

In this case, the exploitation of a natural resource, the ladybird, seems to be ecologically as well as economically beneficial. The cost of the use of these wild-collected ladybirds is on a par with the use of chemical aphidicides, but less damaging. Because *H. convergens* has a relatively catholic diet, it will also eat pests other than aphids, such as scale insects, mealy bugs, caterpillars, white fly and leafhoppers. Furthermore, there may be a number of significant benefits to the ladybirds. Firstly, once harvested, the ladybirds are kept in optimum conditions for survival for the rest of the winter, thereby being protected from climatic caprices. Secondly, they are saved an energy consuming and potentially hazardous spring dispersal flight from the mountains to the lowlands. Thirdly, they will be released when and where aphids are present, hence conditions for reproduction are likely to be good. This, then, is an example of augmentative biocontrol, although here the ladybirds are not mass cultured in captivity, just stored in confinement for part of their adult lives.

Other successes for augmentative biocontrol using ladybirds are sadly few. *Cryptolaemus montrouzieri* is used in California against mealy bugs, particularly *P. citri* infesting citrus. This species was one of those Koebele introduced into California in 1892, and one of the few to establish, although it is only permanently established along a narrow coastal zone in southern California. Inland, it cannot survive the winter. Even in the zone of permanent establishment, *C. montrouzieri* does not attack mealy bug populations sufficiently early in the season to prevent crop damage. However, this ladybird is easy and cheap to culture in captivity on mealy bugs bred on potato sprouts. Mass rearing and release of this species to augment natural populations in the citrus groves began early in the twentieth century (Smith and Armitage, 1920). In 1960, the production of 30 million ladybirds cost just $17 000. An augmentative biological control programme, in which 20 of these ladybirds are released per tree with up to five releases per year, has proved successful (Hodek, 1973), and various other successes using *C. montrouzieri* are outlined by Michaud (2012). A European augmentative biocontrol example involves the use of *R. forestieri* on the island of Chios to control coccids, and *Nephus reunioni*, *Nephus anomus* and *Nephus 4-maculatus* to control mealybugs in Greece (Katsoyannos, 1996).

One of the problems of augmentative biocontrol is that the ladybirds released onto a particular crop have a tendency to fly away if conditions are not precisely right. Vast releases of *H. convergens*, collected from montane dormancy sites, have failed to control *Aphis gossypii* on melon and chrysanthemums, or *Macrosiphum rosae* on roses (Carnes, 1912). Most of the ladybirds simply flew away, and virtually none remained in the release areas three days after release (Hagen, 1962; Flint et al., 1995; Dreistadt and Flint, 1996).

These occurrences do not appear to deter the use of this species in pest control. In the 1990s, 30 million *H. convergens* were imported from America to Holland. These were released onto lime trees, *Tilia* × *europaea*, in towns and cities. The aim was to reduce the dirtying of cars parked along these streets by traffic pollution sticking to the honeydew that had dropped onto cars from aphids in the trees above. The programme was a total failure, as all 30 million ladybirds promptly flew away. One has to wonder about importing a species of ladybird that rarely breeds on trees to control aphids on trees, and one that is adapted to overwintering at high altitude, when Holland is not especially noted for its mountains. More poignantly, one also wonders about the fate of those 30 million misplaced ladybirds.

Another problem is that, with few exceptions, such as the special case of *H. convergens*, ladybirds are not very amenable to augmentative biocontrol. Two difficulties arise. Firstly, most ladybirds are not cheap to mass rear in captivity. Many of the aphidophagous species require aphids in their diet if they are to reproduce at anywhere near their top potential. To grow enough plants to support sufficient numbers of aphids for the mass rearing of ladybirds is simply not cost effective when compared with the costs of insecticides. Various attempts have been made to produce a synthetic food, which will induce rapid reproduction, but no cheap general alternative to aphids has yet been found. Only in the case of *H. axyridis* has an alternative diet, based on industrially cultured eggs of the flour moth, *Ephestia kuehniella*, proved economically viable for mass laboratory rearing (Schanderl et al.,

1988; Berkvens et al., 2008). *Harmonia axyridis* has been continuously reared for over 100 generations on this artificial diet. One side effect of this mass rearing technique, however, is a reduced response of the ladybirds in these cultures to aphids, leading to lower capture efficiency (Ferran et al., 1997). Nonetheless, some success has been achieved with releases of *H. axyridis* larvae against rose aphids, *M. rosae* and the damson-hop aphid, *Phorodon humuli* (Trouve et al., 1997; Ferran et al., 1998). Similarly, mass reared *A. 2-punctata* has been used to control *Dysaphis plantaginea* on roses (Wyss et al., 1999). The predilection of ladybirds to fly away from the crops where they are released has lead to a long-term and ultimately successful programme to produce ladybirds that do not fly. Following artificial mutagenesis and careful selection of flightless strains with high fecundity and normal foraging behaviour, a strain of *H. axyridis* has been produced that differs little in fitness from normal ladybirds (Tourniaire et al., 2000). The flightlessness is a function of reduction in the flight muscles in the meso- and metathoracic segments, rather than reduction in the wings themselves. Flightless *H. axyridis* have been used with some success in open fields against *A. gossypii* (Seko et al., 2008).

Secondly, if substantial numbers of ladybirds are reared in captivity, they must then be stored until needed, in conditions that cause only very low mortality. Work on *S. 11-notata* has shown that high storage survival is achieved if newly hatched adults are fed on aphids for just two days, then given an artificial diet of agar mixed with honey for five days, before being stored at 5–8°C until use (Iperti and Hodek, 1974). However, different species are likely to need different regimens, which will have to be identified in each case, and this feeding and storage will add to the overall costs of augmentative biocontrol programmes.

Husbandry of Native Predator Populations

If ladybirds are only successful as classical biocontrol agents in a limited number of cases, mainly the control of coccids, and if their use in augmentative control programmes is usually uneconomic because of the cost of culture and storage, what of their role in programmes for the conservation of native predators? Here, perhaps, there is more hope.

In his description of the third method of biological control, Luck (1990) writes of the husbanding in the field of native predator populations by appropriate management practices. A startlingly successful example comes from China, where people collect *C. 7-punctata* in huge numbers from wheat. They are then transferred to cotton to control *A. gossypii*. A release density of one and a half to three ladybirds per square metre is reported to reduce aphid abundance by 98% in just two days (Pu, 1976).

This is perhaps an exceptional instance, but the use of ladybirds and other native beneficial insects forms one of the pillars of so-called integrated pest management strategies. The place of ladybirds in such schemes was given prominence in a conference entitled 'The Ecology of Aphidophagous Insects' held in 1965, at Liblice, near Prague, Czechoslovakia (Hodek, 1966). In the introduction to the symposium proceedings, Professor Ivo Hodek summarised the status of biological control of aphids at that time: 'The astonishing success of modern insecticides after World War II temporarily eliminated nearly all attempts at biological control. More recently their adverse side effects have come into the limelight, as for instance, their harm to human health as well as to beneficial animals, stimulation of some pests (e.g. spider mites), and the evolution of insecticide-resistant strains. Consequently, we are witnessing a new wave of interest in natural enemies of insect pests in a different way. We do not rely on biological control alone; the effect of predators, parasites and diseases is incorporated into a complex integrated control programme. We strive, in a longer-term perspective, to keep pests under economic threshold levels without chemical treatments. Such methods have already been rather developed for stable habitats, mainly orchards, where the populations of natural enemies can be gradually increased. The programme for annual crop plants, the fauna of which is re-established every year, is, so far, only at its very beginning.'

Nearly 20 years later, in 1984, a second conference was held in Czechoslovakia (Hodek, 1986), since when the 'Aphidophaga' conference has become a regular event, at two- to three-year intervals.

These symposia encompass all aphidophages, but the Coccinellidae, as the best-researched, although not necessarily the most effective aphid enemies, are the most prominent. It is worth giving some specific examples involving the conservation of native predators in an integrated pest management scheme. A number of researchers have shown that changes in crop management practices can increase populations of ladybirds and other aphidophages. For example, when lucerne meadows were harvested in alternate strips, rather than whole fields at one time, the adult ladybird population rose by a factor of five, with their larvae increasing 20-fold (Schlinger and Dietrick, 1960). Other aphid predators and parasites also increased in density under strip-mowing regimens. Although the aphid *Therioaphis maculata* occurred at similar densities under both regimens early in the season, it rose to damaging densities in fully mown fields, but was reduced to negligible levels in the strip-cut fields.

Hodek (1973) also noted that in central Europe it is common practice to burn old grass on headlands, dykes and other habitats in arable land, and that this is very detrimental to ladybirds, which use such habitats to overwinter. Eradication of this practice would undoubtedly increase ladybird populations.

The same is true of the use of fire to remove stubble left after harvesting grain crops, such as wheat, which must incinerate vast numbers of ladybirds and other small creatures. In Britain, the outlawing of stubble burning in the early 1990s has had a beneficial effect on *C. 7-punctata* and *A. 2-punctata* populations. Maintaining a diverse array of crop types in an area can also be beneficial to aphid predators. Mixed farming of maize and legumes is common in tropical and subtropical regions, and studies have shown that this type of farming supports higher numbers of aphid enemies, in terms of both numbers of species and number of individuals, than do maize monocultures (Coderre, 1988), although the effects of mixed cropping on ladybird populations are not always predictable (Helenius, 1990). The integration of small areas of woodland, hedges and unmown meadowlands into agricultural areas, particularly where mainly annual crops are grown in large acreages, will be highly beneficial in providing overwintering sites for ladybirds and other aphid enemies close to crops, which they may protect the following year. Intercropping, in which two crops are grown together, or leaving weeds to grow freely, may also benefit coccinellid densities (Altieri et al., 1989; Horn, 1991). In contrast, mechanical harvesting of crops using chopper harvesters kills more than 90% of ladybird adults on the crop (Hodek and Honek, 1996). Honek (1982) showed that from 1978 to 1981, the first cut of alfalfa in the year, in late May or early June, killed up to 46% of the total adult population of *C. 7-punctata* on arable land. These examples illustrate that the effects of different agricultural practices should routinely be assessed, and alternatives to the most environmentally detrimental methods should be sought.

One of the prime motivations behind the desire to use coccinellids in aphid control is to reduce the use of insecticides. It is worth noting that use of insecticides has often had unexpectedly detrimental effects on crop yield. Pest species are usually rapid reproducers, and they are often found to evolve resistance to pesticides more quickly than do their predators or parasites. This allows outbreaks of new or old pest species, which had previously been kept under control by natural enemies. The resurgence of *I. purchasi* in California, following the destruction of *R. cardinalis* by DDT and Malathion applications, demonstrated this point at considerable cost (de Bach, 1947; Bartlett and Legace, 1960). Many studies have been conducted on the effects of pesticides (including acaricides, fungicides and herbicides) on ladybirds. Many of these compounds are highly toxic to ladybirds, while others, although they do not kill them, reduce reproductive output. In general, contact insecticides are most harmful, while systemic insecticides are more selective, particularly if applied via the soil.

Hodek (1973) made a number of interesting points about the timing of insecticide applications. Firstly, pesticides should never be used as a routine preventative measure. Rather, selective compounds should be used when pest monitoring indicates the need. Moreover, aphicides should never be used when aphids reach peak densities for a number of reasons: (i) it is too late to prevent loss of crop yield; (ii) the aphid population will decline, often very rapidly, after the peak, even without aphicide treatment; and (iii) the natural enemies are likely to be present in large numbers, so that the destruction of these helpful insects is likely to have a knock-on effect later in the season, or in future seasons. Fagan et al. (2010) found that careful timing of insecticide application can work synergistically with natural enemies in integrated pest management schemes in New Zealand.

Alien versus Native Species

In 1994 I wrote: 'I have yet to meet anyone who actively disliked ladybirds, let alone had a phobia about them'. This is no longer true, and the reason for the change is a single species of ladybird, *H. axyridis*. Over the last few years, this Asian ladybird has received a huge amount of scientific and media attention on both sides of the Atlantic because it is now considered an invasive alien in North America, South America, Africa, Europe and New Zealand. To describe the events that led to this ladybird being featured on the front page of *The Times* newspaper in Britain in 2005, I will describe two case histories, the first concerning the establishment and consequences of *C. 7-punctata* in North America, and the second dealing with *H. axyridis* in North America and Europe.

I have previously written of my concern that we are homogenising the world's fauna and flora (Majerus, 1994a). Increases in the speed of travel and advances in communication technology have already homogenised human culture to a significant extent, the so-called westernisation of Asian and Third World cultures being well documented. In the same way, the increased transport of organisms around the world by man, both intentionally and unintentionally, threatens many native species and ecosystems (McKinney and Lockwood, 1999; Hulme, 2009). It is alarmingly easy to think of examples: the great variety of introductions into Australasia, which have resulted in 'plagues' of species such as rabbits, house mice, cane toads and prickly pear cacti, or the introduction of predators, such as cats and rats into New Zealand, that have pushed species, such as the black robin and the flightless parrot, to the verge of extinction. In addition, great swaths of monoculture crops have been planted in many parts of the world, with great loss of habitat and biodiversity.

Biotic homogenisation now has widely recognised ecological and evolutionary consequences and is considered among the greatest threats to biodiversity

(Olden et al., 2006). The rapid increase in non-native species and the potential that these species have to become invasive (Kajita et al., 2006) contributes very significantly to the problems of biotic homogenisation. Although many accidentally introduced species have had little or no effect on native ecosystems, and many agricultural and biological control species have been beneficial (Williamson, 1999), the impacts of some alien species have been highly undesirable (Williamson, 1996; Vila et al., 2011).

Anthropogenic factors are responsible for the increase in movement and the establishment of alien species around the world. In recognition of the growing threat from invasive alien species, the Convention on Biological Diversity requires Parties 'as far as possible and as appropriate, (to) prevent the introduction of, control or eradicate those alien species which threaten ecosystems, habitats or species' (Article 8(h)), and a new regulation on the prevention and management of the introduction and spread of invasive alien species has recently come into force from the European Commission (EUR-Lex 2014).

Over the last four decades, some biologists and naturalists have become increasingly alarmed by the effect that ladybirds, imported to countries outside their natural range in classical biocontrol programmes, are having on the endemic species of the countries where they are released. There have been reports that introduced species of ladybird, particularly in North America, have harmed native coccinellid populations, non-target aphids and some other insects, such as butterflies. Reports in the 1980s and early 1990s were largely anecdotal, and concerned the impact of introduced *C. 7-punctata* on indigenous North American ladybirds. That reports were not based on strong comparative evidence or long-term monitoring was not surprising as, due to natural fluctuations in populations of ladybirds in general and aphidophagous species in particular, proving the population changes in one species of predator are caused by another predator would require tremendous monitoring effort. Funding was simply not available for such work. The situation was analogous to that of global warming in the 1960s and 1970s. Then, the first warnings of the possibility that the climate of Earth was changing due to emissions of so-called greenhouse gases were treated with great skepticism. Since then, research has given a better understanding of the effect of these gases, deeper theoretical prediction, and exhaustive empirical evidence showing that the Earth is indeed warming up. Today, few people who have examined the evidence in detail would deny that the greenhouse effect is a reality and that we are going to have to face huge changes in the climate of Earth over the next century.

In the case of ladybirds, anecdotal evidence that some non-native species are having a detrimental effect on natives is now being replaced by empirical data.

Coccinella 7-punctata in North America

Coccinella 7-punctata is possibly unique in having established in North America following both intentional and unintentional introductions (Majerus, 1994a). Many attempts were made to introduce this species between 1956 and 1971, from material originating in France, India, Norway and Sweden. All of these attempts apparently failed. However, in 1973, the species was found in New Jersey, the origin of this colony being unknown, and presumably the result of an accidental introduction (Angelet and Jacques, 1975). Subsequent intentional introductions have been more successful, resulting in establishment in Connecticut, Delaware, Maine, New York, Oklahoma and Pennsylvania. In 1959, *C. 7-punctata* was released in New Brunswick, Canada, but failed to establish. However, by the late 1970s it was well established in the state of Quebec, either due to accidental introduction, or having spread north from Maine (Larochelle, 1979). Further releases have been made in virtually every state of the United States and in southern Canada (Gordon and Vandenberg, 1991). In Utah, *C. 7-punctata* first appeared in low numbers in alfalfa fields in 1991. By 1999, 95% of ladybirds collected in these fields in the spring were *C. 7-punctata* (Evans, 2000).

One of the earliest reports that *C. 7-punctata* might be having a detrimental effect on native insects was that of Horn (1991). He reported that, in Ohio, *C. 7-punctata* was rapidly superseding *H. convergens* as the most abundant aphid-eating ladybird. Further, he suggested that increasing numbers of this ladybird might be responsible for declines in populations of two endangered species of lycaenid butterflies, *Lycaena melissa samuelis* and *Incisalia iris*. The butterflies oviposit on lupins in May and June, at a time when *C. 7-punctata* is active on lupins. Horn (1991) showed that *C. 7-punctata* will eat lycaenid eggs as an alternative food, and suggested that this predation may account for the decline in these butterflies, although egg predation has not been observed in the field.

Having a passion for butterflies and moths, this report saddened me. However, it was Horn's comment that *C. 7-punctata* was superceding *H. convergens* that I found even more alarming. Nor is this the only such report, with native ladybirds apparently suffering from *C. 7-punctata* in various states from Maryland to South Dakota. In Maryland, Staines et al. (1990) found that *C. 7-punctata* completely displaced *Coccinella 9-notata* from nursery habitats. In South Dakota, invasion by *C. 7-punctata* led to *A. 2-punctata* and *Coccinella transversoguttata* becoming 20 to 30 times less abundant (Elliot et al., 1996).

The spread and increase of *C. 7-punctata* in the 1980s and early 1990s in North America was rapid, so that by the mid-1990s, it was the most abundant ladybird in many parts of North America. Census data from agricultural systems has shown significant declines in a number of aphidophagous ladybirds, such as *C. 9-notata* and *C. transversoguttata*, coinciding with increases in *C. 7-punctata* (Brown, 2003; Turnock et al., 2003; Alyokhin and Sewell, 2004; Evans, 2004). Its increase and impact on native species probably would have continued had another imported ladybird, *H. axyridis*, not usurped its position as the most abundant coccinellid.

The Case of Harmonia axyridis

'The Ladybird Has Landed. A new ladybird has arrived in Britain. But not just any ladybird: this is *Harmonia axyridis*, the most invasive ladybird on Earth.' This is how I began a press release (Box 10.1), in 2004, in response to the first record of *H. axyridis* in Britain. The reasons for this alarmist release were routed in the history of the establishment, spread, increase and impact of *H. axyridis* in America (Koch, 2003) and more recently in continental Europe (Adriaens et al., 2003; Brown et al., 2008a).

The first established population of *H. axyridis* in North America was recorded from south-eastern Louisiana in 1988 (Chapin and Brou, 1991). It spread rapidly, but unevenly from this population, both through its own natural dispersal, and as a result of intentional redistribution for biological control. In the East, by 1993, it had reached Mississippi, Georgia, Alabama, Virginia, Delaware, New Jersey and Pennsylvania (Chapin and Brou, 1991; Gordon and Vandenberg, 1991; Day et al., 1994; Tedders and Schaefer, 1994). In a ladybird-monitoring programme in Pennsylvania, *H. axyridis* was not recorded from any of the 61 sites sampled in 1993, but was found in 52 of 124 sites by 1995 (Wheeler and Stoops, 1996).

Releases of *H. axyridis* from the south-eastern states were made in California, New Mexico and Texas in the early 1990s (Tedders and Schaefer, 1994). Records of *H. axyridis* in Oregon in 1991 probably also represent anthropogenic transport, although this is uncertain. Now, in 2016, *H. axyridis* is very widespread and abundant over most of the United States, with the exception of the peripheral states of Alaska and Hawaii (Brown et al., 2011b). In Canada, it is abundant in the southern parts of the eastern provinces and in southern British Columbia. It is rare in Alberta and has not been recorded from Labrador or Saskatchewan (Hicks et al., 2010).

The rapid spread and increase in *H. axyridis* in North America is a consequence of its adaptability. In respect of habitat, climate and diet, *H. axyridis* should be considered a generalist (Comont et al., 2012).

Although *H. axyridis* is usually described as an arboreal species (Hodek, 1973; Chapin and Brou, 1991; Brown and Miller, 1998), it can also thrive and breed in many agricultural habitats, including forage crops (LaMana and

Box 10.1 Press Release (September 2004, M. Majerus)

The Ladybird Has Landed

A new ladybird has arrived in Britain. But not just any ladybird: this is *Harmonia axyridis*, the most invasive ladybird on Earth.

Harmonia axyridis, which is variously called the Harlequin ladybird or the Multicoloured ladybug, is a deadly threat to a suite of insects, including butterflies, lacewings and many other ladybirds.

Introduced from Asia into North America for biocontrol of aphids, the Harlequin has swept across the States, becoming by far the most common ladybird in less than a quarter of a century since establishing there, and now Canada is seeing a similar spotted tide. In the last decade, its catastrophic increase in numbers has threatened endemic North American ladybirds and other aphid predators, many of which are plummeting alarmingly as the Harlequins consume their prey. Despite this unwelcome and well-publicised take-over by the Eastern invaders, Harlequin ladybirds are still sold in continental Europe by biocontrol companies, and it now roams across France, Belgium and Holland, with numbers soaring annually.

Now, it is here! On Sunday 19th September, Mr Ian Wright found an 'odd' ladybird in the garden of the White Lion pub, in Sible Hedingham, Essex. The ladybird was identified by Dr Michael Majerus of the Genetics Department, Cambridge University, an international ladybird expert.

Dr Majerus, who admits to 'having an inordinate fondness for ladybirds', said: 'this is without doubt the ladybird I have least wanted to see here. Given its proximity in Holland, I knew it was on its way, but I hoped that it wouldn't be so soon. Now many of our ladybirds will be in direct competition with this aggressively invasive species, and some will simply not cope'.

The Harlequin ladybird not only threatens other insects; in America it is also in conflict with humans. September sees many houses inundated by hundreds of thousands of these beetles seeking places to pass the winter. Harlequins also feed on fruit juices as they fuel up for the winter and fruit-growers are finding that they blemish many soft fruits, reducing the value of the crop. Indeed, so fond are they of grapes, that wineries have reported that the huge numbers of this ladybird among the harvested grapes, taint the vintage because of their acrid defensive chemicals. If this was not enough, reports of Harlequins biting people in the late summer, as they run out of aphid prey, are escalating.

Box 10.1 (cont.)

In North America and continental Europe, it will be difficult to control this invasive species, as numbers are already so great. However, in Britain we may still have time. Dr Majerus urges anyone who finds this ladybird to send it to him with precise details on when and where the ladybird was found. Although highly variable in its colour and pattern, none of the forms is easily mistaken for any British ladybirds. 'It is critical to monitor this ladybird now, before it gets out of control and starts to annihilate our own British ladybirds', he says.

Miller, 1996), corn, soybean, wheat (Colunga-Garcia and Gage, 1998), potato (Jansen and Hautier, 2008), and in conifer woodland (McClure, 1986). For example, in south-western Michigan, where the landscape is one of agricultural fields interspersed with deciduous and coniferous plantations, *H. axyridis* had become a dominant coccinellid, found in all the habitats monitored within four years of its arrival (Colunga-Garcia and Gage, 1998). That a diverse array of habitats appear favourable to *H. axyridis* in Michigan and elsewhere in America is not surprising, for it is a habitat generalist in its native range, occurring in all those habitats mentioned above and in grasslands in Siberia and reedbeds across its range.

The extensive native range of *H. axyridis* in Asia, in terms of both latitude and longitude (Orlova-Bienkowskaja et al., 2015), suggest that it can develop and breed in a broad range of climates. On the latitude scale, it extends from temperate regions to the subtropics, and is well adapted to winter temperatures below freezing and to summer temperatures in excess of 30°C (LaMana and Miller, 1998). On the longitude scale, it is adapted to both the extreme maritime conditions of the Eastern seaboards of Russia, Korea, China, and Japan, and the extreme continental climate of central Russia and Mongolia, with hot summers and very cold winters (Roy et al., 2016).

Although *H. axyridis* feeds most commonly on aphids, it has a greater dietary diversity than most other ladybirds, and can breed and complete development feeding on a variety of other prey, including coccids, psyllids, adelgids, and the immature stages of many other insects.

These habitat, climatic and dietary generalisms of *H. axyridis* suggest that once established in a region, it will rapidly be able to adapt to local conditions; a recent review highlighted the success distribution of *H. axyridis* (Roy et al., 2016). There is considerable evidence of its phenotypic plasticity in the literature. This is shown in both the variation in voltinism and colour pattern of this species.

Harmonia axyridis is a multivoltine species. Although described as bivoltine over much of Asia (Osawa, 2000), and in North America (LaMana and Miller, 1996; Koch and Hutchison, 2003), up to four or five generations of *H. axyridis* per year have been observed (Wang, 1986; Katsoyannos et al., 1997). Many ladybirds from temperate regions require a period of dormancy before becoming reproductively mature, but *H. axyridis* does not. Thus, if appropriate food is available, *H. axyridis* will breed whatever the season as long as temperatures are in the range 12–35°C. Moreover, in more maritime climates, *H. axyridis* is able to breed throughout the summer, and does not require a summer dormancy period as it does in regions that experience very hot dry summers. This reproductive flexibility, coupled with the varied diet, gives introduced populations of *H. axyridis* the chance to survive even when their 'genetic programming' is not perfectly adapted to their new environment.

Harmonia axyridis shows extreme colour pattern polymorphism. The geographic variation in the frequencies of the forms of *H. axyridis* suggests that some forms are particularly suited to specific environments. Thus, for example, in central Russia, f. *axyridis* predominates, while in eastern Russia and Korea, f. *succinea* is the most abundant form, and in central Honshu, Japan, the melanic forms *spectabilis* and *conspicua* are in the majority. Interestingly, in North America, f. *succinea* comprises 99% of the population, with melanics rare and f. *axyridis* absent (Hesler et al., 2004). This is despite documentary evidence that shows that melanics were released in biocontrol schemes in the United States (Krasfur et al., 1997; Koch et al., 2004). Similarly, melanic forms are unknown in South Africa (Roy et al., 2016).

The existence of considerable colour pattern variation will facilitate local adaptation simply because selection has genetic variability to work upon. Moreover, although the main colour pattern morphs of *H. axyridis* are under genetic control, environmental factors also have some effect on elytral patterns. Pupae that experience low temperatures produce forms of the *succinea* complex that have larger spots, which are frequently fused, one into another (Tan and Li, 1934; Majerus et al., 2006; Michie et al., 2010). This increase in the deposition of melanic pigments is probably a consequence of slower adult development in the pupa, and is beneficial in cooler conditions as these dark adults will remain active at lower temperatures and have longer to forage to build up fat bodies for the winter than would less melanised individuals. A survey of the degree of spotting (number and size of spots) of f. *succinea*, both through the year, and more importantly, along temperature gradients (latitudinal or altitudinal), would be very valuable in this regard.

Harmonia axyridis can disperse rapidly and over considerable distances. When breeding, females fly readily between host plants seeking high-density aphid populations, while males fly to find females. In addition, in many parts of Asia and America it flies long distances to and from overwintering sites (Nalepa et al., 2000; I. Zakharov, pers. comm.). Flights to dormancy sites may start as early as late August in Siberia (M. Majerus, pers. obs.) although most take place from late-September through to late-November (Liu and Qin, 1989; Sakurai et al., 1993). Large aggregations form, often on prominent, light-coloured objects, such as rocky outcrops (Tanagishi, 1976; Obata, 1986).

The history of *H. axyridis* in the United States, since its establishment in south-eastern Louisiana, tells of a predator that can adapt to a wide array of environments, is a highly effective predator of a range of plant pest species and can be used alongside other forms of pest control. So why did I write such an alarmist press release? The answer is that, despite the positive aspects of *H. axyridis* from a biocontrol perspective, it also has had a number of negative impacts in North America.

Negative Effects of Harmonia axyridis *on Non-target Prey*

The polyphagy of *H. axyridis* has meant that negative impacts on non-target prey species were inevitable. Unfortunately, little work has been done to assess the effects of *H. axyridis* on the population demography of non-target aphids, coccids and other prey species away from crop systems.

Boettner et al. (2000) stressed the need to examine the potential negative adverse effects of *H. axyridis* on non-pest flora and fauna and the lack of information on this subject. Since then, Koch et al. (2003) have identified *H. axyridis* as a predator of immature stages of the monarch butterfly, *Danaus plexippus*, an aposematic species that contains defensive chemicals.

Negative Effects of Harmonia axyridis *on Other Aphidophages*

Since *H. axyridis* became established in America, data suggesting that it is adversely affecting indigenous aphidophages has been accumulating. Brown and Miller (1998) observed a decrease in the abundance of native ladybirds in apple orchards in West Virginia over 13 years following the establishment of *C. 7-punctata* and *H. axyridis*. Colunga-Garcia and Gage (1998), from a nine-year study in Michigan, showed that populations of *Brachiacantha ursina*, *Cycloneda munda* and *Chilocorus stigma* had all declined following the establishment of *H. axyridis*. Michaud (2002) reported a decline in *Cycloneda sanguinea* inversely correlated with an increase in *H. axyridis* in citrus groves over five years.

Harmonia axyridis is likely to have negative effects on other aphidophages in any of three ways: resource competition, intraguild predation (IGP) and intra-specific competition. The diverse diet of *H. axyridis*, coupled with its mobility, its multivoltinism and high fecundity, gives this species the potential to significantly reduce prey populations. This may be beneficial in crop and horticultural systems, but not where negative effects may manifest as a reduction in biodiversity and declines in native beneficial predators, parasitoids and pathogens of aphids and coccids. For example, Nakata (1995) has shown that *H. axyridis* may have a direct impact on aphid parasitoids because both adult and larval *H. axyridis* feed on parasitised aphids that have not yet mummified. Larvae fed on parasitised aphids reached similar adult weights and had similar survival and development rates as those fed on unparasitised aphids (Takizawa et al., 2000). Moreover, Takizawa et al. (2000) suggested that the presence of *H. axyridis* larvae on an aphid colony may reduce parasitoid oviposition rate and thereby parasitoid numbers. Roy et al. (2008a) have also shown that *H. axyridis* will feed on dead aphids infected with pathogenic fungi more readily than do other ladybirds.

Harmonia axyridis is one of the top predators within aphidophagous and coccidophagous guilds (Yasuda and Ohnuma, 1999). For instance, in Japan, *H. axyridis* repeatedly arrived in alfalfa fields a short time after a number of other ladybirds, allowing *H. axyridis* to feed on the pre-pupae and pupae of other coccinellids (Takahashi, 1989). Lucas et al. (1998) proposed that *H. axyridis'* aggressive nature and the shape of its mandibles make it a dominant intraguild predator, but do not provide supporting empirical evidence.

Harmonia axyridis larvae and adults frequently feed on the immature stages of other aphidophagous insects (reviewed in Koch, 2003). The outcome of interactions between *H. axyridis* larvae and those of other aphidophagous coccinellids tends to be in favour of the *H. axyridis* larvae (Snyder et al., 2004; Yasuda et al., 2004). For example, *H. axyridis* preys on immature stages of three common generalist aphidophagous coccinellids in Europe, *C. 7-punctata* (Yasuda et al., 2001), *A. 2-punctata* (Burgio et al., 2002) and *P. 14-punctata* (Lynch et al., 2001). In all these cases, IGP is strongly biased in favour of *H. axyridis*. Indeed, few observations of other coccinellids successfully attacking *H. axyridis* have been made, and evidence suggests that the immature stages of this ladybird are resistant to attacks by most ladybirds. The large size of *H. axyridis* may have some impact in this regard, for Majerus, (1994a) noted that in IGP interactions between coccinellid larvae, the larger usually eats the smaller if both are mobile. However, this cannot be the only factor involved, for in interactions between *H. axyridis* and *C. 7-punctata*, which are of similar size, the former invariably won. Yasuda et al. (2001) attributed the *H. axyridis'* success here to its higher attack rates and greater escape ability. Further work showed similar results when *H. axyridis* larvae were put in competition with two native North American coccinellids, *H. convergens* and *C. transversoguttata* (Yasuda et al., 2004).

Ware et al. (2008), investigating IGP between *H. axyridis* and a variety of ladybirds both from its native range and its introduced range, found that

H. axyridis readily ate the eggs of most other species. From observations of this IGP, they concluded that defensive chemistry was the most influential factor in levels and direction of IGP of ladybird eggs. This is in line with previous studies showing that the larvae of many species of ladybird find *H. axyridis* eggs unpalatable (Hemptinne et al., 2000c; Alam et al., 2002). Only the eggs of three species, *Coccinula crotchi, Eocaria muiri* and *Calvia 14-guttata,* are well defended against predation by *H. axyridis* larvae. Moreover, species that are sympatric with *H. axyridis* are not more strongly defended than non-sympatric species (Ware et al., 2008). In the case of IGP between larvae, Ware and Majerus (2008) found that IGP was in favour of *H. axyridis* in all cases except with *Harmonia 4-punctata* and *C. 14-guttata,* where IGP was bidirectional, and *Anatis ocellata,* where it favoured *A. ocellata.* Observations of predation suggested that the differences in susceptibilities of larvae to attack by *H. axyridis* were due to variation in larval defence structures. *Harmonia axyridis* larvae have broad and robust spikes (Fig. 10.2), as do larvae of *A. ocellata, H. 4-punctata* and *C. 14-guttata*; the other species tested all possess fine hairs only (Fig. 10.3). Finally, IGP of pre-pupae and pupae by fourth instar larvae was unidirectional or significantly biased towards *H. axyridis* for most species (Fig. 10.4). Only in tests involving *H. axyridis, H. 4-punctata* and *A. ocellata* was bidirectional IGP observed. Similar results were obtained by Katsanis et al. (2013). Ware and Majerus (2008) concluded that variation in vulnerability was largely due to the thickness and integrity of the pupal casing. They also noted that pupae of all species 'flicked' in response to larval investigation, but that this proved ineffective against *H. axyridis* larvae and seemed to facilitate ventral attack (Fig. 10.5).

The results of this laboratory study, particularly with regard to active larvae, should be viewed in the light of tests carried out by Sato et al. (2005),

Fig. 10.2. Robust spikes of a *Harmonia axyridis* larva. (© Gilles San Martin.)

Fig. 10.3. Fine hairs of a *Coccinella 7-punctata* larva. (© Gilles San Martin.)

Fig. 10.4. *Harmonia axyridis* larva attacking *Coccinella 7-punctata* larva.

in which they looked at the reaction of first instar larvae of *A. 2-punctata* and *C. 7-punctata* to attack by fourth instar *H. axyridis*. Experiments were carried out on plants. They found that while nearly half the *C. 7-punctata* larvae dropped off the plants, thereby escaping, none of the *A. 2-punctata* did. The difference in the reactions of the larvae is not surprising. *Adalia 2-punctata* larvae often occur on trees, and should first instar larvae drop from a tree, they may starve to death before they find their way back into the tree to find prey. Conversely, *C. 7-punctata* larvae live on low-growing plants, and would be much more likely to find their way back to prey after dropping. This means that while the threat to *A. 2-punctata* and *C. 7-punctata* from IGP by

Fig. 10.5. *Harmonia axyridis* larva attacking ladybird pupa ventrally.

H. axyridis appears similar from the laboratory tests of Ware et al. (2008), in the field, *C. 7-punctata* larvae may suffer less because they have an effective escape.

The levels of intraguild predation by *H. axyridis* are inversely correlated to aphid or coccid density (Yasuda and Shinya, 1997; Burgio et al., 2002). However, neonate *H. axyridis* larvae frequently attack and consume the eggs of other coccinellid species when they encounter them, even when aphids are plentiful (M. Majerus, pers. obs.).

Intraguild predation is an important factor influencing the structure of aphidophagous guilds (Kajita et al., 2000). Experimental work on IGP, together with field observations from North America and latterly from Europe, support the view that *H. axyridis* is an aggressive coccinellid with a tendency for IGP, so that it has the potential to disrupt native guilds wherever it is introduced. Certainly, it is likely to seriously affect the abundance of native coccinellids and significantly reduce their available niches in the predator complex (Elliott et al., 1996; Brown et al., 2011a; Roy et al., 2012).

Negative Effects of Harmonia axyridis *on Humans*

In addition to the negative effects that *H. axyridis* is likely to have on other insects, *H. axyridis* has also become problematic for humans more directly. In autumn, when conditions become unfavourable, *H. axyridis* adults seek over-wintering sites (Obata, 1986; Huelsman et al., 2002). Due to their hypostatic behaviour, they migrate towards prominent, light-coloured objects on the horizon (Obata, 1986). In North America and Europe, buildings have often become the targets of these migrations, with huge numbers of *H. axyridis* descending on some buildings (Kidd et al., 1995).

Central heating or increases in temperature in the spring may cause *H. axyridis* to increase activity inside homes and offices (Huelsman et al., 2002), and people have reported that having hundreds of ladybirds walking or flying around inside is a considerable nuisance. Moreover, the defensive chemicals in the reflex blood, which the ladybirds release when aggravated, produces a foul odour in buildings and can leave yellow stains on soft furnishing. When food is unavailable, as in the winter, *H. axyridis* sometimes bites people in its search for food. Such bites cause a stinging sensation and a small bump as predigestive enzymes are injected into the skin. However, there have also been a few cases in which people have suffered from a hyperallergic reaction to *H. axyridis*, manifesting as allergic rhinoconjunctivitis (Yarbrough et al., 1999; Huelsman et al., 2002; Magnan et al., 2002; Davis, R.S., et al., 2006).

The tendency of *H. axyridis* to feed on alternative foods and to swarm in late summer has led to *H. axyridis* being designated as a fruit production and processing pest in North America. To build up their body fat for winter, the beetles feed on sweet ripe fruits, such as apples, pears, various berries and grapes, blemishing the fruit and reducing its value (Koch et al., 2004). In some vineyards, *H. axyridis* is attracted in very large numbers to bunches of ripe grapes. These are accidentally harvested with the crop, and when the crop is crushed, the bitter-tasting defensive chemicals of the ladybirds are released and seriously taint the vintage (Ratcliffe, 2002; Ejbich, 2003). This issue has caused some concern in North America (Galvan et al., 2008) and in South Africa, where substantial *H. axyridis* populations have been recorded in important wine production areas such as Western Cape Province (Stals, 2010).

I have personal experience of the spread of introduced ladybirds in North America, having visited the western United States in 1979, worked at Stanford University in 1981, travelled in New York State and down to Missouri in 1987, and driven across North America, from San Francisco to Orlando, by way of Seattle, Edmonton, Winnipeg and Nashville, in 2000. In 1979 and 1981, I recorded neither *C. 7-punctata* nor *H. axyridis*. In 1987, I have just two records of *C. 7-punctata* from New York State. On these visits, many other species, such as *H. convergens*, *Coleomegilla maculata*, *C. munda* (Fig. 10.6), *A. 2-punctata*, *Coccinella californica* and *Hippodamia sinuata* were common. In 2000, the most common ladybird I found, almost wherever I went in the United States, was *H. axyridis*, followed usually by *C. 7-punctata*. In Canada, *C. 7-punctata* was usually the most common species, with *H. axyridis* being recorded only in British Columbia. Nowhere did I find an native aphidophagous coccinellid the most abundant species present except in specialised habitats such as reedbeds, and many of the species I had encountered frequently before were generally scarce or absent. I believe, therefore, that the coccinellid fauna of

Fig. 10.6. *Cycloneda munda*.

North America has suffered a revolution over the last 30 years, to the considerable detriment of indigenous species. Despite my fondness for *C. 7-punctata*, Our Lady's Bird and the most common British species, and for *H. axyridis*, which I have worked on for many years, I am saddened to hear of the success of these aliens outside their native ranges.

Harmonia axyridis *in the United Kingdom*

It was against this backdrop that I received an unusual ladybird, found in Essex on 19 September 2004. The resulting press release (Box 10.1) had the desired effect. Through national newspapers, television and radio, the public was alerted to the arrival of *H. axyridis*. Funding was rapidly sought and gained to monitor its spread and increase in the UK and to assess its impacts on other invertebrates. This led to the launch, less than six months later (15 March 2005) of two websites (www.harlequin-survey.org and www.ladybird-survey.org) to monitor coccinellids in the UK. These surveys use innovative approaches, including a smartphone app, allowing online recording of all Britain's ladybirds (using the biological recording software application *iRecord* www.brc.ac.uk/iRecord), with digital image or specimen verification (Roy and Brown, 2015).

The level of public participation in the UK surveys over a ten-year period has been impressive; for example, over 30 000 *H. axyridis* records have been verified. These data have been used to show a very clear and rapid spread of the non-native ladybird across most of England and Wales (Brown et al., 2008b; Roy et al., 2011; Roy and Brown, 2015). However, *H. axyridis* appears to be climatically limited in Scotland, where the species has only sporadically been recorded.

Harmonia axyridis: *A High-Risk, Invasive Alien Species*

Harmonia axyridis has the ability to spread very rapidly across new environments, colonising a diverse array of habitats and adapting rapidly to local conditions. It is a voracious, generalist predator that dominates in intraguild interactions. There is a growing body of evidence to suggest that the adverse impacts of *H. axyridis* significantly outweigh the benefits that it may bring to agricultural and horticultural communities through pest control.

Owing to the increasing problems of invasive alien species, attempts have been made to predict the probability that an exotic species will become invasive (van Lenteren et al., 2003). Most attention has been paid to classical biological control agents (Hokkanen et al., 2003) and protocols for risk assessment have been suggested for the regulation of such agents (van Lenteren et al., 2003). Criteria used in risk assessment include establishment probability, dispersal, host range, effects on non-target herbivores, intraguild predation and competition with native natural enemies, to produce numerical risk index values. Using this assessment procedure, van Lenteren et al. (2003) allocated their highest risk index to *H. axyridis*.

The evidence of habitat homogenisation is obvious. Travel around lowland America, Europe, Japan or Australia and the similarity in urban, suburban, agricultural and forestry landscapes is inescapable. The dominant generalist species associated with these homogeneous habitats tend to be the same over large geographic ranges, with perhaps the exception of Australia. In contrast, more specialised species have retreated to habitat refuges not open to exploitation for a variety of reasons. If we are to maintain even a small part of the diversity of life on Earth, we need to take this seriously and stop moving species around the world until full risk assessments have been conducted. Moreover, once risks have been assessed, findings should be acted upon, with legislation if necessary. The new Europe-wide regulation on the prevention and management of the introduction and spread of invasive alien species provides opportunities in this regard (EUR-Lex 2014).

Conservation

Why Conserve Ladybirds?

I began this book by pointing out that ladybirds are probably the most popular insects. They are attractive to look at and usually beneficial. Thus far, this chapter has dealt essentially with what ladybirds can do for us. Now I turn to what we can do for ladybirds, or, more to the point, what we must stop doing.

Admittedly, few people consider it necessary to expend time or energy conserving abundant species, such as *H. convergens* or *C. maculata*, *C. 7-punctata* or *H. axyridis*. Two decades ago, I wrote that I thought this view shortsighted, arguing that there was a strong case for conservation of common as well as more scarce coccinellids (Majerus, 1994a). Since that time, the stresses have increased: more pollution, more habitat destruction, more mechanisation of farming methods, and simply more people. Conservation arguments rarely win over development and industrial concerns. Since 1994, many of my concerns have been vindicated, particularly in the United States, where the coccinellid fauna has been changing rapidly and radically, with common and rare native species suffering alike from invasive non-native species.

Although, as noted earlier, many insects are not very attractive, they have proved useful as pollinators, biocontrol agents, and model systems for researchers interested in everything from toxicology to aerodynamics and from genetics to evolution. Yet it is difficult to put a monetary value on these uses (although see Losey and Vaughan, 2006). However, the economic value of the common ladybirds can be used as justification for their conservation.

Much of our modern agriculture nowadays comprises huge monoculture fields of annual crops. Studies in Switzerland showed that if agricultural land is interspersed with hedges, perennial meadowland and woodlands, economic benefit accrues. The studies have considered the role of ladybirds and other aphid predators and parasites in the population dynamics of some important aphid species infesting field crops. The aphid species concerned overwinter away from the annual crops, in hedgerows, woodlands or natural meadows. Most of the yearly cycle is spent on non-crop plants, with only a relatively small proportion on the annual crops. It would seem intuitively obvious that the availability of non-crop overwintering sites and spring breeding sites for the aphids, and good conditions for aphid reproduction and development in the spring, on these non-crop plants, would lead to high aphid numbers on the crop plants later in the year. However, the reverse was found to be the case (Suter and Keller, 1977; Keller and Suter, 1980). During a long-term study, it was discovered that good early aphid development on non-crop perennials allowed aphid predators and parasitoids to build up. As aphids colonised the annual crop plants, these predators and parasitoids followed them, actively or passively, in sufficient numbers to check the increase in aphid numbers, and keep aphid population densities below the damage threshold. In years with low spring aphid development, although fewer aphids initially founded colonies on annual crops, high and damaging infestations resulted, because aphid predators and parasitoids were not present in sufficient numbers to control the aphid population increases on these crops to a significant degree.

In Switzerland, Suter and Keller (1977) noted that hay meadows were of particular importance, as they encouraged a large number of plant species, and therefore a diverse array of insects, including ladybirds. They advised that cutting hay only after aphids had migrated to annual crops would increase the efficiency of biological control using naturally occurring enemies of aphids. This would encourage aphid predators and parasites to migrate to the crops where they would be of most use. They concluded that hedgerows, ditches, streams, river banks, perennial meadows and woodlands should not be considered as agriculturally useless land. Such habitats have an economic value and their protection could be beneficial, particularly in regions characterised by large, intensively controlled, monoculture agriculture. The benefits accruing from increased numbers of natural pest control agents would coincide with more general conservation requirements, with such 'natural habitats' having potential to act as wildlife corridors, and possibly giving refuge and home to rare and endangered species in an otherwise unsympathetic landscape. Many subsequent studies in many different parts of the world have shown that wide field margins, alternative prey or other alternative food (e.g. nectar, pollen) and overwintering areas can be highly beneficial by maintaining populations of aphid predators and parasitoids (Hickman and Wratten, 1996; Powell, 2005; Wratten et al., 2005).

Now, in the early twenty-first century, the emphasis for conservation must be towards communities rather than individual species, towards agricultural fragmentation rather than merger monoculture, and towards long-term continuity of land usage and management. On this large scale, it is evident that conservation of more or less common and habitat generalist ladybird species, through maintenance of an array of different habitats, will be beneficial both to agriculture and to other organisms. On a smaller scale, individuals can apply the same strategy to their own gardens.

Ladybirds in the Garden

I remember visiting a retired couple who lived in a terraced house in south London some years ago. Their garden was their pride and joy, and hence they were distressed when aphids ravaged their pelargoniums, roses and vegetables. The couple did not want to use chemical sprays, but wanted the aphid problem solved. An intensive search of the aphid colonies revealed a few small parasitoid wasps, half a dozen hoverfly larvae and one ladybird. I asked what the garden looked like in the winter, and was shown a photograph of the most pristine lawn and flowerbeds taken in January, with all the bedding plants removed, most of the perennials cut neatly to within a few centimetres of the soil, and not a dead leaf in sight. The problem was that there was nowhere for ladybirds, or other aphid enemies, to pass the winter. I advised them to leave a bit more mess around the garden, to do the big clear up of the herbaceous border in the spring, not the autumn, and to leave dead leaves around the perennial rosettes until the New Year's growth began to show. I remember their reaction to my suggestions was close to horror. Nonetheless, the advice still stands, and is as relevant in California as in Cambridge, in Louisiana as in London, or in Brisbane as in Birmingham. I practise it myself, and have little problem with aphids of any sort.

In Box 10.2 I have set out several recommendations that will encourage, not only ladybirds into one's garden, but also a considerable range of other wildlife, from butterflies to birds, lacewings to dragonflies, and harvest-men to hedgehogs. The constraints of space, situation and family may not permit all these points to be put into practice for a particular garden, but each will have an impact and their effects are largely additive; the more one does, the greater the benefit.

Many urban areas now feature private gardens or community parks. Over the past decade or so, the importance of these green areas has slowly been recognised. Because of individual tastes and preferences, a large group of gardens often provides an extraordinarily diverse and plant species-rich habitat. It is true that many of the plants are of foreign origin, and that generally native species will support a greater diversity of fauna. However, even the foreign species may provide nectar or shelter, and some of the more generalist species adapt to them quickly.

Human activities may then have beneficial effects on wildlife. Certainly, we can all help the more common and generalist species of ladybird with a little thought and without any significant expense or physical effort. The rewards will come in terms of a diverse and sometimes surprising array of wildlife encounters. A garden that is maintained, but not over-managed, with a range of plant types, perhaps a deciduous hedge, a fruit tree, one or two mature native trees, and a small area of grass that is only cut two or three times a year, will support a diverse array of ladybird species. In my own garden on the outskirts of Cambridge, I have recorded over half the British species of ladybirds.

The Conservation of Rare Species

Habitat generalist ladybirds will benefit most from maintaining a diversity of plants and a certain degree of untidiness, promoting what Darwin (1859) called an 'entangled bank'. However, for some habitat specialists, specific measures may need to be taken to protect a species. Highly specialised species are recognised as being more vulnerable to extinction than generalists, because

Box 10.2 Gardening Hints to Encourage Ladybirds (Adapted from Majerus, 1996)

1. Do not use general chemical pesticides. If pesticides must be used, read the label and use brands that at least claim only to harm target pests.
2. Plant a diversity of species. Note, many species of aphid-feeding coccinellid move from host plant to host plant as aphid populations wax and wane.
3. Favour native species of plant and tree rather than imported species. Native species usually support a greater diversity of insects.
4. Leave a wide 'weedy' area, or more preferably two, somewhere in the garden. One may be on previously cultivated, and nutrient-rich soil; the other may be on an area of grass that gets at least some sun and which is only cut once a year, preferably in July. Alternatively, a wild area may be created by digging over a patch of earth and allowing natural plant colonisation of this patch. Best results are obtained if the soil has not been fertilised in the recent past. Such an area needs little maintenance save for the removal of vigorous competitive species that will overwhelm it.
5. Leave low rosettes of dead leaves on perennial plants throughout the winter, and clear them in the spring. These provide overwintering sites for many coccinellid species. In the same vein, disturb ground cover plants as little as possible in the winter, as these too may be used as winter sheltering sites by many ladybirds.
6. Plant some species particularly for the use of ladybirds in the winter. The tight protective tussocks formed by pampas and some other large grasses attract many species that overwinter close to the ground. Evergreen plants, such as *Ilex* spp. and various needled conifers are especially notable in providing overwintering sites for ladybirds. The evergreen foliage of these shrubs and trees offer many sheltered situations for coccinellids that overwinter higher up.
7. Some ladybirds, particularly *Adalia 2-punctata* and *Harmonia axyridis*, often appear on the insides of windows on sunny days in early spring. These should be put outside, preferably late in the morning when the temperature is above 4°C.
8. Because many coccinellids augment their diet with nectar in the late summer, late-flowering nectar plants are important. Buddleia, the so-called 'butterfly bush', is useful in this respect and will attract several species of ladybird to its flowers (as well as many butterflies).

Box 10.2 (cont.)

To prolong the nectar supply, buddleia is best dead-headed every second or third day.

9. A pond is always a joy in a garden. Depending on size, reedy plants may be planted. These may attract reed-bed specialists, such as *Macronaemia* spp., *Anisosticta* spp. and *Coccidula* spp.
10. Watch and learn. Every garden is different and the ecological balance in a garden will vary from place to place and year to year. Try to watch ladybirds in your own garden, determine their requirements and act accordingly. Note which plants they breed upon, and where they pass the winter. Know their early stages so that eggs, larvae and pupae are not destroyed accidentally. In addition, teach your family.

they are less adaptable to changing conditions. With such species, perhaps the greatest difficulty is in understanding the complexities of the ecological interactions that make some habitats favourable and others unfavourable. Who, for example, would have predicted that the introduction of a biological control agent, in the form of a virus, introduced into France to control rabbits, would, a quarter of a century later, lead to the extinction of a butterfly in Britain?

Myxoma is a naturally occurring virus of South and Central American rabbits. It causes illness, but not death. However, this virus was used to great effect to control European rabbits in Australia. Following this success, it was introduced into a rabbit population on an estate near Versailles, on the outskirts of Paris, in 1952. The myxoma virus spread rapidly in France, reaching the north coast and crossing the English Channel in 1953, probably in a mosquito, whereupon it spread across Britain from the south-east, decimating wild rabbit populations. The crash in numbers of wild rabbits coupled with changes in sheep grazing practices dramatically reduced grazing on chalk grasslands, allowing scrub to invade. The character of many chalk grassland habitats thus changed. The large blue butterfly, *Maculinea arion*, lived on chalk grassland in Britain, where it had a strong association with ants of the genus *Myrmica*. The butterfly larvae were carried by the ants from thyme plants, on which the eggs had been laid, into the underground galleries of the ants' nests, where they preyed on ant grubs. This highly specialised butterfly depends for its survival on the presence of both wild thyme and the ants. These only occur together on short-cropped chalk grassland, and the changes in these habitats following myxomatosis resulted in a decline in both, so that *M. arion* disappeared from Britain. The understanding of this system and the requirements of the butterfly have led to its successful reintroduction from European stock, but the habitats in which colonies of the butterfly now reside have to be carefully managed (Thomas et al., 2009).

The same understanding and endeavour may be needed in the case of some of the more specialised coccinellids. An obvious parallel with the case of *M. arion* would be the myrmecophilous ladybirds, such as *Coccinella magnifica* and *Platynaspis luteorubra*. The close distributional association of these species with colonies of appropriate ants strongly suggests that conservation of these species will depend on the conservation of the ants with which they live.

In Britain, the decline in *Formica rufa*, as a result of loss of woodland and heathland, and reduction in coppicing, has reduced the number of suitable sites available to *C. magnifica*. To reverse this decline, heathlands should be managed to promote a diversity of successional stages, and woodlands should be managed to encourage a range of vegetation structures, including a rotation of open glades and wide rides. Wood ants are important in woodland and heathland ecosystems, and are fascinating insects in their own right. Yet when nests are discovered, they are often severely disturbed 'to watch the ants get angry' as one youngster told me. Perhaps the link with a scarce ladybird will give wood ants a better press, and save them from unnecessary disturbance, as well as aiding *C. magnifica*.

Other specialist species may also need some precise element of their ecology, or a set of elements, to be maintained. However, the conservation of most species suffers from the difficulty that in most cases, we do not know what the critical factors are. Future research into the conservation of rare ladybirds needs to be aimed at identifying these critical factors.

The Best That We Can Do

At the root of most conservation issues, whether global warming, increases in a wide array of pollutants, habitat destruction or changes in land usage, is one fact: there are too many of us. This is not the place for a detailed discussion of the problems of human population increase. However, it is worth recognising the enormity of the problem. In 1963, when I was nine, my geography teacher told me that there were about three billion people on Earth. Now there are well over twice that number. In my view, the greatest single contribution that humanity could make to the conservation of ladybirds, and most other organisms on Earth, would be to stop multiplying faster than we are dying.

Ladybirds may perhaps be viewed as conservation flag-bearers for the invertebrates. They are popular and are seen as beneficial to man. As one publicity slogan goes: 'Everyone loves a ladybird'. Well, if we do, it is surely time to give them a little thought and care. A nineteenth century 'Ladybird, ladybird' children's nursery rhyme from Dorset runs:

> Laedy-bird! Leady-bird! Vlee away home,
> Your house is a-vire, your children will burn.

In some ways, the whole world is on fire. By our manipulation of our environment, our destruction of natural habitats and our ever-increasing need for food and space for our multiplying numbers, we are giving ladybirds nowhere else to go. No doubt, some of the generalists will survive, and some specialists will hang on in those extreme environments that we find difficult to colonise. Despite this, I see little long-term hope for these beautiful beetles, or most of the other species on Earth, unless we, as a species, start to take responsibility for the planet we have taken under our control.

References

Ables, J.R., Jones, S.L. and McCommas, D.W., Jr. 1978. Response of selected predator species to different densities of *Aphis gossyppii* and *Heliothis virescens* eggs. *Environmental Entomology*, **7**, 402–4.

Acorn, J. 2007. *Ladybugs of Alberta: Finding the Spots and Connecting the Dots*. Edmonton, AB: University of Alberta.

Adriaens, T., Branquart, E. and Maes, D. 2003. The multicoloured asian ladybird *Harmonia axyridis* Pallas (Coleoptera: Coccinellidae), a threat for native aphid predators in Belgium? *Belgian Journal of Zoology*, **133**, 195–6.

Agarwala, B.K. 1991. Why do ladybirds (Coleoptera, Coccinellidae) cannibalize? *Journal of Bioscience*, **16**, 103–9.

Agarwala, B.K. and Dixon, A.F.G. 1992. Laboratory study of cannibalism and interspecific predation in ladybirds. *Ecological Entomology*, **17**, 303–9.

Agarwala, B.K. and Dixon, A.F.G. 1993. Kin recognition: egg and larval cannibalism in *Adalia bipunctata* (Coleoptera: Coccinellidae). *European Journal of Entomology*, **90**, 45–50.

Al Abassi, S., Birkett, M.A., Pettersson, J., et al. 2000. Response of the seven-spot ladybird to an aphid alarm pheromone and an alarm pheromone inhibitor is mediated by paired olfactory cells. *Journal of Chemical Ecology*, **26**, 1765–71.

Alam, N., Choi, I.S., Song, K.S., et al. 2002. A new alkaloid from two coccinellid beetles *Harmonia axyridis* and *Aiolocaria haexapilota*. *Bulletin of the Korean Chemical Society*, **23**, 497–9.

Alhmedi, A., Haubruge, E. and Francis, F. 2010. Intraguild interactions and aphid predators: biological efficiency of *Harmonia axyridis* and *Episyrphus balteatus*. *Journal of Applied Entomology*, **134**, 34–44.

Allen, D.C., Knight, F.B. and Foltz, J.L. 1970. Invertebrate predators of the jack-pine budworm, *Choristoneura pinus*, in Michigan. *Annals of the Entomological Society of America*, **63**, 59–64.

Altieri, M.A., Glaser, D.L. and Schmidt, L.L. 1989. Diversification of agroecosystems for insect pest regulation: experiments with collards. In Gliessman, S.R., ed. *Agroecology: Researching the Ecological Basis for Sustainable Agriculture*. Ecological Studies 78. Berlin: Springer-Verlag; pp. 70–82.

Alyokhin, A. and Sewell, G. 2004. Changes in a lady beetle community following the establishment of three alien species. *Biological Invasions*, **6**, 463–71.

Anderson, J.M.E. 1982. Seasonal habitat utilization and food of the ladybirds *Scymnodes lividigaster* (Mulsant) and *Leptothea galbula* (Mulsant) (Coleoptera: Coccinellidae). *Australian Journal of Zoology*, **30**, 59–70.

Anderson, J.M.E. and Hales, D.F. 1986. *Coccinella repanda*. Diapause? In Hodek, I., ed. *Ecology of Aphidophaga*. Prague: Academia and Dordrecht: Dr. W. Junk; pp. 233–8.

Anderson, J.M.E., Hales, D.F. and van Brunschot, K.A. 1986. Parasitisation of coccinellids in Australia. In Hodek, I., ed. *Ecology of Aphidophaga*. Prague: Academia and Dordrecht: Dr. W. Junk; pp. 519–24.

Andow, D.A. and Risch, S.J. 1985. Predation in diversified agroecosystems: relationships between a coccinellid predator *Coleomegilla maculata* and its food. *Journal of Applied Ecology*, **22**, 357–72.

Angelet, G.W. and Jacques, R.L. 1975. The establishment of *Coccinella septempunctata* L. in continental United States. *United States Department of Agriculture – Cooperative Economic Insect Report*, **25**, 883–4.

Aoki, S., Akimoto, S. and Yamane, S. 1981. Observations on *Pseudoregma alexanderi* (Homoptera, Pemphigidae), an aphid species producing pseudoscorpion-like soldiers on bamboos. *Kontyû*, **49**, 355–66.

Arakaki, N. 1988. Egg protection with faeces in the lady-beetle, *Pseudoscymnus kurohime* (Miyatake) (Coleoptera: Coccinellidae). *Applied Entomology and Zoology*, **23**, 495–7.

Arakaki, N. 1992. Predators of the sugar cane woolly aphid, *Ceratovacuna lunigera* (Homoptera: Aphididae) in Okinava and predator avoidance of defensive attack by the aphid. *Applied Entomology and Zoology*, **27**, 159–61.

Arnaud, L., Spinneux, Y. and Haubruge, E. 2003. Preliminary observations of sperm storage in *Adalia bipunctata* (Coleoptera: Coccinellidae): sperm size and number. *Applied Entomology and Zoology*, **3**, 301–4.

Attygalle, A.B., Xu, S.-C., McCormick, K.D. and Meinwald, J. 1993. Alkaloids of the Mexican bean beetle, *Epilachna varivestis* (Coccinellidae), *Tetrahedron*, **49**, 9333–42.

Ayr, W.A. and Browne, L.M. 1977. The ladybug alkaloids including synthesis and biosynthesis. *Heterocycles*, **7**, 685–707.

Azam, K.M. and Ali, M.H. 1970. A study of the factors affecting the dissemination of the predatory beetle, *Coccinella septempunctata* L. *Final Technical Report, Department of Entomology*. Hyderabad, India: Andhra Pradesh Agricultural University.

Bach, C.E. 1991. Direct and indirect interactions between ants (*Pheidole megacephala*), scales (*Coccus viridis*) and plants (*Pluchea indica*). *Oecologia*, **87**, 233–9.

Bailey, C.L. and Chada, H.L. 1968. Spider populations in grain sorghums. *Annals of the Entomological Society of America*, **61**, 567–72.

Balduf, W.V. 1935. *The Bionomics of Entomophagous Coleoptera. (13 Coccinellidae – Lady Beetles).* Chicago, IL: John Swift.

Banks, C.J. 1955. An ecological study of Coccinellidae associated with *Aphis fabae* Scop. on *Vicia faba*. *Bulletin of Entomological Research*, **46**, 561–87.

Banks, C.J. 1956. Observations on the behaviour and early mortality of coccinellid larvae before dispersal from egg shells. *Proceedings of the Royal Entomological Society, London A*, **31**, 56–61.

Banks, C.J. 1957. The behaviour of individual coccinellid larvae on plants. *British Journal of Animal Behaviour*, **5**, 12–24.

Banks, C.J. 1962. Effects of the ant *Lasius niger* (L.) on insects preying on small populations of *Aphis fabae* (Scop.) on bean plants. *Annals of Applied Biology*, **50**, 669–79.

Banks, C.J. and Macaulay, E.D.M. 1967. Effects of *Aphis fabae* Scop. and of its attendant ants and insect predators on yields of field beans (*Vicia faba* L.). *Annals of Applied Biology*, **60**, 445–53.

Barczak, T. 1991. Coccinellid beetles associated with colonies of the *Aphis fabae* complex in Poland: preliminary results. In Polgar, L., Chambers R.J., Dixon, A.F.G. and Hodek, I., eds. *Behaviour and Impact of Aphidophaga*. The Hague: SPB Academic Publishing; pp. 103–5.

Barron, A. and Wilson, K. 1998. Overwintering survival in the seven spot ladybird, *Coccinellidae septempunctata* (Coleoptera: Coccinellidae). *European Journal of Entomology*, **95**, 639–42.

Bartlett, B.R. 1978. Coccidae. In Clausen, C.P., ed. *Introduced Parasites and Predators of Arthropod Pests and Weeds: A World Review*. Agricultural Research Service Handbook No. 480, USDA.

Bartlett, B.R. and Legace, C.F. 1960. Interference with the biological control of cottony-cushion scale by insecticides and attempts to re-establish a favourable natural balance. *Journal of Economic Entomology*, **53**, 1055–8.

Bates, H.W. 1862. Contributions to an insect fauna of the Amazon Valley. Lepidoptera: Heliconidae. *Transactions of the Linnean Society of London*, **23**, 495–566.

Baungaard, J. and Hämäläinen, M. 1984. Notes on egg-batch size in *Adalia bipunctata* (Coleoptera, Coccinellidae). *Annales Entomologici Fennici*, **47**, 25–7.

Belcher, D.W. and Thurston, R. 1982. Inhibition of movement of larvae of the convergent lady beetle by leaf trichomes of tobacco. *Environmental Entomology*, **11**, 91–4.

Belicek, J. 1976. Coccinellidae of Western Canada and Alaska with analyses of the transmontane zoogeographic relationships between the fauna of British Columbia and Alberta (Insecta: Coleoptera: Coccinellidae). *Questiones Entomologicae*, **12**, 283–409.

Belovsky, G.E. 1978. Diet optimization in a generalist herbivore: the moose. *Theoretical Population Biology*, **14**, 105–34.

Belshaw, R. 1993. *Tachinid Flies, Diptera: Tachinidae. Handbooks for the Identification of British Insects*, vol. 10, part 4a. London: Royal Entomological Society.

Benham, B.R., Lonsdale, D. and Muggleton, J. 1974. Is polymorphism in the two-spot ladybird an example of non-industrial melanism? *Nature*, **249**, 179–80.

Benton, A.H. and Crump, A.J. 1981. Observations on the spring and summer behavior of the 12-spotted ladybird beetle, *Coleomegilla maculata* (DeGeer) (Coleoptera: Coccinellidae) *Journal of the New York Entomological Society*, **89**, 102–8.

Berkvens, N., Bale, J.S., Berkvens, D., Tirry, L. and De Clercq, P. 2010. Cold tolerance of the harlequin ladybird *Harmonia axyridis* in Europe. *Journal of Insect Physiology*, **56**, 438–44.

Berkvens, N., Bonte, J., Berkvens, D., et al. 2008. Pollen as an alternative food for *Harmonia axyridis*. *BioControl*, **53**, 201–10.

Berti, N., Boulard, M. and Duverger, C. 1983. Fourmis et Coccinelles. Revue bibliographique et observations nouvelles. *Bulletin Société Entomologique de France*, **88**, 271–5.

Betts, M.M. 1955. The food of titmice in oak woodland. *Journal of Animal Ecology*, **24**, 282–323.

Bezzerides, A.L., McGraw, K.J., Parker, R.S. and Husseini, J. 2007. Elytra color as a signal of chemical defense in the Asian ladybird beetle *Harmonia axyridis*. *Behavioral Ecology and Sociobiology*, **61**, 1401–8.

Bhatkar, A.P. 1982. Orientation and defense of ladybeetles (Col., Coccinellidae), following ant trail in search of aphids. *Folia Entomologia Mexicana*, **53**, 75–85.

Bielawski, R. 1959. *Biedronki-Coccinellidae. Klucze do oznaczania owadow Polski*. Part 19. Tom 76. Warszawa: Panstowe Wydawnictwo Naukowe.

Bielawski, R. 1961. *Die in einem Krautpflanzenverein und in einer Kieferschonung in Warszawa/Bielany auftretenden Coccinellidae (Coleoptera)*. Warszawa: Fragmenta Faunistica.

Birch, A.N., Geoghegan, I.E., Majerus, M.E.N., et al. 1999. Tri-trophic interactions involving pest aphids, predatory 2-spot ladybirds and transgenic potatoes expressing snowdrop lectin for aphid resistance. *Molecular Breeding*, **5**, 75–83.

Bishop, J.A., Cook, L.M. and Muggleton, J. 1978. The response of two species of moths to industrialization in northwest England. I. Polymorphism for melanism. *Philosophical Transactions of the Royal Society of London B*, **281**, 489–515.

Blackman, R.L. 1965. Studies on specificity in Coccinellidae. *Annals of Applied Biology*, **56**, 336–8.

Blackman, R.L. 1967. The effects of different aphid foods on *Adalia bipunctata* L. and *Coccinella 7-punctata* L. *Annals of Applied Biology*, **59**, 207–19.

Bodenheimer, F.S. 1943. Studies on the life-history and ecology of Coccinellidae. I. The life-history of *Coccinella septempunctata* L. in four different zoogeographical regions. *Bulletin de la Société Fouad 1er d'Entomologie*, **27**, 1–28.

Boettner, G.H., Elkinton, J.S. and Boettner C.J. 2000. Effects of a biological control introduction on three nontarget native species of saturniid moths. *Conservation Biology*, **14**, 1798–806.

Booth, R.G., Cox, M.L. and Madge, R.B. 1990. *IIE Guides to Insects of Importance to Man. 3. Coleoptera*. Cambridge: Cambridge University Press.

Brackenbury, J.H. 1992. *Insects in Flight*. London: Blandford.

Braconnier, M.F., Braekman, J.C., Daloze, D. and Pasteels, J.M. 1985. (Z)-1, 17-diaminooctadec-9-ene, a novel aliphatic diamine from Coccinellidae. *Experimentia*, **41**, 519–20.

Bradley, G.A. 1973. Effect of *Formica obscuripes* (Hymenoptera: Formicidae) on the predator–prey relationship between *Hyperaspis congressis* (Coleoptera: Coccinellidae) and *Toumeyella numismaticum* (Homoptera: Coccidae). *Canadian Entomologist*, **105**, 1113–18.

Bradley G.A. and Hinks, J.D. 1968. Ants, aphids and jack pine in Manitoba. *Canadian Entomologist*, **100**, 40–50.

Brakefield, P.M. 1984a. Ecological studies on the polymorphic ladybird *Adalia bipunctata* in the Netherlands. II. Population dynamics, differential timing of reproduction and thermal melanism. *Journal of Animal Ecology*, **53**, 775–90.

Brakefield, P.M. 1984b. Ecological studies on the polymorphic ladybird *Adalia bipunctata* in the Netherlands. I. Population biology and geographic variation of melanism. *Journal of Animal Ecology*, **53**, 761–74.

Brakefield, P.M. 1985a. Polymorphic Müllerian mimicry and interactions with thermal melanism in ladybirds and a soldier beetle: a hypothesis. *Biological Journal of the Linnean Society*, **26**, 243–67.

Brakefield, P.M. 1985b. Differential winter mortality and seasonal selection in the polymorphic ladybird *Adalia bipunctata* (L) in The Netherlands. *Biological Journal of the Linnean Society*, **24**, 189–206.

Brakefield, P.M. and Lees, D.R. 1987. Melanism in *Adalia* ladybirds and declining air pollution in Birmingham. *Heredity*, **59**, 273–7.

Brakefield, P.M. and Willmer, P.G. 1985. The basis of thermal melanism in the ladybird *Adalia bipunctata*: differences in reflectance and thermal properties between morphs. *Heredity*, **54**, 9–14.

Briggs, D.J., Cooke, G.R. and Gilbertson, D.D. 1985. The chronology and environmental framework for early man in the Upper Thames Valley. *British Archaeological Reports*, **137**, 1–176.

Bristow, C.M. 1984. Differential benefits from ant attendance to two species of Homoptera on New York ironweed. *Journal of Animal Ecology*, **53**, 715–26.

Brown, M.W. 2003. Intraguild responses of aphid predators on apple to the invasion of an exotic species, *Harmonia axyridis*. *BioControl*, **48**, 141–53.

Brown, M.W. and Miller, S.S. 1998. Coccinellidae (Coleoptera) in apple orchards of eastern West Virginia and the impact of invasion by *Harmonia axyridis*. *Entomological News*, **109**, 143–51.

Brown, N.R. and Clark, R.C. 1959. Studies of predators of the balsam wooly aphid, *Adelges picea* (Ratz.) (Homoptera: Adelgidae). VI. *Aphidecta obliterata* (L.) (Coleoptera: Coccinellidae), an introduced predator in Eastern Canada. *Canadian Entomologist*, **91**, 596–9.

Brown, P.M.J., Adriaens, T., Bathon, H., et al. 2008a. *Harmonia axyridis* in Europe: spread and distribution of a non-native coccinellid. *BioControl*, **53**, 5–21.

Brown, P.M.J., Frost, R., Doberski, J., et al. 2011a. Decline in native ladybirds in response to the arrival of *Harmonia axyridis*: early evidence from England. *Ecological Entomology*, **36**, 231–40.

Brown, P.M.J., Ingels, B., Wheatley, A., et al. 2015. Intraguild predation by *Harmonia axyridis* (Coleoptera: Coccinellidae) on native insects in Europe: molecular detection from field samples. *Entomological Science*, **18**, 130–3.

Brown, P.M.J., Roy, H.E., Rothery, P., et al. 2008b. *Harmonia axyridis* in Great Britain: analysis of the spread and distribution of a non-native coccinellid. *BioControl*, **53**, 55–67.

Brown, P.M.J., Thomas, C., Lombaert, E., et al. 2011b. The global spread of *Harmonia axyridis* (Coleoptera: Coccinellidae): distribution, dispersal and routes of invasion. *BioControl*, **56**, 623–41.

Burgio, G., Santi, F. and Maini, S. 2002. On intra-guild predation and cannibalism in *Harmonia axyridis* (Pallas) and *Adalia bipunctata* L. (Coleoptera: Coccinellidae). *Biological Control*, **24**, 110–16.

Cabral, S., Soares, A.O., Moura, R. and Garcia, P. 2005. Evaluation of the quality of *Aphis fabae, Myzus persicae* (Homoptera:Aphididae) and *Aleyrodes protella* (Homoptera: Aleyrodidae) as prey for *Coccinella undecimpunctata* (Coleoptera: Coccinellidae). *Proceedings of International Symposium on Biological Control of Aphids and Coccids*, Yamagata Unversity; pp. 52–57.

Camargo, F. 1937. Notas taxonomicas e biologicas sobre alguns Coccinellideos do genera *Psyllobora chevrolat* (Col. Coccinellidae). *Revista de Entomologia (Rio de Janeiro)*, **7**, 362–77.

Capra, F. 1947. Note sui Coccinellidi (Col.). III. La larva ed il regime pollinivoro di *Bulaea schovi* Hummel. *Memoire della Societa Entomologica Italiana*, **26**, 80–6.

Carnes, E.K. 1912. Collecting ladybirds (Coccinellidae) by the ton. *Monthly Bulletin of the State Commission of Horticulture*, **1**, 71–81.

Carroll, C.R. and Janzen, D.H. 1973. Ecology of foraging by ants. *Annual Review of Ecology and Systematics*, **4**, 231–57.

Carroll, L. 1962. *Through the Looking-glass and What Alice Found There*. London: Macmillan and Co Ltd.

Carter, M.C. and Dixon, A.F.G. 1982. Habitat quality and the foraging behaviour of coccinellid larvae. *Journal of Animal Ecology*, **51**, 865–78.

Carter, M.C. and Dixon, A.F.G. 1984. Honeydew: an arrestant stimulus for coccinellids. *Ecological Entomology*, **9**, 383–7.

Carter, M.C., Sutherland, D. and Dixon, A.F.G. 1984. Plant structure and searching efficiency of coccinellid larvae. *Oecologia*, **63**, 394–7.

Cartwright, B., Eikenbary, R.D. and Angelet, G.W. 1982. Parasitism by *Perilitus coccinellae* (Hym.: Braconidae) of indigenous coccinellid hosts and the introduced *Coccinella septempunctata* (Col.: Coccinellidae), with notes on winter mortality. *Entomophaga*, **27**, 237–44.

Ceryngier, P. and Godeau, J. 2013. Predominance of *Vibidia duodecimguttata* (Poda, 1761) in the assemblages of ladybird beetles (Coleoptera: Coccinellidae) overwintering in floodplain forests. *Baltic Journal of Coleopterology*, **13**, 41–50.

Ceryngier, P. and Hodek, I. 1996. Enemies of Coccinellidae. In Hodek, I. and Honek, A., eds. *Ecology of Coccinellidae*. Dordrecht: Kluwer Academic Publishers; pp. 319–50.

Ceryngier, P., Roy, H.E. and Poland, R.L. 2012. Natural enemies of ladybird beetles. In Hodek, I., van Emden, H.F. and Honek, A., eds. *Ecology and Behaviour of the Ladybird Beetles (Coccinellidae)*. Chichester: Wiley-Blackwell; pp. 375–443.

Chapin, B. and Brou, V.A. 1991. *Harmonia axyridis* (Pallas), the third species of the genus to be found in the United States (Coleoptera: Coccinellidae). *Proceedings of the Entomological Society of Washington*, **93**, 630–5.

Chapin, E.A. 1966. A new species of myrmecophilous Coccinellidae, with notes on the other Hyperaspini (Coleoptera). *Psyche*, **73**, 278–83.

Chapman, J.A., Romer, J.I. and Stark, J. 1955. Ladybird and army cutworm adults as food for grizzly bears in Montana. *Ecology*, **36**, 156–8.

Chapman, R.F. 1969. *The Insects: Structure and Function*. London: The English Universities Press.

Chapuis, F. 1876. *Histoire Naturelle des Insectes. Genera des Coleopteres, Paris*, **12**, 1–424.

Chazeau, J., Fürsch, H. and Sasaji, H. 1989. Taxonomy of Coccinellids. Coccinella (Passau), **1**, 6–8.

Chudek, J.A., Hunter, G., MacKay, R.L., et al. 1998. MRM, an alternative approach to the study of host/parasitoid relationships in insects. In Blümich, B., Blümler, P., Botto, R. and Fukushima, E., eds. *Spatially Resolved Magnetic Resonance: Methods and Applications in Materials Science, Agriculture and Biomedicine*. Weinheim: Wiley-Vch; pp. 467–71.

Clausen, C.P. 1940. *Entomophagous Insects*. New York, NY: McGraw-Hill.

Coderre, D. 1988. The numerical response of predators to aphid availability in maize: why coccinellids fail. In Niemczyk, E. and Dixon, A.F.G., eds. *Ecology and Effectiveness of Aphidophaga*. The Hague: SPB Academic Publishing; pp. 219–23.

Colburn, R. and Asquith, D. 1970. A cage used to study the finding of a host by the ladybird *Stethorus punctum*. *Journal of Economic Entomology*, **63**, 1376–7.

Colunga-Garcia, M. and Gage, S. H. 1998. Arrival, establishment, and habitat use of the multicolored Asian lady beetle (Coleoptera: Coccinellidae) in a Michigan landscape. *Environmental Entomology*, **27**, 1574–80.

Comont, R.F., Purse, B.V., Phillips, W., et al. 2014. Escape from parasitism by the invasive alien ladybird, *Harmonia axyridis*. *Insect Conservation and Diversity*, **7**, 334–42.

Comont, R.F., Roy, H.E., Lewis, O.T., et al. 2012. Using biological traits to explain ladybird distribution patterns. *Journal of Biogeography*, **39**, 1772–81.

Conrad, M.S. 1959. The spotted lady beetle, *Coleomegilla maculata* (De Geer), as a predator of European corn borer eggs. *Journal of Economic Entomology*, **52**, 843–7.

Constantine, B. and Majerus, M.E.N. 1994. *Cryptoleamus montrouzieri* (Coleoptera: Coccinellidae) in Britain. *Entomologists' Monthly Magazine*, **130**, 45–6.

Cooke, P. 1987. Aphid galls and ladybirds. *Cecidology*, **2**, 20.

Coope, G.R. 1970. Interpretation of quaternary insect fossils. *Annual Review of Entomology*, **15**, 97–120.

Coope, G.R. and Angus, R.B. 1975. An ecological study of a temperate interlude of the middle of the last glaciation based on fossil Coleoptera from Isleworth, Middlesex. *Journal of Animal Ecology*, **44**, 365–91.

Coope, G.R. and Sands, C.H.S. 1966. Insect faunas of the last glaciation from the Tame Valley, Warwickshire. *Proceedings of the Royal Society of London B*, **165**, 389–412.

Corbara, B., Dejean, A. and Cerdan, P. 1999. Une coccinelle myrmecophile associée à la fourmi arboricole *Dolichoderus bidens* (Dolichoderinae). *Actes College Insectes Sociaux*, **12**, 171–9.

Corry, A. 1995. Adalia 2-punctata *Mating and Oviposition Strategies in the Wild*. Unpublished Undergraduate Project Report, University of Cambridge.

Creed, E.R. 1966. Geographical variation in the two-spot ladybird in England and Wales. *Heredity*, **21**, 57–72.

Creed, E.R. 1971a. Melanism in the two-spot ladybird, *Adalia bipunctata*, in Great Britain. In Creed, E.R., ed. *Ecological Genetics and Evolution*. Oxford: Blackwell.

Creed, E.R. 1971b. Industrial melanism and smoke abatement. *Evolution*, **25**, 290–3.

Crotch, G.R. 1874. *A Revision of the Coleopterous Family Coccinellidae*. London: Janson.

Crowson, R.A. 1955. *The Natural Classification of the Families of Coleoptera*. London: Nathaniel Lloyd.

Crowson, R.A. 1981. *The Biology of the Coleoptera*. London: Academic Press.

Daloze, D., Braekman, J.-C. and Pasteels, J.M. 1995. Ladybird defence alkaloids: structural, chemotaxonomic and biosynthetic aspects (Col.: Coccinellidae). *Chemoecology*, **5/6**, 173–83.

Darwin, C. 1859. *On the Origin of Species by Means of Natural Selection, or the Preservation of Favoured Races in the Struggle for Life*. London: John Murray.

Darwin, C. 1871. *The Descent of Man and Selection in Relation to Sex*. London: John Murray.

Darwin, C. 1874. *The Descent of Man and Selection in Relation to Sex*. 2nd edn. London: John Murray.

Darwin, C. 1887. *The Autobiography of Charles Darwin 1809–1882*. 1958 edn, Barlow, N., ed. London: Collins.

Dauguet, P. 1949. *Les Coccinellini de France*. Paris.

Davey, K.G. 1959. Spermatophore production in *Rhodnius prolixus*. *Quarterly Journal of Microscopical Science*, **100**, 221–30.

David, M.H. and Wilde, G. 1973. Susceptibility of the convergens lady beetle to parasitism by *Perilitus coccinellae* (Shrank) (Hymenoptera: Braconidae). *Journal of the Kansas Entomological Society*, **46**, 359–62.

Davis, D.S., Stewart, S.L., Manica, A. and Majerus, M.E.N. 2006. Adaptive preferential selection of female coccinellid hosts by the parasitoid wasp *Dinocampus coccinellae* (Hymenoptera: Braconidae). *European Journal of Entomology*, **103**, 41–5.

Davis, R.S., Vandewalker, M.L., Hutcheson, P.S. and Slavin, R.G. 2006. Facial angioedema in children due to ladybug (*Harmonia axyridis*) contact: 2 case reports. *Annals of Allergy, Asthma and Immunology*, **97**, 440–2.

Day, W.H., Prokrym, D.R., Ellis, D.R. and Chianese, R.J. 1994. The known distribution of the predator *Propylea quatuordecimpunctata* (Coleoptera: Coccinellidae) in the United States, and thoughts on the origin of this species and five other exotic lady beetles in eastern North America. *Entomology News*, **105**, 244–56.

De Bach, P. 1947. Cottony-cushion scale, vedalia and DDT in central California. *California Citrograph*, **32**, 406–7.

De Bach P., ed. 1964. *Biological Control of Insect Pests and Weeds*. London: Chapman and Hall.

De Bach, P., Fleschner, C.A. and Dietrick, E.J. 1951. A biological check method for evaluating the effectiveness of entomophagous insects. *Journal of Economic Entomology*, **44**, 763–6.

Dechene, R. 1970. Studies of some behavioural patterns of *Iridomyrmex humilis* Mayr (Formicidae, Dolichoderinae). *Wasmann Journal of Biology*, **28**, 175–84.

De Jong, P.W., Holloway, G.J., Brakefield, P.M. and de Vos, H. 1991. Chemical defence in ladybird beetles (Coccinellidae). II. Amount of reflex fluid, the alkaloid adaline and individual variation in defence in 2-spot ladybirds (*Adalia bipunctata*). *Chemoecology*, **2**, 15–19.

De Jong, P.W., Verhoog, M.D. and Brakefield, P.M. 1993. Sperm competition and melanic polymorphism in the two-spot ladybird, *Adalia bipunctata* (Coleoptera, Coccinellidae). *Heredity*, **70**, 172–8.

Delbecque, J.-P., Weidner, K. and Hoffmann, K.H. 1990. Alternative sites for ecdysteroid production in insects. *Invertebrate Reproductive Development*, **18**, 29–42.

Delucchi, V. 1953. *Aphidecta obliterata* L. (Coleoptera, Coccinellidae) als Rauber von *Dreyfusia* (Adelges) *piceae* Ratz. *Pflanzenschutzberichte*, **11**, 73–83.

Denlinger, D., Yocum, G. and Rinehart, J. 2004. Hormonal control of diapause. *Comprehensive Insect Molecular Science*, **3**, 615–50.

Disney, R.H.L. 1979. Natural history notes on some British Phoridae (Diptera) with comments on a changing picture. *Entomologist's Gazette*, **30**, 141–50.

Disney, R.H.L., Majerus, M.E.N. and Walpole, M. 1994. Phoridea (Diptera) parasitising Coccinellidae (Coleoptera). *Entomologist*, **113**, 28–42.

Dixon, A.G.H. 1958. The escape response shown by certain aphids to the presence of the coccinellid beetle *Adalia decempunctata* (L.). *Transactions of the Royal Entomological Society of London*, **110**, 319–34.

Dixon, A.G.H. 1959. An experimental study of the searching behaviour of the predatory coccinellid *Adalia decempunctata* (L.). *Journal of Animal Ecology*, **39**, 739–51.

Dixon, A.G.F. 2000. *Insect Predator–Prey Dynamics: Ladybird Beetles and Biological Control*. Cambridge: Cambridge University Press.

Dixon, A.G.F. and Guo, Y. 1993. Egg and cluster size in ladybird beetles (Coleoptera: Coccinellidae): the direct and indirect effects of aphid abundance. *European Journal of Entomology*, **90**, 457–63.

Dixon, A.G.F., Hemptinne, J.-L. and Kindlmann, P. 1997. Effectiveness of ladybirds as biological control agents: patterns and processes. *Entomophaga*, **42**, 71–83.

Dobrzhanskii, F.G. (Dobzhansky, T.) 1922a. Imaginal diapause in Coccinellidae. *Izvestiia Otdel Prikladnaia Entomologia*, **2**, 103–24.

Dobrzhanskii, F.G. (Dobzhansky, T.) 1922b. Mass aggregations and migrations in Coccinellidae. *Izvestiia Otdel Prikladnaia Entomologia*, **2**, 229–34.

Dobzhansky, T. 1933. Geographical variation in lady-beetles. *American Naturalist*, **67**, 97–126.

Dobzhansky, T. 1951. *Genetics and the Origin of Species*, 3rd edn. New York, NY: Columbia University Press.

Dobzhansky, T. and Sivertzev-Dobzhansky, N.P. 1927. Die geographische Variabilitat von *Coccinella septempunctata*. *Biologicheskii Zhurnal*, **47**, 556–69.

Dolenska, M., Nedvěd, O., Vesely, P., Tesarova, M. and Fuchs, R. 2009. What constitutes optical warning signals of ladybirds (Coleoptera: Coccinellidae) towards bird predators: colour, pattern or general look? *Biological Journal of the Linnean Society*, **98**, 234–42.

Donisthorpe, H.StJ.K. 1919–1920. The myrmecophilous ladybird *Coccinella distincta*, Fald., its life-history and association with ants. *Entomologist's Record and Journal of Variation*, **32**, 1–3.

Doumbia, M., Hemptinne, J.-L. and Dixon, A.F.G. 1998. Assessment of patch quality by ladybirds: role of larval tracks. *Oecologia*, **113**, 197–202.

Dreistadt, S.H. and Flint, M.L. 1996. Melon aphid (Homoptera: Aphididae) control by inundative convergent lady beetle (Coleoptera: Coccinellidae) release on chrysanthemum. *Environmental Entomology*, **25**, 688–97.

Duverger, Ch. 1989. Contribution a l'étude des Hyperaspinae. L'ère note (Coleoptera, Coccinellidae). *Bulletin de la Societé Linéenne de Bordeaux*, **17**, 143–57.

Dyadechko, N.P. 1954. *Coccinellids of the Ukranian SSR*. Kiev.

Eastop, V.F. and Pope, R.D. 1966. Notes on the ecology and phenology of some British Coccinellidae. *Entomologist*, **99**, 287–9.

Eastop, V.F. and Pope, R.D. 1969. Notes on the biology of some British Coccinellidae. *Entomologist*, **102**, 162–4.

Eberhard, W.G. 1996. *Female Control: Sexual Selection by Cryptic Female Choice*. Princeton, NJ: Princeton University Press.

Edwards, A.W.F. 1998. Natural selection and the sex ratio: Fisher's sources. *American Naturalist*, **151**, 564–9.

Edwards, J.S. 1966. Defense by smear: supercooling in the cornicle wax of aphids. *Nature*, **211**, 73–4.

Eisner, T. and Eisner, M. 1992. Operation and defensive role of "gin traps" in a coccinellid pupa (*Cycloneda sanguinea*). *Psyche*, **99**, 265–74.

Eisner, T., Goetz, M., Aneshansley, D., Ferstandig-Arnold, G. and Meinwald, J. 1986. Defensive alkaloid in blood of Mexican bean beetle (*Epilachna varivestis*). *Experientia*, **42**, 204–7.

Eisner, T., Hicks, K. and Eisner, M. 1978. "Wolf-in-sheep's-clothing" strategy of a predaceous insect larva. *Science*, **199**, 790–4.

Eisner, T., Ziegler, R., McCormick, J.L., et al. 1994. Defensive use of an acquired substance (carminic acid) by predaceous insect larvae. *Experimentia*, **50**, 610–15.

Ejbich, K. 2003. Producers in Ontario and northern U.S. bugged by bad odors in wine. *Wine Spectator*, 15 May, 16.

Elliot, N., Kieckhefer, R. and Kauffman, W. 1996. Effects of invading coccinellid on native coccinellids in an agricultural landscape. *Oecologia*, **105**, 537–44.

Elnagdy, S. E., Majerus, M. E. N., Lawson Handley, L.-J. 2011. The value of an egg: resource reallocation in ladybirds (Coleoptera: Coccinellidae) infected with male-killing bacteria. *Evolutionary Biology*, **24**, 2164–72.

El-Ziady, S. and Kennedy, J.S. 1956. Beneficial effects of the common garden ant *Lasius niger* L., on the black bean aphid, *Aphis fabae* Scopoli. *Proceedings of the Royal Entomological Society of London A*, **31**, 61–5.

Enders, D. and Bartzen, D. 1991. Enantioselective total synthesis of harmonine, a defence alkaloid of ladybugs (Coleoptera: Coccinellidae). *Liebigs Annalen der Chemie*, (B), 569–74.

Ettifouri, M. and Ferran, A. 1993. Influence of larval rearing diet on intensive searching behaviour of *Harmonia axyridis* (Col., Coccinellidae) larvae. *Entomophaga*, **38**, 51–9.

Evans, E.W. 2000. Morphology of invasion: body size patterns with establishment of *Coccinella septempunctata* (Coleoptera: Coccinellidae) in western North America. *European Journal of Entomology*, **97**, 469–74.

Evans, E.W. 2004. Habitat displacement of North American ladybirds by an introduced species. *Ecology*, **85**, 637–47.

Evans, E.W. and Dixon, A.G.F. 1986. Cues for oviposition by ladybird beetles (Coccinellidae): response to aphids. *Journal of Animal Ecology*, **55**, 1027–34.

Ewert, M.A. and Chiang, H.C. 1966. Effects of some environmental factors on the distribution of three species of Coccinellidae in their microhabitat. In Hodek, I., ed. *Ecology of Aphidophagous Insects*. Prague: Academia and The Hague: Dr W. Junk; pp. 195–219.

Exell, A.W. 1991. *The History of the Ladybird*, 2nd edn. Shotesham, Norfolk: Erskine Press.

Fabres, G., Nenon, J.P., Kiyindou, A. and Biassangama, A. 1991. Reflexions sur l'acclimatation d'entomophages exotique pour la regulation des populations de la cochenille du marioc au Congo. *Bulletin de la Société Zoologique de France*, **114**, 43–8.

Fagan, L.L., McLachlan, A., Till, C.M. and Walker, M.K. 2010. Synergy between chemical and biological control in the IPM of currant-lettuce aphid (*Nasonovia ribisnigri*) in Canterbury, New Zealand. *Bulletin of Entomological Research*, **100**, 217–23.

Fain, A., Hurst, G.D.D., Tweddle, J.C., et al. 1995. Description and observations of two new species of Hemisarcoptidae from deutonymphs phoretic on Coccinellidae (Coleoptera) in Britain. *International Journal of Acarology*, **21**, 1–8.

Fain, A., Hurst, G.D.D., Fassotte, C., et al. 1997. New observations on the mites of the family Hemisarcoptidae (Acari: Astigmata) phoretic on Coccinellidae (Coleoptera). *Entomologie*, **67**, 89–94.

Ferran, A., Gambier, J., Parent, S., et al. 1997. The effect of rearing the ladybird *Harmonia axyridis* on *Ephestia kuehniella* eggs on the response of its larvae to aphid tracks. *Journal of Insect Behavior*, **10**, 129–43.

Ferran, A., Giuge, L., Tourniaure, R., Gambier, J. and Fournier, D. 1998. An artificial non-flying mutation to improve the efficiency of the ladybird *Harmonia axyridis* in biological control of aphids. *BioControl*, **43**, 53–64.

Filatova, I.T. 1974. The parasites of Coccinellidae (Coleoptera) in West Siberia. In Kolomyietz, N.G., ed. *The Fauna and Ecology of Insects from Siberia*. Novosibirsk: Nauka, Siberian Branch; pp. 173–185.

Fisher, R.A. 1930. *The Genetical Theory of Natural Selection*. Oxford: Oxford University Press.

Fisher, T.W. 1959. Occurrence of spermatophore in certain species of *Chilocorus*. *Pan-Pacific Entomology*, **35**, 205–8.

Flint, M.L., Dreistadt, S.H., Rentner, J. and Parella, M.P. 1995. Lady beetle release controls aphids on potted plants. *California Agriculture*, **49**, 5–8.

Forbes, S.A. 1883. The food relations of the Carabidae and Coccinellidae. *Bulletin of the Illinois State Laboratory of Natural History*, **1**, 33–64.

Ford, E.B. 1940. Polymorphism and taxonomy. In Huxley, J., ed. *The New Systematics*. Oxford: Clarendon Press.

Ford, E.B. 1945. *Butterflies*. New Naturalist Series No. 1. London: Collins.

Ford, E.B. 1955. *Moths*. New Naturalist Series No. 30. London: Collins.

Ford, E.B. 1964. *Ecological Genetics*. London: Methuen.

Ford, R.L.E. 1979. Dipterous parasites of ladybirds. *Proceedings of the Isle of Wight Natural History and Archaeological Society*, **7**, 471976.

Fowler, S.V. 2004. Biological control of an exotic scale, *Orthezia insignis* Browne (Homoptera: Ortheziidae), saves the endemic gumwood tree, *Commidendrum robustum* (Roxb.) DC. (Asteraceae) on the island of St. Helena. *Biological Control*, **29**, 367–74.

Fowles, A.P. 1990. Observations on the over-wintering behaviour of the orange ladybird. *Dyfed Invertebrate Group Newsletter*, **17**, 13–18.

Frazer, B.D. and Ives, P.M. 1976. *Homalotylus californicus* (Hymenoptera: Encyrtidae), a parasite of *Coccinella californica* (Coleoptera: Coccinellidae) in British Columbia. *Journal of the Entomological Society of British Columbia*, **73**, 6–7.

Frazer, B.D. and McGregor, R.R. 1994. Searching behaviour of adult female Coccinellidae (Coleoptera) on stem and leaf models. *Canadian Entomologist*, **126**, 389–99.

Frazer, J.F.D. and Rothschild, M. 1960. Defence mechanisms in warningly coloured moths and other insects. *Proceedings of the International Congress of Entomology*, **11**, 249–56.

Fürsch, H. 1990. Taxonomy of Coccinellids, corrected version. Coccinella *(Passau)*, **2**, 4–6.

Gagné, W.C. and Martin, J.L. 1968. The insect ecology of red pine plantations in central Ontario. V. The Coccinellidae (Coleoptera). *Canadian Entomologist*, **100**, 835–46.

Galvan, T.L., Koch, R.L. and Hutchison, W.D. 2008. Impact of fruit feeding on overwintering survival of the multicolored Asian lady beetle, and the ability of this insect and paper wasps to injure wine grape berries. *Entomologia Experimentalis et Applicata*, **128**, 429–36.

Ganglbauer, L. 1899. *Die Käfer von Mitteleuropa*. Band 3. Wien.

Geoghegan, I.E., Chudek, J.A., MacKay, R.L., et al. 2000. Study of anatomical changes in *Coccinella septempunctata* (Coleoptera: Coccinellidae) induced by diet and by infection with the larvae of *Dinocampus coccinellae* (Hymenoptera: Braconidae) using magnetic resonance microimaging. *European Journal of Entomology*, **97**, 457–61.

Geoghegan, I.E., Majerus, T.M.O. and Majerus, M.E.N. 1998a. A record of a rare male of the parthenogenetic parasitoid *Dinocampus coccinellae* (Schrank) (Hym., Braconidae). *Entomologist's Record and Journal of Variation*, **110**, 171–2.

Geoghegan, I.E., Majerus, T.M.O. and Majerus, M.E.N. 1998b. Differential parasitisation of adult and pre-imaginal *Coccinella 7-punctata* (Coleoptera: Coccinellidae) by *Dinocampus coccinellae* (Hymenoptera: Braconidae). *European Journal of Entomology*, **95**, 571–9.

Geoghegan, I.E., Thomas, W.P. and Majerus, M.E.N. 1997. Notes on the Coccinellid parasitoid *Dinocampus coccinellae* (Schrank) (Hymenoptera: Braconidae) in Scotland. *Entomologist*, **116**, 179–84.

George, K.S. 1957. Preliminary investigations on the biology and ecology of the parasites and predators of *Brevicoryne brassicae* (L.). *Bulletin of Entomological Research*, **48**, 619–29.

Gerson, U., O'Connor, B.A. and Houck, M.A. 1990. Acari. In Rosen, D., ed. *Armored Scale Insects: Their Biology, Natural Enemies and Control*. Amsterdam: Elsevier; pp. 77–97.

Geyer, J.W.C. 1947. A study of the biology and ecology of *Exochomus flavipes* Thunb. (Coccinellidae, Coleoptera). Parts 1 and 2. *Journal of the Entomological Society of South Africa*, **9**, 219–34 and **10**, 64–109.

Giroux, S., Côté, J.-C., Vincent, C., Martel, P. and Coderre, D. 1994. Bacteriological insecticide M-One effects on predation efficiency and mortality of adult *Coleomegilla maculata lengi* (Coleoptera: Coccinellidae). *Journal of Economic Entomology*, **87**, 39–43.

Godeau, J.-F. 1997. Adaptations à la cohabitation avec des fourmis: le cas de Coccinella magnifica Redtenbacher. Mémoire de D.E.A., Faculté des Sciences Agronomiques de Gembloux.

Godeau, J.-F. 2000. Coccinelles amies des fourmis. 2/2/ Groupe de Travail Coccinulla. *Feuille de Contact*, **2**, 10–15.

Godeau, J.-F., Hemptinne, J.-L. and Verhaeghe, J.-C. 2003. Ant trail: a highway for Coccinella magnifica Redtenbacher (Coleoptera: Coccinellidae). In Soares, A.O., Ventura, M.A., Garcia, V. and Hemptinne, J.-L., eds. *Proceedings of the 8th International Symposium on Ecology of Aphidophaga: Biology, Ecology and Behaviour of Aphidophagous Insects*. Arquipélago, Life and Marine Sciences, Supplement 5; pp. 79–83.

Godeau, J.-F., Hemptinne, J.-L., Dixon, A.F.G. and Verhaeghe, J.-C. 2009. Reaction of ants to, and feeding biology of, a congeneric myrmecophilous and non-myrmecophilous ladybird. *Journal of Insect Behavior*, **22**, 173–85.

Gordon, R.D. 1971. A generic review of the Cryptognathini, new tribe, with a Description of a new genus (Coleoptera: Coccinellidae). *Acta Zoologica Lilloana*, **26**, 181–96.

Gordon, R.D. 1985. The Coccinellidae (Coleoptera) of America North of Mexico. *Journal of the New York Entomological Society*, **93**, 1–912.

Gordon, R.D. and Vandenberg, N.J. 1991. Field guide to recently introduced species of Coccinellidae (Coleoptera) in North America, with a revised key to North American genera of Coccinellini. *Proceedings of the Entomological Society of Washington*, **93**, 845–64.

Gordon, R.D. and Vandenberg, N.J. 1993. Larval systematics of North American *Cycloneda* Crotch (Coleoptera: Coccinellidae). *Entomologica Scandinavica*, **24**, 301–12.

Gordon, R.D. and Vandenberg, N.J. 1995. Larval systematics of North American *Coccinella* L. (Coleoptera: Coccinellidae). *Entomologica Scandinavica*, **26**, 67–86.

Gould, J.S. 1980. *The Panda's Thumb*. New York, NY: W.W. Norton.

Graf, P. and Kriegl, M. 1968. Methoden zur Massensammlung europaischerr Adelgidenrauber und Hinweise auf ihre okologie. *Anzeiger für Schädlingskunde und Pflanzenschutz*, **41**, 151–5.

Greathead, D.J. 1989. Biological control as an introduction phenomenon: a preliminary examination of programmes against Homoptera. *Entomologist*, **108**, 28–37.

Grevstad, F.S. and Klepetka, B.W. 1992. The influence of plant architecture on the foraging efficiencies of a suite of ladybird beetles feeding on aphids. *Oecologia*, **92**, 399–404.

Grill, C.P. and Moore, A.J. 1998. Effects of a larval antipredator response and larval diet on adult phenotype in an aposematic ladybird beetle. *Oecologia*, **114**, 274–82.

Groden, E., Drummond, F.A., Casagrande, R.A. and Haynes, D.L. 1990. *Coleomegilla maculata* (Coleoptera: Coccinellidae): its predation upon the Colorado potato beetle (Coleoptera: Chrysomelidae) and its incidence in potatoes and surrounding crops. *Journal of Economic Entomology*, **83**, 1306–15.

Guan, X. and Chen, E. 1986. Corpus allatum activity in the female *Coccinella septempunctata* L. adults. *Acta Entomologia Sinica*, **29**, 10–14.

Gurney, B. and Hussey, N.W. 1970. Evaluation of some Coccinellids species for the biological control of aphids in protected cropping. *Annals of Applied Biology*, **65**, 451–8.

Haddrill, P.R. 2001. *The Development and Use of Molecular Genetic Markers to Study Sexual Selection and Population Genetics in the 2-spot Ladybird,* Adalia bipunctata (L.). Unpublished PhD thesis, University of Cambridge.

Haddrill, P.R., Majerus, M.E.N. and Mayes, S. 2002. Isolation and characterization of highly polymorphic microsatellite loci in the 2-spot ladybird, *Adalia bipunctata*. *Molecular Ecology Notes*, **2**, 316–19.

Haddrill, P.R., Shuker, D.M., Amos, W., Majerus, M.E.N. and Mayes, S. 2008. Female multiple mating in wild and laboratory populations of the two-spot ladybird, *Adalia bipunctata*. *Molecular Ecology*, **17**, 3189–97.

Hagedorn, H.H. 1985. The role of ecdysteroids in reproduction. In Kerkut, G.A. and Gilbert, L.I., eds. *Comprehensive Insect Physiology, Biochemistry and Pharmacology*. Vol. **8**. Oxford: Pergamon Press; pp. 205–62.

Hagen, K.S. 1962. Biology and ecology of predaceous Coccinellidae. *Annual Review of Entomology*, **7**, 289–326.

Hagen, K.S. 1974. The significance of predaceous Coccinellidae in biological and integrated control of insects. *Entomophaga*, **7**, 25–44.

Hamilton, W.D. 1967. Extraordinary sex ratios. *Science*, **156**, 477–88.

Hamilton, W.D. and Zuk, M. 1982. Heritable true fitness and bright birds: a role for parasites? *Science*, **218**, 384–7.

Happ, G.M. and Eisner, T. 1961. Hemorrage in a Coccinellid beetle and its repellent effects on ants. *Science*, **134**, 329–31.

Hariri, G. 1966a. Laboratory studies on the reproduction of *Adalia bipunctata* (Coleoptera, Coccinellidae). *Entomologie Experimentalis et Applicata*, **9**, 200–4.

Hariri, G. 1966b. Changes in metabolic reserves of three species of aphidophagous Coccinellidae (Coleoptera) during metamorphosis. *Entomologie Experimentalis et Applicata*, **9**, 349–58.

Hariri, G. 1966c. Studies on the physiology of hibernating Coccinellidae (Coleoptera): changes in the metabolic reserves and gonads. *Proceedings of the Royal Entomological Society of London A*, **41**, 133–44.

Harmon, J.P., Losey, J.E. and Ives, A.R. 1998. The role of vision and color in the close proximity foraging behaviour of four coccinellid species. *Oecologia*, **115**, 287–92.

Harper, A.M. and Lilly, C.E. 1982. Aggregations and winter survival in Southern Alberta of *Hippodamia quinquesignata* (Coleoptera, Coccinellidae), a predator of pea aphid (Homoptera: Aphididae). *Canadian Entomologist*, **114**, 303–9.

Harris, R.H.T.P. 1921. A note on *Ortalia pallens* Muls. *South African Journal of Science*, **17**, 170–1.

Hattingh, V. and Samways, M.J. 1992. Prey choice and substitution in *Chilocorus* spp. (Coleoptera: Coccinellidae). *Bulletin of Entomological Research*, **82**, 327–34.

Hattingh, V. and Samways, M.J. 1995. Visual and olfactory location of biotypes, prey patches, and individual prey by the ladybeetle *Chilocorus nigritus*. *Entomologia Experimentalis et Applicata*, **75**, 87–98.

Hautier, L., San Martin y Gomez, G., Callier, P., de Biseau, J. and Gregoire, J. 2011. Alkaloids provide evidence of intraguild predation on native Coccinellids by *Harmonia axyridis* in the field. *Biological Invasions*, **13**, 1805–14.

Hawkes, O.A.M. 1920. Observations on the life-history, biology and genetics of the ladybird beetle *Adalia bipunctata* (Mulsant). *Proceedings of the Zoological Society of London*, **90**, 475–90.

Hawkins, R.D. 2000. *Ladybirds of Surrey*. Woking: Surrey Wildlife Trust.

Hazzard, R.V. and Ferro, D.N. 1991. Feeding responses of adult *Coleomegilla maculata* (Coleoptera: Coccinellidae) to eggs of Colorado potato beetle (Coleoptera: Chrysomelidae) and green peach aphids (Homoptera: Aphididae). *Environmental Entomology*, **20**, 644–51.

Heidari, M. and Copland, M.J.W. 1992. Host finding by *Cryptolaemus montrouzieri* (Col., Coccinellidae), a predator of mealybugs (Hom., Pseudococcidae). *Entomophaga*, **37**, 621–5.

Heinz, K.M. and Parrella, P.M. 1994. Poinsettia (*Euphoria pulcherrima* Willd. ex Koltz.) cultivar-mediated differences in performance of five natural enemies of *Bemisia argentifollii* Bellows and Perring. n. sp. (Homoptera: Aleyrodidae). *Biological Control*, **4**, 305–8.

Helenius, J. 1990. Effect of epigeal predators on infestation by the aphid *Rhopalosiphum padi* and on grain yield of oats in monocrops and mixed intercrops. *Entomologia Experimentalis et Applicata*, **54**, 225–36.

Hemptinne, J.-L. and Desprets, A. 1986. Pollen as a spring food for *Adalia bipunctata*. In Hodek, I., ed. *Ecology of Aphidophaga*. Prague: Academia and Dordrecht: Dr. W. Junk.

Hemptinne, J.-L. and Dixon, A.F.G. 1991. Why ladybirds have generally been so ineffective in biological control? In Polgar, L., Chambers R.J., Dixon, A.F.G. and Hodek, I., eds. *Behaviour and Impact of Aphidophaga*. The Hague: SPB Academic Publishing; pp. 149–57.

Hemptinne, J.-L., Dixon, A.F.G. and Coffin, J. 1992. Attack strategy of ladybird beetles (Coccinellidae): factors shaping their numerical response. *Oecologia*, **90**, 238–45.

Hemptinne, J.-L., Dixon, A.F.G., Doucet, J.L. and Petersen, J.E. 1993. Optimal foraging by hoverflies (Diptera: Syrphidae) and ladybirds (Coleoptera: Coccinellidae): mechanisms. *European Journal of Entomology*, **90**, 451–5.

Hemptinne, J.-L., Dixon, A.F.G. and Gauthier, C. 2000c. Nutritive cost of intraguild predation on eggs of *Coccinella septempunctata* and *Adalia bipunctata* (Coleoptera: Coccinellidae). *European Journal of Entomology*, **97**, 559–62.

Hemptinne, J.-L., Dixon, A.F.G. and Lognay, G. 1996. Searching behaviour and mate recognition by males of the two-spot ladybird beetle, *Adalia bipunctata*. *Ecological Entomology*, **21**, 165–70.

Hemptinne, J.-L., Doumbia, M. and Dixon, A.F.G. 2000b. Assessment of patch quality by ladybirds role of aphids and plant phenology. *Journal of Insect Behavior*, **13**, 353–9.

Hemptinne, J.-L., Gaudin, M., Dixon, A.F.G. and Lognay, G. 2000a. Social feeding in ladybird beetles: adaptive significance and mechanism. *Chemoecology*, **10**, 149–52.

Hemptinne, J.-L., Lognay, G., Doumbia, M. and Dixon, A.F.G. 2001. Chemical nature and persistence of the oviposition deterring pheromone in the tracks of larvae of the two-spot ladybird, *Adalia bipunctata* (Coleoptera: Coccinellidae). *Chemoecology*, **11**, 43–7.

Henson, R.D., Thompson, A.C., Hedin, P.A., Nichols, P.R. and Neel, W.W. 1975. Identification of precoccinellin in the ladybird beetle, *Coleomegilla maculata*. *Experientia*, **31**, 145.

Hesler, L.S., Kieckhefer, R.W. and Catangui, M.A. 2004. Surveys and field observations of *Harmonia axyridis* and other Coccinellidae (Coleoptera) in eastern and central South Dakota. *Transactions of the American Entomological Society*, **130**, 113–33.

Hickman, J.M. and Wratten, S.D. 1996. Use of *Phacelia tanacetifolia* strips to enhance biological control of aphids by hoverfly larvae in cereal fields. *Journal of Economic Entomology*, **89**, 832–40.

Hicks, B., Majka, C.G. and Moores, S.P. 2010. *Harmonia axyridis* (Pallas) (Coleoptera: Coccinellidae) found in Newfoundland. *Coleopterists Bulletin*, **64**, 50.

Hills, L.D. 1969. *Biological Pest Control: Report 3*. Essex: Henry Doubleday Research Association.

Hippa, H., Koponen, S. and Laine, T. 1978. On the feeding biology of *Coccinella hieroglyphica* L. (Col., Coccinellidae). *Reports from the Kevo Subarctic Research Station*, **14**, 18–20.

Hippa, H., Koponen, S. and Roine, R. 1984. Larval growth of *Coccinella hoieroglyphica* (Col., Coccinellidae) fed on aphids and pre-imaginal stages of *Galerucella sagittariae* (Col., Chrysomelidae). *Report from the Kevo Subarctic Research Station*, **19**, 67–70.

Hodek, I. 1956. The influence of *Aphis sambuci* L. as prey of the ladybird beetle *Coccinella septempunctata* L. *Vestnik Ceskoslovenske Zoologiscke Spolecnosti*, **20**, 62–74.

Hodek, I. 1957. The influence of *Aphis sambuci* L. as food for *Coccinella septempunctata* L. II. *Casopis Ceskoslovenske Spolecnosti Entomologicke*, **54**, 10–17.

Hodek, I. 1958. Influence of temperature, relative humidity and photoperiodicity on the speed of development of *Coccinella septempunctata* L. *Casopis Ceskoslovenske Spolecnosti Entomologicke*, **55**, 121–41.

Hodek, I. 1959. Ecology of *Aphidophagous Coccinellidae. International Conference on Insect Pathology and Biological Control*, Prague 1958, 543–7.

Hodek, I. 1960. The influence of various aphid species as food for two ladybirds *Coccinella 7-punctata* L. and *Adalia bipunctata* L. In *The Ontogeny of Insects. Symposium Proceedings*, Praha 1959. Prague: Academia; pp. 314–16.

Hodek, I. 1962. Experimental influencing of the imaginal diapause in *Coccinella septempunctata* L. (Col., Coccinellidae), 2nd Part. *Casopis Ceskoslovenske Spolecnosti Entomologicke*, **59**, 297–313.

Hodek, I. 1966. *Ecology of Aphidophagous Insects*. Prague: Academia and The Hague: Dr W. Junk.

Hodek, I. 1973. *Biology of Coccinellidae*. Prague: Academia and The Hague: Dr W. Junk.

Hodek, I. 1986. *Ecology of Aphidophaga*. Prague: Academia and Dordrecht: Dr W. Junk.

Hodek, I. 1996a. Food relationships. In Hodek, I and Honek, A., eds. *Ecology of Coccinellidae*. Dordrecht: Kluwer Academic Publishers; pp. 143–238.

Hodek, I. 1996b. Dormancy. In Hodek, I. and Honek, A., eds. *Ecology of Coccinellidae*. Dordrecht: Kluwer Academic Publishers; pp. 239–318.

Hodek, I. 2012. Diapause/dormancy. In Hodek, I., van Emden, H.F. and Honek, A., eds. *Ecology and Behaviour of the Ladybird Beetles (Coccinellidae)*. Chichester: Wiley-Blackwell; pp. 275–342.

Hodek, I. and Cerkasov, J. 1961. Prevention and artificial induction of imaginal diapause in *Coccinella septempunctata* L. (Col., Coccinellidae). *Entomologia Experimentalis et Applicata*, **4**, 179–90.

Hodek, I. and Cerkasov, J. 1963. Imaginal dormancy in *Semiadalia undecimnotata* Schneid. (Coccinellidae, Col.). II. Changes in water, fat and glycogen content. *Vestnik Ceskoslovenske Zoologiscke Spolecnosti*, **27**, 298–318.

Hodek, I. and Evans, E.W. 2012. Food relationships. In Hodek, I., van Emden, H.F. and Honek, A., eds. *Ecology and Behaviour of the Ladybird Beetles (Coccinellidae)*. Chichester: Wiley-Blackwell; pp. 141–274.

Hodek, I., Hodkova, M. and Sem'Yanov, V.P. 1989. Physiological state of *Coccinella septempunctata* adults from northern Greece sampled in mid-hibernation. *Acta Entomologica Bohemoslovaca*, **86**, 241–51.

Hodek, I., Holman, J., Novak, K., et al. 1966. The present possibilities and prospects of integrated control of *Aphis fabae* Scop. In Hodek, I., ed. *Ecology of Aphidophagous Insects*. Prague: Academia and The Hague: Dr W. Junk; pp. 331–5.

Hodek, I. and Honek, A., eds. 1996. *Ecology of Coccinellidae*. Dordrecht: Kluwer Academic Publishers.

Hodek, I. and Landa, V. 1971. Anatomical and histological changes during dormancy in two Coccinellidae. *Entomophaga*, **16**, 239–51.

Hodek, I. and Michaud, J.P. 2008. Why is *Coccinella septempunctata* so successful? (A point-of-view). *European Journal of Entomology*, **105**, 1–12.

Hodek, I. and Okuda, T. 1993. A weak tendency to "obligatory" diapause in *Coccinella septempunctata* from southern Spain. *Entomophaga*, **38**, 139–42.

Hodek, I., Ruzicka, Z. and Hodkova, M. 1978. Pollinivorie et aphidiphagie chez *Coleomegilla maculata*. *Annales Zoologie-Écologie Animale*, **10**, 453–9.

Hodková, M. 1992. Storage of the photoperiodic 'information' within the implanted neuroendocrine complex of the linden bug *Pyrrhocoris apterus* (L.) (Heteroptera). *Journal of Insect Physiology*, **38**, 357–63.

Hodková, M. 1996. Physiological mechanisms of adult diapause. In Hodek, I. and Honek, A., eds. *Ecology of Coccinellidae*. Dordrecht: Kluwer Academic Publishers; pp. 315–18.

Hodson, A.C. 1937. Some aspects of the role of water in insect hibernation. *Ecological Monographs*, **7**, 271–315.

Hoelmer, K.A., Osborne, L.S. and Yokomi, R.K. 1994. Interactions of the whitefly predator *Delphastus pusillus* (Coleoptera: Coccinellidae) with parasitized sweetpotato whitefly (Homoptera: Aleyrodidae). *Environmental Entomology*, **23**, 136–9.

Hoffmann, A.A. and Turelli, M. 1997. Cytoplasmic incompatibility in insects. In O'Neill, S.L., Hoffmann, A.A. and Werren, J.H., eds. *Influential Passengers*. Oxford: Oxford University Press; pp. 42–80.

Hokkanen, H.M.T., Bigler, F., Burgio, G., van Lenteren, J.C. and Thomas, M.B. 2003. Ecological risk assessment framework for biological control. In Hokkanen, H.M.T. and Hajek, A.E., eds. *Environmental Impacts of Microbial Insecticides: Need and Methods for Risk Assessment*. Dordrecht: Kluwer; pp. 1–14.

Hölldobler, B. and Wilson, E.O. 1990. *The Ants*. Berlin: Springer Verlag.

Holloway, G.J., de Jong, P.W., Brakefield, P.M. and de Vos, H. 1991. Chemical defense in ladybird beetles (Coccinellidae). I. Distribution of coccinelline and individual variation in defense in 7-spot ladybirds (*Coccinella septempunctata*). *Chemoecology*, **2**, 7–14.

Holloway, G.J., de Jong, P.W. and Ottenheim, M. 1993. The genetics and cost of chemical defence in the 2-spot ladybird (*Adalia bipunctata* L.). *Evolution*, **47**, 1229–39.

Honek, A. 1979. Plant density and occurrence of *Coccinella septempunctata* and *Propylea quatuordecimpunctata* (Coleoptera, Coccinellidae) in cereals. *Acta Entomologica Bohemoslovaca*, **76**, 308–12.

Honek, A. 1982. Factors which determine the composition of field communities of adult aphidophagous Coccinellidae (Coleoptera). *Zeitschrilt für Angewandte Entomologie*, **94**, 157–68.

Honek, A. 1985. Habitat preferences of aphidophagous Coccinellids [Coleoptera]. *Entomophaga*, **30**, 253–64.

Honek, A. 1997. Factors determining winter survival in *Coccinella septempunctata* (Col.: Coccinellidae). *Entomophaga*, **42**, 119–24.

Honek, A. and Martinkova, Z. 1991. Competition between maize and barnyard grass *Echinochola crus-galli* and its effect on aphids and their predators. *Acta Oecologia*, **12**, 741–51.

Honek, A., Martinkova, Z. and Pekar, S. 2005. Aggregation of coccinellid species at hibernation sites. *Proceedings of International Symposium on Biological Control of Aphids and Coccids*, Yamagata University; pp. 62–3.

Honek, A., Martinkova, Z. and Pekar, S. 2007. Aggregation characteristics of three species of Coccinellidae (Coleoptera) at hibernation sites. *European Journal of Entomology*, **104**, 51–6.

Horn, D.J. 1991. Potential impact of *Coccinella septempunctata* on endangered Lycaenidae (Lepidoptera) in Northwestern Ohio, USA. In Polgar, L., Chambers R.J., Dixon, A.F.G. and Hodek, I., eds. *Behaviour and Impact of Aphidophaga*. The Hague: SPB Academic Publishing; pp. 159–62.

Hosino, Y. 1936. Genetical study of the lady-bird beetle, *Harmonia axyridis Pallas Rep. II. Japanese Journal of Genetics*, **12**, 307–20.

Houck, M. A. and O'Connor, B.M. 1991. Ecological and evolutionary significance of phoresy in the Astigmata. *Annual Review of Entomology*, **36**, 611–36.

Howard, N.F. and Landis, B.J. 1936. Parasites and predators of the Mexican bean beetle in the United States. *USDA Circular*, **418**, 1–11.

Huelsman, MF., Kovach, J., Jasinski, J., Young, C. and Eisley, B. 2002. Multicolored Asian lady beetle *(Harmonia axyridis)* as a nuisance pest in households in Ohio. In Jones, S.C., Zhai, J. and Robinson, W.H., eds. *Proceedings of 4th International Conference on Urban Pests*; pp. 243–50.

Hukusima, S. and Kamei, M. 1970. Effects of various species of aphids as food on development, fecundity and longevity of *Harmonia axyridis* Pallas (Coleoptera: Coccinellidae). *Research Bulletin of the Faculty of Agriculture, Gifu University*, **29**, 53–66.

Hulme, P.E. 2009. Trade, transport and trouble: managing invasive species pathways in an era of globalization. *Journal of Applied Ecology*, **46**, 10–18.

Hurej, M. 1988. Natural reduction of *Aphis fabae* Scop. by Coccinellidae on sugar beet crops. In Niemczyk, E. and Dixon, A.F.G., eds. *Ecology and Effectiveness of Aphidophaga*. The Hague: SPB Academic Publishing; pp. 225–9.

Hurst, G.D.D. 1993. *Studies of Biased Sex-ratios in* Adalia bipunctata *L.* Unpublished PhD thesis, University of Cambridge.

Hurst, G.D.D., Hurst, L.D. and Majerus, M.E.N. 1997b. Cytoplasmic sex-ratio distorters. In O'Neill S.L., Hoffmann, A.A. and Werren, J.H., eds. *Influential Passengers*. Oxford: Oxford University Press; pp. 125–54.

Hurst, G.D.D., Jiggins, F.M. and Majerus, M.E.N. 2003. Inherited microorganisms that selectively kill male hosts: the hidden players of insect evolution. In Bourtzis, K. and Miller, T., eds. *Insect Symbiosis*. Boca Raton, FL: CRC Press; pp. 177–97.

Hurst, G.D.D., Jiggins, F.M., von der Schulenberg, J.H.G., et al. 1999b. Male-killing *Wolbachia* in two species of insect. *Proceedings of the Royal Society of London B*, **266**, 735–40.

Hurst, G.D.D., Majerus, M.E.N. and Walker, L.E. 1992. Cytoplasmic male-killing elements in *Adalia bipunctata* (Linnaeus) (Coleoptera: Coccinellidae). *Heredity*, **69**, 84–91.

Hurst, G.D.D., Majerus, M.E.N. and Fain, A. 1997a. Coccinellidae (Coleoptera) as vectors of mites. *European Journal of Entomology*, **94**, 317–19.

Hurst, G.D.D., McMeechan, F. and Majerus, M.E.N. 1998. Phoridae (Diptera) parasitizing *Coccinella septempunctata* (Coleoptera: Coccinellidae) select older prepupal hosts. *European Journal of Entomology*, **95**, 179–81.

Hurst, G.D.D., Sharpe, R.G., Broomfield, A.H., et al. 1995. Sexually transmitted disease in a promiscuous insect, *Adalia bipunctata*. *Ecological Entomology*, **20**, 230–6.

Hurst, G.D.D., von der Schulenberg, J.H.G., Majerus, T.M.O., et al. 1999a. Invasion of one insect species, *Adalia bipunctata*, by two different male-killing bacteria. *Insect Molecular Biology*, **8**, 133–9.

Hurst, L.D. 1991. The incidences and evolution of cytoplasmic male killers. *Proceedings of the Royal Society of London B*, **244**, 91–9.

Husband, R.W. 1981. The African species of *Coccipolipus* with a description of all stages of *Coccipolipus solanophilae* (Acarina: Podapolipidae). *Revue de Zoologie Africaine*, **95**, 283–99.

Ichimori, T., Ohtomo, R., Suzuki, K. and Kurihara, M. 1990. Specific protein related to adult diapause in the leaf beetle, *Gastrophysa atrocyanea*. *Journal of Insect Physiology*, **36**, 85–91.

Iperti, G. 1964. Les parasites des Coccinelles aphidiphages dans les Basses-Alpes et les Alpes-Maritimes. *Entomophaga*, **9**, 153–80.

Iperti, G. 1965. Contribution a etude de la specificite chez les principales Coccinelles aphidiphages des Alpes-Maritimes et Basses-Alpes. *Entomophaga*, **10**, 159–78.

Iperti, G. 1966a. Protection of coccinellids against mycosis. In Hodek, I., ed. *Ecology of Aphidophagous Coccinellidae*. Prague: Academia and The Hague: Dr W. Junk; pp. 189–90.

Iperti, G. 1966b. The natural enemies of aphidophagous Coccinellidae. In Hodek, I., ed. *Ecology of Aphidophagous Coccinellidae*. Prague: Academia and The Hague: Dr W. Junk; pp. 185–7.

Iperti, G. and Hodek, I. 1974. Introduction alimentaire de la dormance imaginale chez *Semiadalia undecimnotata* Schn. (Coleoptera: Coccinellidae) pour aider a la conservation des coccinellis elevees au laboratoire avant une utilization ulterieure. *Annales de Zoologie-Écologie Animale*, **6**, 41–51.

Iperti, G. and Prudent, P. 1986. Effect of photoperiod on egg-laying in *Adalia bipunctata*. In Hodek, I., ed. *Ecology of Aphidophaga*. Prague: Academia and Dordrecht: Dr. W. Junk; pp. 245–6.

Iperti, G. and van Waerebeke, D. 1968. Description, biologie et importance d'une nouvelle espèce d'Allantonematidae (Nématode) parasite des coccelles aphidiphages: *Parasitilenchus coccinellinae*, n. sp. *Entomophaga*, **13**, 107–19.

Ireland, H., Kearns, P.W.E. and Majerus, M.E.N. 1986. Interspecific hybridisation in the coccinellids: some observations on an old controversy. *Entomologist's Record and Journal of Variation*, **98**, 181–5.

Isaac, N.J.B., van Strien, A.J., August, T.J., de Zeeuw, M.P. and Roy, D.B. 2014. Statistics for citizen science: extracting signals of change from noisy ecological data. *Methods in Ecology and Evolution*, **5**, 1052–60.

Itioka, T. and Inoue, T. 1996. The role of predators and attendant ants in the regulation of a population of the citrus mealybug *Pseudococcus citriculus* in a Satsuma orange orchard. *Applied Entomology and Zoology*, **31**, 195–202.

Iwata, K. 1932. On the biology of two large lady-birds in Japan. *Transactions of the Kansas Entomological Society*, **3**, 13–26.

Izraylevich, S. and Gerson, U. 1993. Mite parasitisation on armoured scale insects: host suitability. *Experimental Applied Acarology*, **17**, 877–88.

Jaeger, B. 1859. *The Life of North American Insects*. New York, NY.

Jalali, S.K. and Singh, S.P. 1989. Biotic potential of three coccinellid predators on various Diaspine hosts. *Journal of Biological Control*, **3**, 20–3.

James, R.R. and Lighthart, B. 1992. The effect of temperature, diet, and larval instar on the susceptibility of an aphid predator, *Hippodamia convergens* (Coleoptera: Coccinellidae), to the weak bacterial pathogen *Pseudomonas fluorescens*. *Journal of Invertebrate Pathology*, **60**, 215–18.

Jansen, J.P. and Hautier, L. 2008. Ladybird population dynamics in potato: comparison of native species with an invasive species, *Harmonia axyridis*. *BioControl*, **53**, 223–33.

Jeffries, D.L., Chapman, J., Roy, H.E., et al. 2013. Characteristics and drivers of high-altitude ladybird flight: insights from vertical-looking entomological radar. *Plos One* **8**, e82278.

Jeffries, M.J. and Lawton, J.H. 1984. Enemy-free space and the structure of ecological communities. *Biological Journal of the Linnean Society*, **23**, 269–86.

Jenions, M.D. 1997. Female promiscuity and genetic incompatibility. *Trends in Ecology and Evolution*, **12**, 251–3.

Ji, L., Gerson, U. and Izraylevich, S. 1994. The mite *Hemisarcoptes* sp. (Astigmata: Hemisarcoptidae) parasitizing willow oyster scale (Homoptera: Diaspididae) on poplars in northern China. *Experimental Applied Acarology*, **18**, 623–7.

Jiggins, C., Majerus, M.E.N. and Gough, U. 1993. Ant defence of colonies of *Aphis fabae* Scopoli (Hemiptera: Aphididae), against predation by ladybirds. *British Journal of Entomology and Natural History*, **6**, 129–38.

Jiggins, F.M., Hurst, G.D.D. and Majerus, M.E.N. 1998. Sex-ratio distortion in *Acraea encedon* (Lepidopera: Nymphalidae) is caused by a male-killing bacterium. *Heredity*, **81**, 87–91.

Jiggins, F.M., Hurst, G.D.D. and Majerus, M.E.N. 1999. Sex ratio distorting *Wolbachia* causes sex role reversal in its butterfly host. *Proceedings of the Royal Society of London B*, **267**, 69–73.

Jiggins, F.M., Hurst, G.D.D., Dolman, C.E. and Majerus, M.E.N. 2000a. High prevalence male-killing *Wolbachia* in the butterfly *Acraea encedana*. *Journal of Evolutionary Biology*, **13**, 495–501.

Jiggins, F.M., Hurst, G.D.D., von der Schulenburg, J.H.G. and Majerus, M.E.N. 2000b. The butterfly *Danaus chrysippus* is infected with a male-killing *Spiroplasma* bacterium. *Parasitology*, **120**, 439–46.

Johnson, C.G. 1969. *Migration and Dispersal of Insects by Flight*. London: Methuen.

Johnstone, R.A. and Hurst, G.D.D. 1996. Maternally inherited male-killing microorganisms may confound interpretation of mtDNA variation in insects. *Biological Journal of the Linnean Society*, **58**, 453–70.

Jones, T.H. and Blum, M.S. 1983. Arthropod alkaloids: distribution, functions, and chemistry. In Pelletier, S.W., ed. *Alkaloids Vol. 1, Chemical and Biological Perspectives*. New York, NY: John Wiley; pp. 33–84.

Jonson, B. 1599. *The Fountaine of Self-love, or Cynthia's Revels*. Reprinted 1908. London: Nutt.

Kadono-Okuda, K., Sakurai, H., Takeda, S. and Okuda, T. 1995. Synchronous growth of a parasitoid, *Perilitus coccinellae*, and teratocytes with the development of the host, *Coccinella septempunctata*. *Entomologia Experimentalis et Applicata*, **75**, 145–9.

Kairo, M.T.K. and Murphy, S.T. 1995. The life history of *Rodolia iceryae* Janson (Col., Coccinellidae) and the potential for use in innoculative releases against *Icerya pattersoni* Newstead (Hom., Margarodidae) on coffee. *Journal of Applied Entomology*, **119**, 487–91.

Kajita, Y., Takano, F., Yasuda, H. and Agarwala, B.K. 2000. Effects of indigenous ladybird species (Coleoptera: Coccinellidae) on the survival of an exotic species in relation to prey abundance. *Applied Entomology and Zoology*, **35**, 473–9.

Kajita, Y., Takano, F., Yasuda, H. and Evans, E.W. 2006. Interactions between introduced and native predatory ladybirds (Coleoptera, Coccinellidae): factors influencing the success of species introductions. *Ecological Entomology*, **31**, 58–67.

Kalushkov, P. 1998. Ten species (Sternorrhyncha: Aphididae) as prey for *Adalia bipunctata* (Coleoptera: Coccinellidae). *European Journal of Entomology*, **95**, 343–9.

Kalushkov, P. 1999. The effect of aphid prey quality of searching behaviour of *Adalia bipunctata* and its susceptibility to insecticides. *Entomologia Experimentalis et Applicata*, **92**, 277–82.

Kaneko, S. 2002. Aphid-attending ants increase the number of emerging adults of the aphid's primary parasitoid and hyperparasitoids by repelling intraguild predators (behavior and ecology). *Entomological Science*, **5**, 131–46.

Kaneko, S. 2007. Larvae of two ladybirds, *Phymatosternus lewisii* and *Scymnus posticalis* (Coleoptera: Coccinellidae), exploiting colonies of the brown citrus aphid *Toxoptera citricidus* (Homoptera: Aphididae) attended by the ant *Pristomyrmex pungens* (Hymenoptera: Formicidae). *Applied Entomology and Zoology*, **42**, 181–7.

Kanervo, V. 1940. Beobachtungen und Versuche zur Ermittlung der Nahrung einiger Coccinelliden. *Annales Entomologici Fennici*, **6**, 89–110.

Kareiva, P. and Sahakian, R. 1990. Tritrophic effects of a simple architectural mutation in pea plants. *Nature*, **345**, 433–4.

Karpenko, A.V., Timchenko, G.A. and Dubitskaja, S.L. 1969. Birds eat lady beetles. *Zashchita rastenii*, **9**, 53.

Katakura, H. 1985. Sperm transfer in the potato ladybird *Henosepilachna vigintioctomaculata* (Coleoptera, Coccinellidae, Epilachninae). *Kontyû*, **53**, 652–7.

Katakura, H., Nakano, S., Hosogai, T. and Kahono, S. 1994. Female internal reproductive organs, modes of sperm transfer, and phylogeny of Asian Epilachninae (Coleoptera: Coccinellidae). *Journal of Natural History*, **28**, 577–83.

Katsanis, A., Babendreier, D., Nentwig, W. and Kenis, M. 2013. Intraguild predation between the invasive ladybird *Harmonia axyridis* and non-target European Coccinellid species. *BioControl*, **58**, 73–83.

Katsoyannos, P. 1984. The establishment of *Rhyzobius forestieri* (Col.: Coccinellidae) in Greece and its efficiency as an auxilliary control agent against heavy infestation of *Saissetia oleae* (Hom.: Coccidae). *Entomophaga*, **29**, 387–97.

Katsoyannos, P. 1996. *Integrated Insect Pest Management for Citrus in Northern Mediterranean Countries*. Athens: Benaki Phytopathological Institute.

Katsoyannos, P., Kontodimas, D.C., Stathas, G.J. and Tsartsalis, C.T. 1997. Establishment of *Harmonia axyridis* on citrus and some data on its phenology in Greece. *Phytoparasitica*, **25**, 183–91.

Katsoyannos, P., Kontodimas, D.C. and Stathas, G. 2005. Summer diapause and winter quiescence of Hippodamia (Semiadalia) undecimnotata (Coleoptera: Coccinellidae) in central Greece. *European Journal of Entomology*, **102**, 453–7.

Kaufmann, T. 1996. Dynamics of sperm transfer, mixing and fertilization in *Cryptolaemus montrouzieri* (Coleoptera: Coccinellidae) in Kenya. *Annals of the Entomological Society of America*, **89**, 238–42.

Kearns, P.W.E., Tomlinson, I.P.M., O'Donald, P. and Veltman, C.J. 1990. Non-random mating in the two-spot ladybird (*Adalia bipunctata*). I. A reassessment of the evidence. *Heredity*, **65**, 229–40.

Kearns, P.W.E., Tomlinson, I.P.M., Veltman, C.J. and O'Donald, P. 1992. Non-random mating in the two-spot ladybird (*Adalia bipunctata*). II. Further tests for female mating preference. *Heredity*, **68**, 385–9.

Keller, S. and Suter, H. 1980. Epizootiologische Untersuchungen uber das Entomophthora-Auftreten bei feldbaulich wichtigen Blattlausarten. *Acta Oecologia Applicata*, **1**, 63–81.

Kesten, U. 1969. Zur Morphologie und Biologie von *Anatis ocellata* (L.) (Coleoptera, Coccinellidae). *Zeitschrilt für Angewandte Entomologie*, **63**, 412–45.

Kettlewell, H.B.D. 1973. *The Evolution of Melanism*. Oxford: Clarendon Press.

Khalil, S.H., Shah, M.A. and Baloch, U.K. 1985. Optical orientation in predatory coccinellids. *Pakistan Journal of Agricultural Research*, **6**, 40–4.

Kidd, K.A., Nalepa, C.A. and Waldvogel, M.G. 1995. Distribution of *Harmonia axyridis* (Pallas) (Coleoptera: Coccinellidae) in North Carolina and Virginia. *Proceedings of the Entomological Society of Washington*, **97**, 729–31.

Kirby, W. and Spence, W. 1815. *An Introduction to Entomology*. London: Longman, Brown, Green and Longman.

Klausnitzer, B. 1968. Zur Biologie von *Myrrha octodecimguttata* (L.). (Col. Coccinellidae). *Entomologische Nachrichten*, **12**, 102–4.

Klausnitzer, B. 1969. Zur Kenntnis der Entomoparasiten mitteleuropaischer Coccinellidae. *Abhandlung und Berichte des Naturkundemuseums, Gorlitz*, **44**, 1–15.

Klausnitzer, B. 1992. Coccinelliden als Prädatoren der Holunderblattlaus (*Aphis sambuci* L.) im Wärmefrühjahr 1992. *Entomologische Nachrichten und Berichte*, **36**, 185–90.

Klausnitzer, B. and Klausnitzer, H. 1972. *Marienkäfer*. Wittenberg: Die Neue Brehm Bucherei.

Klausnitzer, B. and Klausnitzer, H. 1986. *Marienkäfer (Coccinellidae)*. 3. Überard. Lutherstadt Wittenberg: Aufl., A. Ziemsen Verlag.

Koch, R.L. 2003. The multicoloured Asian lady beetle, *Harmonia axyridis*: a review of its biology, uses in biological control and non-target impacts. *Journal of Insect Science*, **3**, 32.

Koch, R.L., Burkness, E.C., Wold Burkess, S.J. and Hutchison, W.D. 2004. Phytophagous preferences of the multicolored Asian lady beetle (Coleoptera: Coccinellidae) to autumn ripening fruit. *Journal of Economic Entomology*, **97**, 539–44.

Koch, R.L. and Hutchison, W.D. 2003. Phenology and blacklight trapping of the multicoloured Asian lady beetle (Coleoptera: Coccinellidae) in a Minnesota agricultural landscape. *Journal of Entomological Science*, **38**, 477–80.

Koch, R.L., Hutchison, W.D., Venette, R.C. and Heimpel, G.E. 2003. Susceptibility of immature monarch butterfly, *Danaus plexippus* (Lepidoptera: Nymphalidae: Danainae), to predation by *Harmonia axyridis* (Coleoptera: Coccinellidae). *Biological Control*, **28**, 265–70.

Komai, T. 1956. Genetics of ladybeetles. *Advances in Genetics*, **8**, 155–88.

Korschefsky, R. 1931. *Coleopterorum Catalogus*. Pars **118**. Coccinellidae. I. Berlin.

Korschefsky, R. 1932. *Coleopterorum Catalogus*. Pars **120**. Coccinellidae. II. Berlin.

Kovár, I. 1996. Phylogeny. In Hodek, I. and Honek, A., eds. *Ecology of Coccinellidae*. Dordrecht: Kluwer Academic Publishers; pp. 19–31.

Krafsur, E.S., Kring, T.J., Miller, J.C., et al. 1997. Gene flow in the colonizing ladybeetle *Harmonia axyridis* in North America. *Biological Control*, **8**, 207–14.

Kristín, A. 1984. The diet and trophic ecology of the tree sparrow (*Passer montanus*) in the Bratislava area. *Folia Zoologika*, **33**, 143–57.

Kristín, A. 1986. Heteroptera, Coccinea, Coccinellidae and Syrphidae in the food of *Passer montanus* L. and *Pica pica* L. *Biologia (Bratislava)*, **41**, 143–50.

Kristín, A., Lebedeva, N. and Pinowski, J. 1995. The diet of nestling tree sparrows (*Passer montanus*). *Preliminary report. International Studies on Sparrows*, **20**, 3–20.

Kulman, H.M. 1971. Parasitism of *Anatis quindecimpunctata* by *Homalotylus terminalis*. *Annals of the Entomological Society of America*, **64**, 953.

Kuznetsov, V.N. 1997. *Lady Beetles of the Russian Far East*. Memoir Series No. 1. Gainesville, FL: Center for Systematic Entomology.

LaMana, M.L. and Miller, J.C. 1996. Field observations on *Harmonia axyridis* Pallas (Coleoptera: Coccinellidae) in Oregon. *Biological Control*, **6**, 232–7.

LaMana, M.L. and Miller, J.C. 1998. Temperature-dependent development in an Oregon population of *Harmonia axyridis* (Coleoptera: Coccinellidae). *Environmental Entomology*, **27**, 1001–5.

Lane, C. and Rothschild, M. 1960. Notes on wasps visiting a mercury vapour trap, together with some observations on their behaviour towards their prey. *Entomologists' Monthly Magazine*, **95**, 277–9.

Larochelle, A. 1979. Les Coleopteres Coccinellidae du Quebec. *Cordulia*, Supplement **10**, 1–111.

Laudého, Y., Ormiéres, R. and Iperti, G. 1969. Les entomophages de Parlatoria blanchardi Targ. Dans les palmeraies de l'Adrar Mauritanien. II Étude d'un parasite de Coccinellidae "Gregarina Katherina" Watson. *Annales Zoologica Écologica Animales*, **1**, 395–406.

Laurent, P., Braekman, J. and Daloze, D. 2005. Insect chemical defense. *Topics in Current Chemistry*, **240**, 167–229.

Lawrence, J.F. and Newton, A.F. Jr. 1995. Families and subfamilies of Coleoptera (within selected genera, notes, references and data on family-group names). In Pakaluk, J. and Slipinski, S.A., eds. *Biology, Phylogeny and Classification of Coleoptera; Papers Celebrating the 80th Birthday of Roy A. Crowson*. Warsaw: Museum I Instytut Zoologii PAN; pp. 779–1006.

Lee, R.E. Jr. 1980. Aggregation of lady beetles on the shores of lakes (Coleoptera: Coccinellidae). *American Midland Naturalist*, **104**, 295–304.

Lees, D.R. 1981. Industrial melanism: genetic adaptation of animals to air pollution. In Bishop, J.A. and Cook, L.M., eds. *Genetic Consequences of Man-Made Change*. London: Academic Press; pp. 129–76.

Le Monnier, Y. and Livory, A. 2003. *Atlas des Coccinelles de La Manche*. Basse-Normandie, France: Manche-Nature.

Le Pelley, R.H. 1959. *Agricultural Insects of East Africa*. Nairobi: E. A. High Commission.

Lesage, L. 1991. Coccinellidae (Cucujoidea), the lady beetles, ladybirds. In Stehr, F.W., ed. *Immature Insects*. Vol. 2. Dubuque, IA: Kendall/Hunt; pp. 485–94.

Liere, H. and Perfecto, I. 2008. Cheating on a mutualism: indirect benefits of ant attendance to a coccidophagous coccinellid. *Environmental Entomology*, **37**, 143–9.

Lipa, J.J. and Sem'Yanov, V.P. 1967. The parasites of the ladybirds in the Leningrad region. *Entomologicheskoe Obozrenie*, **46**, 75–80.

Liu, H. and Qin, L. 1989. The population fluctuations of some dominant species of ladybird beetles in Eastern Hebei Province. *Chinese Journal of Biological Control*, **5**, 92.

Losey, J.E. and Vaughan, M. 2006. The economic value of ecological services provided by insects. *Bioscience*, **56**, 311–23.

Lucas, E. 2005. Intraguild predation among aphidophagous predators. *European Journal of Entomology*, **102**, 351–63.

Lucas, E., Coderre D. and Brodeur, J. 1998. Intraguild predation among aphid predators: characterization and influence of extraguild prey density. *Ecology*, **79**, 1084–92.

Luck, R.F. 1990. Evaluation of natural enemies for biological control: a behavioural approach. *Trends in Ecology and Evolution*, **5**, 196–9.

Lus, Ya.Ya. (Lusis, J.J.) 1928. On the inheritance of colour and pattern in lady beetles *Adalia bipunctata* L. and *Adalia decempunctata* L. *Izvestie Byuro Genetiki, Leningrad*, **6**, 89–163.

Lus, Ya.Ya. (Lusis, J.J.) 1932. An analysis of the dominance phenomenon in the inheritance of the elytra and pronotum colour in *Adalia bipunctata*. *Trudy Laboratorii Genetika, Leningrad*, **9**, 135–62.

Lusis, J.J. 1947a. Some rules of reproduction in populations of *Adalia bipunctata*: heterozygosity of lethal alleles in populations. *Doklady Akademii Nauk SSSR*, **57**, 825–8.

Lusis, J.J. 1947b. Some aspects of the population increase in *Adalia bipunctata*. 2. The strains without males. *Doklady Akademii Nauk SSSR*, **57**, 951–4.

Lusis, J.J. 1961. On the biological meaning of colour polymorphism of ladybeetle *Adalia bipunctata* L. *Latvian Entomologist*, **4**, 3–29.

Lusis, J.J. 1973. Taxonomic relations and geographical distribution of forms in beetles of the genus *Adalia* Mulsant. *The Problems of Genetics and Evolution*, **1**, 5–128.

Lynch, L.D., Hokkanen, H.M.T., Babendreier, D., et al. 2001. Insect biological control and non-target effects: a European perspective. In Wajnberg, E., Scott, J.K. and Quinby, P.C., eds. *Evaluating Indirect Ecological Effects of Biological Control*. Wallingford: CABI Publishing; pp. 99–125.

MacKay, W.P. 1983. Beetles associated with the harvester ants, *Pogonomyrmex montanus, P. subnitidus* and *P. rugosus* (Hymenoptera: Formicidae). *Coleopterists' Bulletin*, **37**, 239–46.

Mader, L. 1926–1937. *Evidenz der Palaarktischen Coccinelliden und ihrer Aberationen in Wort und Bild.* Vienna.

Maeta, Y. 1969. Biological studies on the natural enemies of some Coccinellid beetles. I. On *Perilitus coccinellae* (Schrank). *Kontyû*, **37**, 147–66.

Magnan, E.M., Sanchez, H., Luskin, A.T. and Bush, R.K. 2002. Multicolored Asian lady beetle *(Harmonia axyridis)* sensitivity. *Journal of Allergy and Clinical Immunology*, **109**, 205.

Magro, A., Dixon, A.F.G., Bastin, N. and Hemptinne, J.-L. 2005. Assessment of patch quality in ladybirds' guild: role for larval tracks. *Proceedings of International Symposium on Biological Control of Aphids and Coccids*, Yamagata Unversity; p. 219.

Magro, A., Lecompte, E., Magne, F., Hemptinne, J.L. and Crouau-Roy, B. 2010. Phylogeny of ladybirds (Coleoptera: Coccinellidae): are the subfamilies monophyletic? *Molecular Phylogenetics and Evolution*, **54**, 833–48.

Majerus, M.E.N. 1986. The genetics and evolution of female choice. *Trends in Ecology and Evolution*, **1**, 1–7.

Majerus, M.E.N. 1988. Some notes on the 18-spot ladybird (*Myrrha 18-guttata* L.) (Coleoptera: Coccinellidae). *British Journal of Entomology and Natural History*, **1**, 11–13.

Majerus, M.E.N. 1989. *Coccinella magnifica* (Redtenbacher) – a myrmecophilous ladybird. *British Journal of Entomology and Natural History*, **2**, 97–106.

Majerus, M.E.N. 1991a. Habitat and host plant preferences of British ladybirds (Col., Coccinellidae). *Entomologists' Monthly Magazine*, **127**, 167–75.

Majerus, M.E.N. 1991b Predation of the eyed ladybird by house martins. *Entomologist*, **110**, 75.

Majerus, M.E.N. 1991c. Notes on the behaviour of the ladybird parasitoid *Perilitus coccinellae* (Schrank), from an unusual source. *Bulletin of the Amateur Entomologists' Society*, **50**, 37–40.

Majerus, M.E.N. 1991d. A rare melanic form of *Tytthaspis sedecimpunctata* L. (Col., Coccinellidae). *Entomologists' Monthly Magazine*, **127**, 176.

Majerus, M.E.N. 1992. Aspects of the overwintering biology of ladybirds in Britain, Parts I and II. *Entomologist's Record and Journal of Variation*, **104**, 173–84.

Majerus, M.E.N. 1993. Notes on the inheritance of a scarce form of the striped ladybird, *Myzia oblongoguttata* Linnaeus (Coleoptera: Coccinellidae). *Entomologist's Record and Journal of Variation*, **105**, 271–8.

Majerus, M.E.N. 1994a. *Ladybirds*. New Naturalist Series No. 81. London: HarperCollins.

Majerus, M.E.N. 1994b. Female promiscuity maintains high fertility in ladybirds (Col., Coccinellidae). *Entomologists' Monthly Magazine*, **130**, 205–9.

Majerus, M.E.N. 1994c. An aberrant form of the cream-spot ladybird, *Calvia 14-guttata* (Col: Coccinellidae). *Entomologist*, **113**, 85.

Majerus, M.E.N. 1995. *The Current Status of Ladybirds in Britain: Final Report of the Cambridge Ladybird Survey: 1984–1994*. Cambridge: Department of Genetics, University of Cambridge.

Majerus, M.E.N. 1996. Ladybird, ladybird fly to my home! (or how to encourage ladybirds to your garden). *Bulletin of the Amateur Entomologists' Society*, **55**, 83–90.

Majerus, M.E.N. 1997a. Interspecific hybridisation in the Coccinellidae. *Entomologist's Record and Journal of Variation*, **109**, 11–23.

Majerus, M.E.N. 1997b. How is *Adalia bipunctata* (Linn.) (Coleoptera: Coccinellidae) attracted to overwintering sites? *Entomologist*, **116**, 212–17.

Majerus, M.E.N. 1997c. Parasitisation of British ladybirds by *Dinocampus coccinellae* (Schrank) (Hymenoptera: Braconidae). *British Journal of Entomology and Natural History*, **10**, 15–24.

Majerus, M.E.N. 1998a. Predation of ladybirds (Coccinellidae) by other beetles. *Entomologist's Record and Journal of Variation*, **110**, 27–30.

Majerus, M.E.N. 1998b. *Melanism: Evolution in Action*. Oxford: Oxford University Press.

Majerus, M.E.N. 1999. Simbiontes hereditarios causantes de efectos deletéreos en los artrópodos/ Deleterious endosymbionts of Arthropods. In Melic, A., de Haro, J.J., Méndez, M. and Ribera, I., eds. *The Evolution and Ecology of Arthropods*. (In Spanish and English.) Zaragosa, Spain: Sociedad Entomologica Aragonera; pp. 777–806.

Majerus, M.E.N. 2002. *Moths*. New Naturalist Series No. 90. London: HarperCollins.

Majerus, M.E.N. 2003. *Sex Wars: Genes, Bacteria, and Sex Ratios*. Princeton, NJ: Princeton University Press.

Majerus, M.E.N. 2006a. Male ladybirds under attack down under. *Genetics Society News*, July, 44–6.

Majerus, M.E.N. 2006b. The impact of male-killing bacteria on the evolution of aphidophagous coccinellids. *European Journal of Entomology*, **103**, 1–7.

Majerus, M.E.N., Amos, W. and Hurst, G.D.D. 1996. *Evolution: The Four Billion Year War*. Harlow: Longmans.

Majerus, M.E.N., Bayne, E., Betts, H. and Haddrill, P. 2000a. Survival of ladybird hosts infected with phorid parasitoids. *Entomologist's Record and Journal of Variation*, **112**, 123–4.

Majerus, M.E.N. and Fowles, A.P. 1989. The rediscovery of the 5-spot ladybird *(Coccinella 5-punctata* L.) (Col., Coccinellidae) in Britain. *Entomologists' Monthly Magazine*, **125**, 177–81.

Majerus, M.E.N., Geoghegan, I.E. and Majerus, T.M.O. 2000b. Adaptive preferential selection of young hosts by the parasitoid wasp *Dinocampus coccinellae* (Schrank) (Hymenoptera: Braconidae). *European Journal of Entomology*, **97**, 161–4.

Majerus, M.E.N. and Hurst, G.D.D. 1996. Extension to the genome. *The Genetical Society Newsletter*, **32**, 6.

Majerus, M.E.N. and Hurst, G.D.D. 1997. Ladybirds as a model system for the study of male-killing symbionts. *Entomophaga*, **42**, 13–20.

Majerus, M.E.N., Ireland, H. and Kearns, P.W.E. 1987. Description of a new form of *Adalia bipunctata* with notes on its inheritance. *Entomologist's Record and Journal of Variation*, **99**, 255–7.

Majerus, M.E.N. and Kearns, P.W.E. 1989. *Ladybirds*. Naturalists' Handbooks No. 10. Slough: Richmond Publishing.

Majerus, M.E.N. and Majerus, T.M.O. 1996. Ladybird population explosions. *British Journal of Entomology and Natural History*, **9**, 65–76.

Majerus, M.E.N. and Majerus, T.M.O. 1997a. Cannibalism among ladybirds. *Bulletin of the Amateur Entomologists' Society*, **56**, 235–48.

Majerus, M.E.N. and Majerus, T.M.O. 1997b. Predation of ladybirds by birds in the wild. *Entomologists' Monthly Magazine*, **133**, 55–61.

Majerus, M.E.N. and Majerus, T.M.O. 1998. Mimicry in ladybirds. *Bulletin of the Amateur Entomologists' Society*, **57**, 126–40.

Majerus, M.E.N. and Majerus, T.M.O. 2000. Female-biased sex ratio due to male-killing in the Japanese ladybird *Coccinula sinensis*. *Ecological Entomology*, **25**, 234–8.

Majerus, M.E.N., Majerus, T.M.O. and Cronin, A. 1999. The inheritance of a melanic form of the 7-spot ladybird, *Coccinella septempunctata* L. (Coleoptera: Coccinellidae). *British Journal of Entomology and Natural History*, **11**, 180–4.

Majerus, M.E.N., O'Donald, P. and Wier, J. 1982a. Evidence for preferential mating in *Adalia bipunctata*. *Heredity*, **49**, 37–49.

Majerus, M.E.N., O'Donald, P. and Wier, J. 1982b. Female mating preference is genetic. *Nature*, **300**, 521–3.

Majerus, M.E.N., O'Donald, P., Kearns, P.W.E. and Ireland, H. 1986. The genetics and evolution of female choice. *Nature*, **321**, 164–7.

Majerus, M.E.N., Sloggett, J.J., Godeau, J.-F. and Hemptinne, J.-L. 2007. Interactions between ants and aphidophagous and coccidophagous ladybirds. *Population Ecology*, **49**, 15–27.

Majerus, M.E.N., Strawson, V. and Roy, H.E. 2006. The potential impacts of the arrival of the harlequin ladybird, *Harmonia axyridis* (Pallas) (Coleoptera: Coccinellidae), in Britain. *Ecological Entomology*, **31**, 207–15.

Majerus, M.E.N., von der Schulenburg, J.H.G. and Zakharov, I.A. 2000c. Multiple causes of male-killing in a single sample of the 2-spot ladybird, *Adalia bipunctata* (Coleoptera: Coccinellidae) from Moscow. *Heredity*, **84**, 605–9.

Majerus, M.E.N. and Williams, Z. 1989. The distribution and life history of the orange ladybird, *Halyzia 16-guttata* (L.) (Coleoptera: Coccinellidae) in Britain. *Entomologist's Gazette*, **40**, 71–8.

Majerus, M.E.N. and Zakharov, I.A. 2000. Does thermal melanism maintain melanic polymorphism in the two-spot ladybird, *Adalia bipunctata* (Coleoptera: Coccinellidae)? *Journal Obshchei Biologia*, **61**, 381–92.

Majerus, T. M. O. and Majerus, M. E. N. 2010. Intergenomic arms races: detection of a nuclear rescue gene of male-killing in a ladybird. *PLOS Pathology* **6**, e1000987.

Malausa, J.C., Franco, E. and Brun, P. 1988. Acclimatation sur la Cote d'Azur et en Corse de *Serangium parcesetosum* (Col.: Coccinellidae) predateur de l'aleurode des citrus *Dialeurodes citri* (Hom.: Aleyrodidae). *Entomophaga*, **33**, 517–19.

Malcolm, S.B. 1990. Chemical defence in chewing and sucking insect herbivores: plant-derived cardenolides in the monarch butterfly and oleander aphid. *Chemoecology*, **1**, 12–21.

Mann, W.M. 1911. On some northwestern ants and their guests. *Psyche*, **18**, 102–9.

Mariau, D. and Julia, J.F. 1977. Nouvelles rechérches sur la cochenille du cocotier *Aspidotus destructor* (Sign.) *Oléagineux*, **32**, 217–24.

Marples, N.M. 1993. Do wild birds use size to distinguish palatable and unpalatable prey types? *Animal Behaviour*, **46**, 347–54.

Marples, N.M., Brakefield, P.M. and Cowie, R.J. 1989. Differences between the 7-spot and 2-spot ladybird beetles (Coccinellidae) in their toxic effects on a bird predator. *Ecological Entomology*, **14**, 79–84.

Marples, N.M., de Jong, P.W., Ottenheim, M.M., Verhoog, M.D. and Brakefield, P.M. 1993. The inheritance of a wingless character in the 2-spot ladybird (*Adalia bipunctata*). *Entomologia Experimentalis et Applicata*, **69**, 69–73.

Martelli, G. 1914. Notizie su due Coccinellidi micofaga. *Bollettino del Laboratorio de Zoologia Generale e Agraria, Portici*, **9**, 151–60.

Masaki, S. 1984. Unity band diversity in insect photoperiodism. In Porter, R. and Collins, G.M., eds. *Photoperiodic Regulation of Insect and Molluscan Hormones. CIBA Foundation Symposium 104*. London: Pitman; pp. 7–25.

Mattson, D.J., Gillin, C.M., Benson, S.A. and Knight, R.R. 1991. Bear feeding activity at alpine insect aggregation sites in the Yellowstone ecosystem. *Canadian Journal of Zoology*, **69**, 2430–5.

Maynard Smith, J. 1976. Sexual selection and the handicap principle. *Journal of Theoretical Biology*, **57**, 239–42.

Maynard Smith, J. 1989. *Evolutionary Genetics*. Oxford: Oxford University Press.

Mayr, E. 1963. *Animal Species and Evolution*. Cambridge, MA: Harvard University Press.

McClure, M.S. 1986. Role of predators in regulation of endemic populations of *Matsucoccus-Matsumurae* (Homoptera: Margarodidae) in Japan. *Environmental Entomology*, **15**, 976–83.

McDaniel, B. and Morrill, W. 1969. A new species of *Tetrapolipus* from *Hippodamia convergens* from South Dakota (Acarina: Podapolipidae). *Annals of the Entomological Society of America*, **62**, 1456–8.

McKinney, M.L. and Lockwood, J.L. 1999. Biotic homogenization: a few winners replacing many losers in the next mass extinction. *Trends in Ecology and Evolution*, **14**, 450–3.

McMullen, R.D. 1967. A field study of diapause in *Coccinella novemnotata* (Coleoptera: Coccinellidae). *Canadian Entomologist*, **99**, 42–9.

McMurtry, J.A., Scriven, G.T. and Malone, R.S. 1974. Factors affecting oviposition of *Stethorus picipes* (Coleoptera: Coccinellidae), with special reference to photoperiod. *Environmental Entomology*, **3**, 123–7.

Meinwald, J., Meinwald, Y.C., Chalmers, A.M. and Eisner, T. 1968. Dihydromatricaria acid: acetylinic acid secreted by soldier beetle. *Science*, **160**, 890–2.

Merlin, J., Lemaitre, O. and Grégoire, J.-C. 1996a. Oviposition in *Cryptolaemus montrouzieri* stimulated by wax filaments of its prey. *Entomologia Experimentalis et Applicta*, **79**, 141–6.

Merlin, J., Lemaitre, O. and Grégoire, J.-C. 1996b. Chemical cues produced by conspecific larvae deter oviposition by the coccidophagous ladybird beetle, *Cryptolaemus montrouzieri*. *Entomologia Experimentalis et Applicta*, **79**, 147–51.

Meyerdirk, D.E. 1983. Biology of *Diomus flavifrons* (Blackburn) (Coleoptera: Coccinellidae), a citrus mealybug predator. *Environmental Entomology*, **12**, 1275–7.

Michaud, J.P. 2002. Invasion of the Florida citrus ecosystem by *Harmonia axyridis* (Coleoptera: Coccinellidae) and asymmetric competition with a native species, *Cycloneda sanguinea*. *Environmental Entomology*, **31**, 827–35.

Michaud, J.P. 2012. Coccinellids in biological control. In Hodek, I., van Emden, H.F. and Honek, A., eds. *Ecology and Behaviour of the Ladybird Beetles (Coccinellidae)*. Chichester: Wiley-Blackwell; pp. 488–519.

Michie, L.J., Mallard, F., Majerus, M.E.N. and Jiggins, F.M. 2010. Melanic through nature or nurture: genetic polymorphism and phenotypic plasticity in *Harmonia axyridis*. *Journal of Evolutionary Biology* **23**, 1699–1707.

Mills, N.J. 1979. Adalia bipunctata *(L.) as a Generalist Predator of Aphids*. Unpublished PhD thesis, University of East Anglia.

Minelli, A. and Pasqual, C. 1977. The mouthparts of ladybirds: structure and function. *Bolletino Zoologia Napoli*, **44**, 183–7.

Montgomery, H.W. Jr. and Goodrich, M.A. 2002. The *Brachiacantha* (Coleoptera: Coccinelllidae) of Illinois. *Transactions of the Illinois State Academy of Science*, **95**, 111–30.

Moore, B.P., Brown, W.V. and Rothschild, M. 1990. Methylalkylpyrazines in aposematic insects, their hostplants and mimics. *Chemoecology*, **1**, 43–51.

Morales, J. and Burandt, C.L. Jr. 1985. Interactions between *Cycloneda sanguinea* and the brown citrus aphid: adult feeding and larval mortality. *Environmental Entomology*, **14**, 520–2.

Morgan, C.L. 1896. *Habit and Instinct*. London.

Morton Jones, F. 1932. Insect coloration and the relative acceptability of insects to birds. *Transactions of the Entomological Society of London*, **80**, 345–85.

Moter, G. 1959. *Untersuchungen zur Biologie von* Stethorus punctillum *Weise*. Unpublished PhD thesis, University of Köln.

Muggleton, J. 1978. Selection against the melanic morphs of *Adalia bipunctata* (two-spot ladybird): a review and some new data. *Heredity*, **40**, 269–80.

Muggleton, J. 1979. Non-random mating in wild populations of polymorphic *Adalia bipunctata*. *Heredity*, **42**, 57–65.

Muggleton, J., Lonsdale, D. and Benham, B.R. 1975. Melanism in *Adalia bipunctata* L. (Col., Coccinellidae) and its relationship to atmospheric pollution. *Journal of Applied Ecology*, **12**, 451–64.

Müller, F. 1878. [Notes on Brazilian entomology]. *Transactions f the Entomological Society of London*, **3**, 211–23.

Mulsant, M.E. 1846. *Histoire naturelle des Coleopteres de France: Sulcicolles-Securipalpes*. Paris.

Mulsant, M.E. 1850. Species de Coleopteres trimeres securipalpes. *Annales des Sciences Physiques et Naturelles, d'Agriculture et d'Industrie (Lyon)*, **2**, 1–1104.

Mulsant, M.E. 1853. Supplement a la monographie des Coleopteres trimeres securipalpes. *Opuscula Entomologica*, **3**, 1–178.

Mulsant, M.E. 1866. Monographie des coccinellides. *Memoire de L'Academie des Science, Belles – Lettres et Arts de Lyon*, **15**, 1–112.

Nakamuta, K. 1984. Visual orientation of the ladybeetle, *Coccinella septempunctata* L. (Coleoptera: Coccinellidae), towards its prey. *Applied Entomology and Zoology*, **19**, 82–6.

Nakata, T. 1995. Population fluctuations of aphids and their natural enemies on potato in Hokkaido. *Japanese Journal of Applied Entomology and Zoology*, **30**, 129–38.

Nalepa, C.A., Kidd, K.A. and Hopkins, D.I. 2000. The multicoloured Asian ladybeetle (Coleoptera: Coccinellidae): orientation to aggregation sites. *Journal of Entomological Science*, **35**, 150–7.

Nalepa, C.A. and Weir, A. 2007. Infection of *Harmonia axyridis* (Coleoptera: Coccinellidae) by *Hesperomyces virescens* (Ascomycetes: Laboulbeniales): role of mating status and aggregation behavior. *Journal of Invertebrate Pathology*, **94**, 196–203.

Nechaev, V.A. and Kuznetsov, V.N. 1973. Avian predation on coccinellids in Primorie Region. *Entomofauna Sovetskogo Dalnego Vostoka. Trudy Biologo – Pochvennyi Institut, Vladivostok*, **9**, 97–9.

Nedvěd, O. 1993. Comparison of cold hardiness in two ladybird beetles (Coleoptera: Coccinellidae) with contrasting hibernation behaviour. *European Journal of Entomology*, **90**, 465–70.

Nedvěd, O. 1995. Cold reacclimation in postdormant *Exochomus quadripustulatus* (Coleoptera: Coccinellidae). *Cryo-Letters*, **16**, 47–50.

Nedvěd, O. and Cihakova, V. 2004. Phylogeny of Coccinellidae – the third cladistic attempt. *Abstracts of Ecology of Aphidophaga 9 Ceske Budejovice*, September 2004, p. 55.

Nedvěd, O. and Kalushkov, P. 2012. Effect of air humidity on sex ratio and development of ladybird *Harmonia axyridis* (Coleoptera: Coccinellidae). *Psyche: A Journal of Entomology*, **2012**, Article ID 173482.

Nedvěd, O. and Kovár, I. 2012. Phylogeny and classification. In: *Ecology and Behaviour of the Ladybird Beetles (Coccinellidae)*. Hodek, I., van Emden, H.F. and Honek, A., eds. Chichester: Wiley-Blackwell; pp. 1–12.

Newell, R.H. 1845. *The Zoology of the English Poets, Corrected by the Writings of Modern Naturalists*. London.

Nishida, R. and Fukami, H. 1989. Host plant iridoid-based chemical defense of an aphid, *Acyrthosiphon nipponicus*, against ladybird beetles. *Journal of Chemical Ecology*, **15**, 1837–45.

Novák, B and Grenarová, A. 1967. Coccinelliden an der Grenze des Feld – und Waldbiotops – Hibernationsversuche mit den Imagines führender Arten. *Konferenz über die Schädlinge der Hackfrüchte III*. Phaha; pp. 49–59.

Obata, S. 1986. Determination of hibernation site in the ladybird beetle, *Harmonia axyridis* Pallas (Coleoptra, Coccinellidae). *Kontyu*, **54**, 218–23.

Obata, S. 1988. Mating behaviour and sperm transfer in the ladybird beetle, *Harmonia axyridis* Pallas (Coleoptera: Coccinellidae). In Niemczyk, E. and Dixon, A.F.G., eds. *Ecology and Effectiveness of Aphidophaga*. The Hague: SPB Academic Publishing; pp. 39–42.

Obata, S. and Hidaki, T. 1987. Ejection and ingestion of the spermatophore by the female beetle, *Harmonia axyridis* Pallas (Coleoptera: Coccinellidae). *Canadian Entomologist*, **119**, 603–4.

Obata, S. and Johki, Y. 1991. Comparative study on copulatory behaviour in four species of aphidophagous ladybirds. In Polgar L., Chambers, R.J., Dixon, A.G.F. and Hodek I., eds. *Behaviour and Impact of Aphidophaga*. The Hague: SPB Academic Publishing; pp. 207–11.

Obatake, H. and Suzuki, H. 1985. On the isolation and identification of canavanine and ethanolamine contained in the young leaves of black locus, *Robinia pseudoacacia*, lethal for the lady beetle *Harmonia axyridis*. *Technical Bulletin of the Faculty of Agriculture, Kagawa University*, **36**, 107–15.

Obrycki, J.J., Tauber, M.J., Tauber, C.A. and Gollands, B. 1983. Environmental control of the seasonal life cycle of *Adalia bipunctata* (Coleoptera: Coccinellidae). *Environmental Entomology*, **12**, 416–21.

Obrycki, J.J., Tauber, M.J. and Tauber, C.A. 1985. *Perilitus coccinellae* (Hymenoptera: Braconidae): parasitization and development in relation to host-stage attacked. *Annals of the Entomological Society of America*, **78**, 852–4.

Oczenascheck, C. 1997. Chemische Ekologie der Entwicklungsstadien des Marienkaefers *Platynaspis luteorubra* Goeze. Unpublished Diploma thesis, University of Bayreuth.

O'Donald, P. 1980. *Genetic Models of Sexual Selection*. Cambridge: Cambridge University Press.

O'Donald, P., Derrick, M., Majerus, M.E.N. and Wier, J. 1984. Population genetic theory of the assortative mating, sexual selection and natural selection of the two-spot ladybird, *Adalia bipunctata*. *Heredity*, **52**, 43–61.

O'Donald, P. and Majerus, M.E.N. 1984. Polymorphism of melanic ladybirds maintained by frequency-dependant sexual selection. *Biological Journal of the Linnean Society*, **23**, 101–11.

O'Donald, P. and Majerus, M.E.N. 1985. Sexual selection and the evolution of preferential mating in ladybirds. I. Selection for high and low lines of female preference. *Heredity*, **55**, 401–12.

O'Donald, P. and Majerus, M.E.N. 1988. Frequency-dependent sexual selection. *Philosophical Transactions of the Royal Society of London B*, **319**, 571–86.

O'Donald, P. and Majerus, M.E.N. 1992. Non-random mating in *Adalia bipunctata* (the two-spot ladybird). III. New evidence of genetic preference. *Heredity*, **61**, 521–6.

O'Donald, P. and Muggleton, J. 1979. Melanic polymorphism in ladybirds maintained by sexual selection. *Heredity*, **43**, 143–8.

Ofuya, T.I. 1991. Aspects of the ecology of predation in two coccinellid species on the cowpea aphid in Nigeria. In Polgar, L., Chambers R.J., Dixon, A.F.G. and Hodek, I., eds. *Behaviour and Impact of Aphidophaga*. The Hague: SPB Academic Publishing; pp. 213–20.

Okamoto, H. 1961. Comparison of ecological characters of the predatory ladybird *Coccinella septempunctata bruckii* fed on the apple grain aphids, *Rhopalosiphum prunifoliae* and the cabbage aphids, *Brevicoryne brassicae*. *Japanese Journal of Applied Entomological Zoology*, **5**, 277–8.

Okuda, T. and Chinzei, Y. 1988. Vitellogenesis in a lady-beetle, *Coccinella septempunctata* in relation to the aestivation-diapause. *Journal of Insect Physiology*, **34**, 393–401.

Okuda, T. and Hodek, I. 1994 .Diapause and photoperiodic response in *Coccinella septempunctata brucki* Mulsant (Coleoptera: Coccinellidae) in Hokkaido, Japan. *Applied Entomological Zoology*, **29**, 549–54.

Olden, J.D., Poff, N.L. and McKinney, M.L. 2006. Forecasting faunal and floral homogenisation associated with human population geography in North America. *Biological Conservation*, **127**, 261–71.

Olszak, R.W. 1986. Suitability of three aphid species as prey for *Propylea quatuordecimpunctata*. In Hodek, I., ed. *Ecology of Aphidophaga*. Prague: Academia and Dordrecht: Dr W. Junk.

Omkar, O. and Mishra, G. 2005. Evolutionary significance of promiscuity in an aphidophagous ladybird, *Propylea dissecta* (Coleoptera: Coccinellidae). *Bulletin of Entomological Research*, **95**, 527–33.

Omkar, O. and Pervez, A. 2003. Ecology and biocontrol potential of a scale-predator, *Chilocorus nigritus*. *Biocontrol Science and Technology*, **13**, 379–90.

O'Neill, S.L., Hoffmann, A.A. and Werren, J.H., eds. 1997. *Influential Passengers*. Oxford: Oxford University Press.

Orivel, J., Servigne, P., Cerdan, Ph., Dejean, A. and Corbara, B., 2004. The ladybird *Thalassa saginata*, an obligatory myrmecophile of *Dolichoderus bidens* ant colonies. *Naturwissenschaften*, **91**, 97–100.

Orlova-Bienkowskaja, M.J., Ukrainsky, A.S. and Brown, P.M.J. 2015. *Harmonia axyridis* (Coleoptera: Coccinellidae) in Asia: a re-examination of the native range and invasion to southeastern Kazakhstan and Kyrgyzstan. *Biological Invasions*, **17**, 1941–8.

Ormond, E.L., Thomas, A.P.M., Pell, J.K., Freeman, S.N. and Roy, H.E. 2011. Avoidance of a generalist entomopathogenic fungus by the ladybird, *Coccinella septempunctata*. *FEMS Microbiology Ecology*, **77**, 229–37.

Ormond, E.L., Thomas, A.P.M., Pugh, P.J.A., Pell, J.K. and Roy, H.E. 2010. A fungal pathogen in time and space: the population dynamics of *Beauveria bassiana* in a conifer forest. *FEMS Microbiology Ecology*, **74**, 146–54.

Osawa, N. 2000. Population field studies on the aphidophagous ladybird beetle *Harmonia axyridis* (Coleoptera: Coccinellidae): resource tracking and population characteristics. *Population Ecology*, **42**, 115–27.

Osawa, N. and Nishida, T. 1992. Seasonal variation in elytral colour polymorphism in *Harmonia axyridis* (the ladybird beetle): the role of non-random mating. *Heredity*, **69**, 297–307.

Osborne, P.J. 1971. The insect faunas of the Wandle Gravels. *Proceedings and Transactions of the Croydon Natural History Society*, **14**, 162–75.

Palmer, M.A. 1911. Some notes on heredity in the coccinellid genus *Adalia* Mulsant. *Annals of the Entomological Society of America*, **4**, 283–302.

Palmer, M.A. 1914. Some notes on life history of lady beetles. *Annals of the Entomological Society of America*, **7**, 13–238.

Palmer, M.A. 1917. Additional notes on heredity and life history in the Coccinellid genus *Adalia* Mulsant. *Annals of the Entomological Society of America*, **10**, 289–302.

Panis, A. 1981. Notes sur quelques insectes auxiliaires régulateurs des populations de Pseudococcidae et de Coccidae (Homoptera, Coccoidea) des agrumes en Provence orientale. *Fruits*, **36**, 49–52.

Pantyukhov, G.A. 1968a. A study of ecology and physiology of the predatory beetle *Chilocorus rubidus* Hope (Coleoptera, Coccinellidae). *Zoologicheskii Zhurnal*, **47**, 376–86.

Pantyukhov, G.A. 1968b. On photoperiodic reaction of *Chilocorus renipustulatus* Scriba (Coleoptera, Coccinellidae). *Entomologicheskoe Obozrenie*, **47**, 376–85.

Park, O. 1930. Studies in the ecology of forest Coleoptera. Seral and seasonal succession of Coleoptera in the Chicago area, with observations on certain phases of hibernation and aggregation. *Annals of the Entomological Society of America*, **23**, 57–80.

Parker, B.L., Whalon, M.E. and Warshaw, M. 1977. Respiration and parasitism in *Coleomegilla maculata lengi* (Coleoptera: Coccinellidae). *Annals of the Entomological Society of America*, **70**, 984–7.

Parker, G.A., Simmons, L.W. and Kirk, H. 1990. Analyzing sperm competition data – simple models for predicting mechanisms. *Behavioural Ecology and Sociobiology*, **27**, 55–65.

Parry, W.H. 1980. Overwintering of *Aphidecta obliterata* L. (Coleoptera: Coccinellidae) in north-east Scotland. *Acta Oecologia Applicata*, **1**, 307–16.

Pasteels, J.M. 1978. Apterous and brachypterous coccinellids at the end of the food chain, *Cionura erecta* (Asclepiadaceae) – *Aphis nerii. Entomologia Experimentalis et Applicata*, **24**, 379–84.

Pasteels, J.M., Deroe, C., Tursch, B., et al. 1973. Distribution et activites des alcaloides défensifs des Coccinellidae. *Journal of Insect Physiology*, **19**, 1771–84.

Pell, J.K., Baverstock, J., Roy, H.E., Ware, R.L. and Majerus, M.E.N. 2008. Intraguild predation involving *Harmonia axyridis*: a review of current knowledge and future perspectives. *BioControl*, **53**, 147–68.

Perry, J.C. and Roitberg, B.D. 2005. Ladybird mothers mitigate offspring starvation risk by laying trophic eggs. *Behavioural Ecology and Sociobiology*, **58**, 578–86.

Perry, J.C., Sharpe, D.M.T. and Rowe, L. 2009. Condition-dependent female remating resistance generates sexual selection on male size in a ladybird beetle. *Animal Behaviour*, **77**, 743–8.

Petersen, J.-E., 1992. *Oviposition Strategy and Larval Oophagy in Coccinellids (Coleoptera: Coccinellidae).* Unpublished Diplomarbeit, University of East Anglia.

Petrie, M. and Kempenaers, B. 1998. Extra-pair paternity in birds: explaining variation between species and populations. *Trends in Ecology and Evolution*, **13**, 52–8.

Pettersson, J. 2012. Coccinellids and semiochemicals. In Hodek, I., van Emden, H.F. and Honek, A. eds. *Ecology and Behaviour of the Ladybird Beetles (Coccinellidae)*. Chichester: Wiley-Blackwell; pp. 444–64.

Phuoc, D.T. and Stehr, F.W. 1974. Morphology and taxonomy of the known pupae of Coccinellidae (Coleoptera) of North America, with a discussion of phylogenetic relationships. *Contributions of the American Entomological Institute*, **10**, 1–125.

Pocock, R.I. 1911. On the palatability of some British insects, with notes on the significance of mimetic resemblances. With notes on the experiments by E.B. Poulton. *Proceedings of the Zoological Society of London*, **1911**, 809–68.

Podoler, H. and Henen, J. 1986. Foraging behaviour of two species of the genus *Chilocorus* (Coccinellidae: Coleoptera): a comparative study. *Phytoparasitica*, **14**, 11–23.

Ponsonby, D.J. and Copland, M.J.W. 1995. Olfactory responses by the scale insect predator *Chilocorus nigritus* (F.) (Coleoptera: Coccinellidae). *Biocontrol Science and Technology*, **5**, 83–93.

Pontin, A.J. 1959. Some records of predators and parasites adapted to attack aphids attended by ants. *Entomologists' Monthly Magazine*, **95**, 154–5.

Pope, R.D. 1979. Wax production by coccinellid larvae (Coleoptera). *Systematic Entomology*, **4**, 171–96.

Poutiers, R. 1930. Sur le comportement du *Novius cardinalis* vis-à-vis le certain alcaloides. *Compte rendu des seances Soc. Biol.*, **103**, 1023–5.

Powell, W. 2005. Using field margins and aphid pheromones for promotion of cereal aphid biocontrol in the UK. *Proceedings of International Symposium on Biological Control of Aphids and Coccids*, Yamagata Unversity; pp. 45–51.

Proksch, P., Wite, L., Wray, V. and Hartmann, T. 1993. Ontogenic variation of defensive alkaloids in the Mexican bean beetle *Epilachna varivestis* (Coleoptera: Coccinellidae). *Entomologia Generalis*, **18**, 1–7.

Provine, W.B. 1986. *Sewall Wright and Evolutionary Biology*. Chicago, IL: Chicago Press.

Pu, C.L. 1976. Biological control of insect pests in China. *Acta Entomologia Sinica*, **19**, 247–52.

Pulliainen, E. 1964. Studies on the humidity and light orientation and the flying activity of *Myrrha 18-guttata* L. (Coleoptera: Coccinellidae). *Annales Entomologici Fennici*, **30**, 117–41.

Purse, B.V., Comont, R., Butler, A., et al. 2015. Landscape and climate determine patterns of spread for all colour morphs of the alien ladybird *Harmonia axyridis*. *Journal of Biogeography*, **42**, 575–88.

Putman, W.L. 1955. Bionomics of *Stethorus punctillum* Weise in Ontario. *Canadian Entomologist*, **87**, 9–33.

Putman, W.L. 1957. Laboratory studies of the foods of some coccinellids (Coleoptera) found in Ontario peach orchards. *Canadian Entomologist*, **89**, 572–9.

Quezada, J.R. and de Bach, P. 1973. Bioecological and population studies of the cottony-cushion scale, *Icerya purchasi* Mask., and its natural enemies, *Rodalia cardinalis* Mul. and *Cryptochaetum icerya* Will., in southern California. *Hilgardia*, **41**, 631–88.

Raak-van den Berg, C.L., Stam, J.M., de Jong, P.W., Hemerik, L. and van Lenteren, J.C. 2012. Winter survival of *Harmonia axyridis* in The Netherlands. *Biological Control*, **60**, 68–76.

Radford, P., Attygalle, A.B., Meinwald, J., Smedley, S.R. and Eisner, T. 1997. Pyrrolidinooxazolidine alkaloids from two species of ladybird beetles. *Journal of Natural Products*, **60**, 755–9.

Ragusa, S. and Russo, A. 1988. Citrus pests in Calabria (Southern Italy). *IOBC/WPRS Bulletin 1988*, **11**, 26–31.

Rana, J.S. and Kakker, J. 2000. Biological studies on 7-spot ladybird beetle, *Coccinella septempunctata* L. with cereal aphid, *Sitobion avenae* (F.) as prey. *Cereal Research Communications*, **28**, 449–54.

Randall, K., Majerus, M.E.N. and Forge, H. 1992. Characteristics for sex determination in British ladybirds (Coleoptera: Coccinellidae). *Entomologist*, **111**, 109–22.

Randerson, J.P., Smith, N.G.C. and Hurst, L.D. 2000. The evolutionary dynamics of male-killers and their hosts. *Heredity*, **84**, 152–60.

Ransford, M.O. 1997. *Sperm Competition in the 2-spot Ladybird,* Adalia bipunctata. Unpublished PhD thesis, University of Cambridge.

Ratcliffe, S. 2002. *National Pest Alert: Multicolored Asian Lady Beetle.* USDA CSREES Regional Integrated Pest Management Program and Pest Management Centers.

Rathcke, B., Hamrum, C.L. and Glass, A.W. 1967. Observations of the interrelationships among ants and aphid predators. *Michigan Entomologist*, **1**, 169–73.

Rees, B.E., Anderson, D.M., Bouk, D. and Gordon, R.D. 1994. Larval key to the genera and selected species of North American Coccinellidae (Coleoptera). *Proceedings of the Entomological Society of Washington*, **96**, 387–412.

Reimer, N.J., Cope, M.-L. and Yasuda, G. 1993. Interference of *Pheidole megacephala* (Hymenoptera: Formicidae) with biological control of *Coccus viridis* (Homoptera: Coccidae) in coffee. *Environmental Entomology*, **22**, 483–8.

Revels, R. and Majerus, M.E.N. 1997. Grouping behaviour in overwintering 16-spot ladybirds (*Tytthaspis 16-punctata*) (Coleoptera: Coccinellidae). *Bulletin of the Amateur Entomologists' Society*, **56**, 143–4.

Rhamhalinghan, M. 1986. Pathologies caused by *Coccinellimermis rubtzov* (Nematoda: Mermithidae) in *Coccinella septempunctata* L. (Coleoptera: Coccinellidae). *Proceedings of the Indian National Science Academy B*, **52**, 228–31.

Ricci, C. 1979. Lapparato boccale pungente succhiante della larva di *Platynaspis luteorubra* Goeze (Col. Coccinellidae). *Bollettino del Laboratorio di Entomologia Agraria Filippo Silvestri di Portici*, **36**, 179–98.

Ricci, C. 1982. Sulla costituzione e funzione delle mandibole della larva di *Tytthaspis sedecimpunctata* (L.) e T. trilineata (Weise). *Frustula Entomologica*, **16**, 205–12.

Ricci, C. 1986. Seasonal food preferences and behaviour of *Rhizobius litura*. In Hodek, I., ed. *Ecology of Aphidophaga*. Prague: Academia and Dordrecht: Dr W. Junk.

Ricci, C., Fiori, G. and Colazza, S. 1983. Regime alimentaire dell'adulto di *Tytthaspis sedecimpunctata* (L.) (Coleoptera Coccinellidae) in ambiente a influenza antropica primaria: prato polifita. *Atti XIII Congrresso Nazionale Italiano Entomologia, Sestriere-Torino*, Torino: Società Entomologica Italiana; pp. 691–8.

Ricci, C. and Stella, I. 1988. Relationship between morphology and function in some Palearctic Coccinellidae. In Niemczyk, E. and Dixon, A.G.F., eds. *Ecology and Effectiveness of Aphidophaga*. The Hague: SPB Academic Publishing; pp. 21–5.

Richards, A.M. 1980a. Sexual selection, guarding and sexual conflict in a species of Coccinellidae (Coleoptera). *Journal of the Australian Entomological Society*, **19**, 26.

Richards, A.M. 1980b. Defence adaptations and behaviour in *Scymnodes lividigaster* (Coleoptera: Coccinellidae). *Journal of Zoology, London*, **192**, 157–68.

Richards, A.M. 1985. Biology and defensive adaptations in *Rodatus major* (Coleoptera: Coccinellidae) and its prey, *Monophlebus pilosior* (Hemiptera: Margarodidae). *Journal of Zoology, London*, **205**, 287–295.

Ridgway, R. 1912. *Color Standards and Color Nomenclature*. Washington, DC.

Ridley, M. 1988. Mating frequency and fecundity in insects. *Biological Reviews*, **63**, 509–49.

Risch, S.J., Wrubel, R. and Andow, D. 1982. Foraging by a predaceous beetle, *Coleomegilla maculata* (Coleoptera: Coccinellidae), in a polyculture: effects of plant density and diversity. *Environmental Entomology*, **11**, 949–50.

Robertson, G.J. 1961. Ovariole numbers in Coleoptera. *Canadian Journal of Zoology*, **39**, 245–63.

Robertson, J.A., Whiting, M.F. and McHugh, J.V. 2008. Searching for natural lineages within the Cerylonid Series (Coleoptera: Cucuioidea). *Molecular Phylogenetics and Evolution*, **46**, 193–205.

Rolley, F., Hodek, I. and Iperti, G. 1974. Influence de la nourriture aphidienne (selon l'âge de la plante-hôte à partir de laquelle les pucerons se multiplient) sur l'induction de la dormance chez *Semiadalia undecimnotata* Schn. (Coleoptera: Coccinellidae). *Annales Zoologique-Écologie Animale*, **6**, 53–60.

Romani, R., Isidoro, N. and Ricci, C. 2004. Morphological evidence of a putative sex pheromone gland in some coccinellid species (Coleoptera). *Abstracts of Ecology of Aphidophaga 9 Ceske Budejovice*, p. 70.

Rosenheim, J.A., Kaya, H.K., Ehler, L.E., Marois, J.J. and Jaffee, B.A. 1995. Intraguild predation among biological-control agents: theory and evidence. *Biological Control*, **5**, 303–35.

Rothschild, M. 1961. Defensive odours and Müllerian mimicry among insects. *Transactions of the Royal Entomological Society of London*, **113**, 101–22.

Rothschild, M. and Reichstein, T. 1976. Some problems associated with the storage of cardiac glycosides by insects. *Nova Acta Leopoldina, Supplement* **7**, 507–50.

Rothschild, M., von Euw, J. and Reichstein, T. 1973. Cardiac glycosides in a scale insect (*Aspidiotus*), a ladybird (*Coccinella*) and a lacewing (*Chrysopa*). *Journal of Entomology, A*, **48**, 89–90.

Roy, H.E., Adriaens, T., Isaac, N., et al. 2012. Invasive alien predator causes rapid declines of native European ladybirds. *Diversity and Distributions*, **18**, 717–25.

Roy, H.E., Baverstock, J., Ware, R.L., et al. 2008a. Intraguild predation of the aphid pathogenic fungus *Pandora neoaphidis* by the invasive coccinellid *Harmonia axyridis*. *Ecological Entomology*, **33**, 175–82.

Roy, H.E., Brown, P.M.J., Rowland, F. and Majerus, M.E.N. 2005. Ecology of *Harmonia axyridis*. *British Wildlife*, **August 2005**, 403–7.

Roy, H.E., Brown, P.M.J., Rothery, P., Ware, R.L. and Majerus, M.E.N. 2008b. Interactions between the fungal pathogen *Beauveria bassiana* and three species of coccinellid: *Harmonia axyridis*, *Coccinella septempunctata* and *Adalia bipunctata*. *BioControl*, **53**, 265–76.

Roy, H.E., Brown, P.M.J., Frost, R. and Poland, R.L. 2011. *Ladybirds (Coccinellidae) of Britain and Ireland*. Shrewsbury: Field Studies Council.

Roy, H.E., Brown, P.M.J., Comont, R., Poland, R.L. and Sloggett, J.J. 2013. *Ladybirds*. Naturalists' Handbooks No. 10. Exeter: Pelagic.

Roy, H.E. and Brown, P.M.J. 2015. Ten years of invasion: *Harmonia axyridis* (Pallas) (Coleoptera: Coccinellidae) in Britain. *Ecological Entomology*, **40**(4), 336–48.

Roy, H.E., Brown, P.M.J., Adriaens, T., et al. 2016. The harlequin ladybird, *Harmonia axyridis*: global perspectives on invasion history and ecology. *Biological Invasions*, **18**(4), 997–1044.

Roy, H.E. and Pell, J.K. 2000. Interactions between entomopathogenic fungi and other natural enemies: implications for biological control. *Biocontrol Science and Technology*, **10**, 737–52.

Rubtsov, I.A. 1954. *Citrus Pests and their Natural Enemies*. Moscow-Leningrad: Izdanii Akademii Nauk SSSR.

Ruzicka, Z. 2003. Perception of oviposition-deterring larval tracks in aphidophagous coccinellids *Cycloneda limbifer* and *Ceratomegilla undecimnotata* (Coleoptera: Coccinellidae). *European Journal of Entomology*, **100**, 345–50.

Ruzicka, Z. and Zemek, R. 2002. Effects of conspecific and heterospecific larval tracks on mobility and searching patterns of *C. limbifer* females (Coleoptera: Coccinellidae). *Proceedings of 8th International Symposium on Ecology of Aphidophaga*, Alphen aan den Rijn: Kluwer Academic Publishing; pp. 85–93.

Sakaguchi, B. and Poulson, D.F. 1963. Interspecific transfer of the "sex ratio" condition from *Drosophila willistoni* to *D. melanogaster*. *Genetics*, **48**, 841–61.

Sakurai, H. 1969. Respiration and glycogen contents in the adult life of the *Coccinella septempunctata* Mulsant and *Epilachna vigintioctopunctata* Fabricius (Coleoptera: Coccinellidae). *Applied Entomology and Zoology*, **4**, 55–7.

Sakurai, H., Goto, K., Mori, Y. and Takeda, S. 1981. Studies on the diapause of *Coccinella septempunctata bruckii* Mulsant. II. Role of corpus allatum related with diapause. *Research Bulletin of the Faculty of Agriculture, Gifu Univiversity*, **48**, 37–45.

Sakurai, H., Kumada, Y. and Takeda, S. 1993. Seasonal prevalence and hibernating-diapause behaviour in the ladybeetle, *Harmonia axyridis*. *Research Bulletin of the Faculty of Agriculture, Gifu University*, **58**, 51–5.

Salt, G. 1920. A contribution to the ethology of the Meliponinae. *Transactions of the Entomological Society of London*, **77**, 431–70.

Samways, M.J., Osborn, R. and Saunders, T.L. 1997. Mandible form relative to the main food type in ladybirds (Coleoptera: Coccinellidae). *Biocontrol Science and Technology*, **7**, 275–86.

Samways, M.J. and Wilson, S.J. 1988. Aspects of the feeding behaviour of *Chilocorus nigritus* (F.) (Col., Coccinellidae) relative to its effectiveness as a biocontrol agent. *Journal of Applied Entomology*, **106**, 177–82.

Sasaji, H. 1968. Phylogeny of the Family Coccinellidae (Coleoptera). *Etizenia*, **35**, 1–37.

Sasaji, H. 1971. *Coccinellidae. Fauna Japonica*. Tokyo: Academic Press of Japan.

Sato, S., Yasuda, H. and Evans, E.W. 2005. Dropping behaviour of larvae of aphidophagous ladybirds and its effect on incidence of intraguild predation: interactions between the intraguild prey, *Adalia bipunctata* (L.) and *Coccinella septempunctata* (L.), and the intraguild predator, *Harmonia axyridis* Pallas. *Ecological Entomology*, **30**, 220–4.

Saunders, D.S. 1982. *Insect Clocks*. 2nd edn. Oxford: Pergamon Press.

Savastano, L. 1918. Talune notizie sul *Novius* e l'*Icerya riguardanti* l'arboricoltore. *Boll. R. Staz. Sper. Agrum. Fruttic.*, **32**, 1–2.

Savoiskaya, G.I. 1965. Biology and perspectives of utilisation of coccinellids in the control of aphids in south-eastern Kazakhstan orchards. *Trudy Institut Zashchity Rastenii, Alma-Ata*, **9**, 128–56.

Savoiskaya, G.I. 1966. The significance of Coccinellidae in the biological control of apple-tree aphids in the Alma-Ata fruit-growing region. In Hodek, I., ed. *Ecology of Aphidophagous Insects*. Prague: Academia and The Hague: Dr W. Junk; pp. 317–19.

Savoiskaya, G.I. 1970. Introduction and acclimatisation of some coccinellids in the Alma-Ata reserve. *Trudy Alma-Atin.skii Gosudarstvennyi Zapovednik*, **9**, 163–7.

Savoiskaya, G.I. 1983. *Kokcinellidy*. Tokyo: Izdatelstvo Nauka Kazachskoi SSRAlma-Ata.

Savoiskaya, G.I. and Klausnitzer, B. 1973. Morphology and taxonomy of the larvae with keys for their identification. In Hodek, I., ed. *Biology of Coccinellidae*. Prague: Academia and The Hague: Dr W. Junk; pp. 36–55.

Scali, V. and Creed, E.R. 1975. The influence of climate on melanism in the two-spot ladybird, *Adalia bipunctata*, in central Italy. *Transactions of the Royal Entomological Society*, **127**, 163–9.

Schaefer, P.W., Dysart, R.J., Flanders, R.V., Burger, T.L. and Ikebe, K. 1983. Mexican bean beetle (Coleoptera: Coccinellidae) larval parasite *Pediobius foveolatus* (Hymenoptera: Eulophidae) from Japan: field release in the United States. *Environmental Entomology*, **12**, 852–4.

Schanderl, H., Ferran, A. and Garcia, V. 1988. L'Elevage de deux coccinelles *Harmonia axyridis* Pallas et *Semiadalia undecimpunctata* Schn. a l'aide d'oeufs d'*Anagasta kuehniella* Zell. tues aux rayons ultraviolets. *Entomologia Experimentalis et Applicata*, **49**, 235–44.

Schlinger, E.I. and Dietrick, E.J. 1960. Biological control of insect pests aided by strip-farming alfalfa in an experimental program. *California Agriculture*, **14**, 9–15.

Schröder, F.C., Farmer, J.J., Attygalle, A.B., et al. 1998. Combinatorial chemistry in insects: a library of defensive macrocyclic polyamines. *Science*, **281**, 428–31.

Schroder, R.F.W. 1982. Effect of infestation with *Coccipolipus epilachnae* Smiley (Acarina: Podapolipidae) on the fecundity and longevity of the Mexican bean beetle. *International Journal of Acarology*, **8**, 81–4.

Seko, T., Yamashita, K. and Miura, K. 2008. Residence period of a flightless strain of the ladybird beetle *Harmonia axyridis* Pallas (Coleoptera: Coccinellidae) in open fields. *Biological Control*, **47**, 194–8.

Sem'Yanov, V.P. 1970. Biological properties of *Adalia bipunctata* L. (Coleoptera, Coccinellidae) in conditions of Leningrad region. *Zashchita Rastenii ot Vreditelei i Boleznei*, **127**, 105–12.

Sem'Yanov, V.P. 1980. Biology of *Calvia quatuordecimguttata* L. (Coleoptera, Coccinellidae). *Entmologicheskoe Obozrenie*, **59**, 757–63.

Sengonca, C., Kotikal, Y.K. and Schade, M. 1995. Olfactory reactions of *Cryptolaemus montrouzieri* Mulsant (Col., Coccinellidae) and *Chrysoperla carnea* Stephens (Neur., Chrysopidae) in relation to period of starvation. *Anzieger Schädlingskunde Pflanzenschutz, Umweltschutz*, **68**, 9–12.

Sengonca, C. and Liu, B. 1994. Responses of the different instar predator, *Coccinella septempunctata* L. (Coleoptera: Coccinellidae), to the kairomones produced by the prey and non-prey insects as well as the predator itself. *Zeitschrift für Pflanzenkrankheiten ind Pflanzenschutz*, **101**, 173–7.

Sergievsky, A.O. and Zakharov, I.A. 1989. Population response on the stress effects: the concept of two-stage response. In *Ontogenesis, Evolution, Biosphere*. Moscow: Nauka Publishing; pp.157–73.

Shah, M.A. 1982. The influence of plant surfaces on the searching behaviour of coccinellid larvae. *Entomologie Experimentalis et Applicata*, **31**, 377–80.

Shaw, M.R., Geoghegan, I.E. and Majerus, M.E.N. 1999. Males of *Dinocampus coccinellae* (Schrank) (Hym.: Braconidae: Euphorinae). *Entomologist's Record and Journal of Variation*, **111**, 195–6.

Shi, X.W., Attygalle, A.B. and Meinwald, J. 1997. Defense mechanisms of arthropods. 149. Synthesis and absolute configuration of two defensive alkaloids from the Mexican bean beetle, *Epilachna varivestis*. *Tetrahedron Letters*, **38**, 6479–82.

Silvestri, F. 1903. Contribuzioni alla conoscenza dei Mirmecophili. I. Osservazioni su alcuni mirmecophili dei dintorni di Portici. *Annuario dell'1st e Museum di Zoologie Universite di Napoli*, **1**, 1–5.

Sivinski, J.M. and Petersson, E. 1997. Mate choice and species isolation in swarming insects. In Choe, J.C. and Crespi, B.J., eds. *Mating Systems in Insects and Arachnids*. Cambridge: Cambridge University Press; pp. 294–309.

Slipinski, S.A. and Pakaluk, J. 1991. Problems in the classification of the Cerylonid series of Cucujoidea (Coleoptera). In Zunino, M., Belles, X. and Blas, M., eds. *Advances in Coleopterology*. Barcelona: European Association of Coleopterology; pp. 79–88.

Sloggett, J.J. 1991. *Provisioning Strategies in* Adalia bipunctata *(L.)*. Unpublished Undergraduate Project Report, University of Cambridge.

Sloggett, J.J. 1998. *Interactions between Coccinellids (Coleoptera) and Ants (Hymenoptera: Formicidae), and the Evolution of Myrmecophily in* Coccinella magnifica *Redtenbacher*. Unpublished PhD thesis, University of Cambridge.

Sloggett, J.J. 2005. Are we studying too few taxa? Insights from aphidophagous ladybird beetles (Coleoptera: Coccinellidae). *European Journal of Entomology*, **102**, 391–8.

Sloggett, J.J., Magro, A., Verheggen, F.J., et al. 2011. The chemical ecology of *Harmonia axyridis*. *BioControl*, **56**, 643–61.

Sloggett, J.J. and Majerus, M.E.N. 2000a. Habitat preferences and diet in the predatory Coccinellidae (Coleoptera): an evolutionary perspective. *Biological Journal of the Linnean Society*, **70**, 63–88.

Sloggett, J.J. and Majerus, M.E.N. 2000b. Aphid-mediated coexistence of ladybirds (Coleoptera: Coccinellidae) and the wood ant *Formica rufa*: seasonal effects, interspecific variability and the evolution of a coccinellid myrmecophile. *Oikos*, **89**, 345–59.

Sloggett, J.J. and Majerus, M.E.N. 2000c. *Myrrha octodecimguttata* (L.) (Coleoptera: Coccinellidae), a newly recorded host of *Dinocampus coccinellae* (Schrank) (Hymenoptera: Braconidae). *British Journal of Entomology and Natural History*, **13**, 126.

Sloggett, J.J. and Majerus, M.E.N. 2003. Adaptations of *Coccinella magnifica*, a myrmecophilous coccinellid to aggression by wood ants (*Formica rufa* Group). II. Larval behaviour, and ladybird oviposition location. *European Journal of Entomology*, **100**, 337–44.

Sloggett, J.J., Völkl, W., Schulze, W., von der Schulenberg, J.H. and Majerus, M.E.N. 2002. The ant-associations and diet of the ladybird *Coccinella magnifica* (Coleoptera: Coccinellidae). *European Journal of Entomology*, **99**, 565–9.

Sloggett, J.J., Wood, R.A. and Majerus, M.E.N. 1998. Adaptations of *Coccinella magnifica* Redtenbacher, a myrmecophilous coccinellid, to aggression by wood ants (*Formica rufa* group). I. Adult behavioral adaptation, its ecological context and evolution. *Journal of Insect Behaviour*, **11**, 889–904.

Smith, B.C. 1960. A technique for rearing coccinellid beetles on dry foods, and influence of various pollens on the development of *Coleomegilla maculata lengi* Timb. (Coleoptera: Coccinellidae). *Canadian Journal of Zoology*, **38**, 1047–9.

Smith, B.C. 1971. Effects of various factors on the local distribution and density of coccinellid adults on corn (Coleoptera, Coccinellidae). *Canadian Entomologist*, **103**, 1115–20.

Smith, H.S. 1919. On some phases of insect control by the biological method. *Journal of Economic Entomology*, **12**, 288–92.

Smith, H.S. and Armitage, H.M. 1920. Biological control of mealybugs in California. *Monthly Bulletin of the California State Department of Agriculture*, **9**, 104–61.

Smith, J.B. 1886. Ants' nests and their inhabitants. *American Naturalist*, **20**, 679–87.

Smith, S.G. 1959. The cytogenetic basis of speciation in Coleoptera. *10th International Conference of Genetics*, **1**, 444–50.

Smith, S.G. 1960. Chromosome numbers of Coleoptera. II. *Canadian Journal of Genetics and Cytology*, **2**, 67–88.

Snyder, W.E., Clevenger, G.M. and Eigenbrode, S.D. 2004. Intraguild predation and successful invasion by introduced ladybird beetles. *Oecologia*, **140**, 559–65.

Solbreck, C. 1974. Maturation of post-hibernation flight behaviour in the coccinellid *Coleomegilla maculata* (DeGeer). *Oecologia*, **17**, 265–75.

Southwood, T.R.E. 1977. Habitat the templet for ecological studies? *Journal of Animal Ecology*, **46**, 337–65.

Stadler, B. 1991. Predation success of *Coccinella septempunctata* when attacking different *Uroleucon* species. In Polgar, L., Chambers R.J., Dixon, A.F.G. and Hodek, I., eds. *Behaviour and Impact of Aphidophaga*. The Hague: SPB Academic Publishing; pp. 265–71.

Staines, C.L. Jr., Rothchild, M.J. and Trumble, R.B. 1990. A survey of the Coccinellidae (Coleoptera) associated with nursery stock in Maryland. *Proceedings of the Entomological Society of Washington*, **92**, 310–13.

Stals, R. 2010. The establishment and rapid spread of an alien invasive lady beetle: *Harmonia axyridis* (Coleoptera: Coccinellidae) in southern Africa, 2001–2009. *IOBC WPRS Bulletin*, **58**, 125–32.

State of Nature partnership. 2013. *The State of Nature Report*. RSPB: Sandy.

Stevens, L.M., Steinhauer, A.L. and Coulson, J.R. 1975. Suppression of Mexican bean beetle on soybeans with annual inoculative releases of *Pediobius foveolatus*. *Environmental Entomology*, **4**, 947–52.

Stewart, J.W., Whitcomb, W.H. and Bell, K.O. 1967. Estivation studies of the convergent lady beetle in Arkansas. *Journal of Economic Entomology*, **60**, 1730–5.

Stewart, L.A., Dixon, A.F.G., Ruzicka, Z. and Iperti, G. 1991b. Clutch and egg size in ladybird beetles. *Entomophaga*, **36**, 329–33.

Stewart, L.A., Hemptinne, J.-L. and Dixon, A.F.G. 1991a. Reproductive tactics of ladybird beetles: relationships between egg size, ovariole number and developmental time. *Functional Ecology*, **5**, 380–5.

Strand, M.R. and Obrycki, J.J. 1996. Host specificity of insect parasitoids and predators. *Bioscience*, **46**, 422–9.

Strong-Gunderson, J.M., Lee, R.E. and Lee, M.R. 1992. Topical application of ice-nucleating-active bacteria decreases insect cold tolerance. *Applied Environmental Microbiology*, **58**, 2711–16.

Stubbs, M. 1980. Another look at prey detection by coccinellids. *Ecological Entomology*, **5**, 179–82.

Sunderland, K.D. 1978. Studies on polyphagous predators of cereal aphids at North Farm. *Glasshouse Crops Research Institute (Littlehampton) Annual Report*, **1978**, 118–19.

Suter, H. and Keller, S. 1977. Richtlinien fur die Durchfuhrung einer mittelfristigen Blattlausprognose in Feldkulturen. *Mitteilungen für die Schweizerische Landwirtschaft*, **25**, 65–9.

Takahashi, K. 1987. Cannibalism by larvae of *Coccinella septempunctata brucki* Mulsant (Coleoptera: Coccinellidae) in mass-rearing experiments. *Japanese Journal of Applied Entomology and Zoology*, **31**, 201–5.

Takahashi, K. 1989. Intra- and interspecific predations of lady beetles in spring alfalfa fields. *Japanese Journal of Entomology*, **57**, 199–203.

Takizawa, T., Yasuda, H. and Agarwala, B.K. 2000. Effect of three species of predatory ladybirds on oviposition of aphid parasitoids. *Entomological Science*, **3**, 465–9.

Tamaki, G., Annis, B. and Weiss, M. 1981. Response of natural enemies to the green peach aphid in different plant cultures. *Environmental Entomology*, **10**, 375–8.

Tan, C.-C. 1949. Seasonal variation of color patterns in *Harmonia axyridis*. *Proceedings of the 8th International Congress of Genetics*. Bonnier, G., ed. Lund, Stokholm: Berlingska Boktryckeriet; pp. 669–70.

Tan, C.-C. and Li, J.-C. 1934. Inheritance of the elytral color patterns in the lady-bird beetle, *Harmonia axyridis* Pallas. *American Naturalist*, **68**, 252–65.

Tanagishi, K. 1976. Hibernation of the lady beetle, *Harmonia axyridis*. *Insectarium*, **13**, 294–8.

Tao, C.C. and Chiu, S.C. 1971. Biological control of citrus, vegetables and tobacco aphids. *Special Publications of the Taiwan Agricultural Research Institute*, **10**, 1–110.

Taylor, H., Taylor, N. and Majerus, M.E.N. 1996. Predation of *Aphidecta obliterata*, L. (Col.: Coccinellidae) by rainbow trout *Salmo irideus*. *Entomologist's Record and Journal of Variation*, **108**, 28–9.

Tedders, W.L. and Schaefer, P.W. 1994. Release and establishment of *Harmonia axyridis* (Col: Cocc.) in the southeastern United States. *Entomological News*, **105**, 228–43.

Telenga, N.A. 1948. *Biological Method of the Insect Pest Control (Predaceous Coccinellids and their Utilisation in USSR)*. Kiev: Izdanii Akademii Nauk SSSR.

Telenga, N.A. and Bogunova, M.V. 1936. The most important predators of coccids and aphids in the Ussuri region of Far East and their utilisation. *Zashchita Rastenii*, 1936, **10**, 75–87.

Thomas, J.A., Simcox, D.J. and Clarke, R.T. 2009. Successful conservation of a threatened Maculinea butterfly. *Science*, **325**, 80–3.

Thompson, W.R. 1951. The specificity of host relations in predaceous insects. *Canadian Entomologist*, **83**, 262–9.

Thornhill, R. 1976. Sexual selection and nuptial feeding behaviour in *Bittacus apicalis* (Insecta: Mecoptera). *American Naturalist*, **110**, 529–48.

Timofeeff-Ressovsky, N.W. 1940. Zur analyse des Polymorphismus bei *Adalia bipunctata* L. *Biologisches Zentralblatt*, **60**, 130–7.

Tinsley, M.C. 2003. *The Ecology and Evolution of Male-killing Bacteria in Ladybirds*. Unpublished PhD thesis, University of Cambridge.

Tourniaire, R., Ferran, A., Gambier, J., Giuge, L. and Bouffault, F. 2000. Locomotory behavior of flightless *Harmonia axyridis* Pallas (Col., Coccinellidae). *Journal of Insect Physiology*, **46**, 721–6.

Tregenza, T. and Wedell, N. 1998. Benefits of multiple mating in the cricket *Gryllus bimaculatus*. *Evolution*, **52**, 1726–30.

Trivers, R.L. 1972. Parental investment and sexual selection. In Campbell, B., ed. *Sexual Selection and the Descent of Man*. London: Heineman; pp. 136–79.

Trouve, C., Ledee, S., Ferran, A. and Brun, J. 1997. Biological control of the damson-hop aphid, *Phorodon humuli* (Hom.: Aphididae), using the ladybeetle *Harmonia axyridis* (Col.: Coccinellidae). *Entomophaga*, **42**, 57–62.

Turian, G. 1969. Coccinelles micromycétophages (Col.). *Mitteilungen für die Schweizerische Landwirtschaft*, **42**, 52–7.

Turnock, W.J., Wise, I.L. and Matheson, F.O. 2003. Abundance of some native coccinellines (Coleoptera: Coccinellidae) before and after the appearance of *Coccinella septempunctata*. *Canadian Entomologist*, **135**, 391–404.

Tursch, B., Braekman, J.C., Daloze, D., et al. 1973. Chemical ecology of arthropods. VI. Adaline, a novel alkaloid from *Adalia bipunctata* L. (Coleoptera: Coccinellidae). *Tetrahedron*, **29**, 201.

Tursch, B., Daloze, D., Dupont, M., et al. 1971. Coccinelline, the defensive alkaloid of the beetle *Coccinella septempunctata*. *Chemia*, **25**, 307.

Tursch, B., Daloze, D., Pasteels, J.M., et al. 1972. Two novel alkaloids from the American ladybug *Hippodamia convergens* (Coleoptera: Coccinellidae). *Bulletin des Société Chimiques Belges*, **81**, 649–50.

Tursch, B., Daloze, D., Braekman, J.C., et al. 1974. Chemical ecology of arthropods. IX. Structure and absolute configuration of hippodamine and convergine, two novel alkaloids from the American ladybug *Hippodamia convergens* (Coleop., Cocc.). *Tetrahedron*, **30**, 409–12.

Tursch, B., Daloze, D., Braekman, J.C., Hootele, C. and Pasteels, J.M. 1975. Chemical ecology of arthropods. X. The structure of myrrhine and the biosynthesis of coccinelline. *Tetrahedron*, **31**, 1541–3.

Ueno, H. 1994. Intraspecific variation of P2 value in a coccinellid beetle, *Harmonia axyridis*. *Journal of Ethology*, **12**, 169–74.

Ueno, H., de Jong, P.W. and Brakefield, P.M. 2004. Genetic basis and fitness consequences of winglessness in the two-spot ladybird beetle, *Adalia bipunctata*. *Heredity*, **93**, 283–9.

Vandenberg, N.J. 2002. Family 93. Coccinellidae Latreille 1807. In Arnett, R.H., Thomas, M.C., Skelley, P.E. and Frank, J.H., eds. *American Beetles, Volume II. Polyphaga: Scarabaeoidea through Curculionoidea*. Boca Raton, FL: CRC Press; pp.371–89.

Van den Meiracker, R.A.F., Hammond, W.N.O. and van Alphen, J.J.M. 1990. The role of kairomones in prey finding by *Diomus* sp. and *Exochomus* sp., two coccinellid predators of the cassava mealybug, *Phenococcus manihoti*. *Entomologia Experimentalis et Applicata*, **56**, 209–17.

Van Lenteren, J.C., Babendreier, D., Bigler, F., et al. 2003. Environmental risk assessment of exotic natural enemies used in inundative biological control. *BioControl*, **48**, 3–38.

Van Valen, L.M. 1973. A new evolutionary law. *Evolutionary Theory*, **1**, 1–30.

Veneti, Z., Bentley, J.K., Koana, T., Braig, H.R. and Hurst, G.D. 2005. A functional dosage compensation complex required for male killing in Drosophila. *Science*, **307**, 1461–3.

Verheggen, F.J., Fagel, Q., Heuskin, S., et al. 2007. Electrophysiological and behavioral responses of the multicolored Asian lady beetle, *Harmonia axyridis* Pallas, to sesquiterpene semiochemicals. *Journal of Chemical Ecology*, **33**, 2148–55.

Vila, M., Espinar, J.L., Hejda, M., et al. 2011. Ecological impacts of invasive alien plants: a meta-analysis of their effects on species, communities and ecosystems. *Ecology Letters*, **14**, 702–8.

Vilcinskas, A., Stoecker, K., Schmidtberg, H., Roehrich, C.R. and Vogel, H. 2013. Invasive harlequin ladybird carries biological weapons against native competitors. *Science*, **340**, 862–3.

Vohland, K. 1996. The influence of plant structure on searching behaviour in the ladybird, *Scymnus nigrinus* (Coleoptera: Coccinellidae). *European Journal of Entomology*, **93**, 151–60.

Völkl, W. 1995. Behavioural and morphological adaptations of the coccinellid, *Platynaspis luteorubra* for exploiting ant-attending resources (Coleoptera: Coccinellidae). *Journal of Insect Behaviour*, **8**, 273–81.

Völkl, W. and Vohland, K. 1996. Wax covers in larvae of two *Scymnus* species: do they enhance coccinellid larval survival? *Oecologia*, **107**, 498–503.

Von der Schulenburg, J.H.G., Habig, M., Sloggett, J.J., et al. 2001b. The incidence of male-killing *Rickettsia* (alpha proteobacteria) in the 10-spot ladybird, *Adalia decempunctata* L. (Coleoptera: Coccinellidae). *Applied Environmental Microbiology*, **67**, 270–7.

Von der Schulenburg, J.H.G., Hancock, J.M., Pagnamenta, A., et al. 2001a. Extreme length and length variation in the first ribosomal internal transcribed spacer of ladybirds (Coleoptera: Coccinellidae). *Molecular Biology and Evolution*, **18**, 648–60.

Wang, L.Y. 1986. Mass rearing and utilization in biological control of the ladybeetle *Leis axyridis* (Pallas). *Acta Entomologica Sinica*, **29**, 104.

Ware, R.L., Evans, N., Malpas, L., et al. 2008. Intraguild predation by the invasive ladybird *Harmonia axyridis*. 1: British and Japanese coccinellid eggs. *Neobiota*, **7**, 263–75.

Ware, R.L. and Majerus, M.E.N. 2008. Intraguild predation of immature stages of British and Japanese coccinellids by the invasive ladybird *Harmonia axyridis*. *BioControl*, **53**, 169–88.

Ware, R.L., Yguel, B. and Majerus, M.E.N. 2009. Effects of competition, cannibalism and intra-guild predation on larval development of the European coccinellid *Adalia bipunctata* and the invasive species *Harmonia axyridis*. *Ecological Entomology*, **34**, 12–19.

Warren, L.O. and Tadic, M. 1967. Biological observations on *Coleomegilla maculata* and its role as a predator of the fall webworm. *Journal of Economic Entomology*, **60**, 1492–6.

Wassmann, E. 1894. *Kritisches Verzeichniss der Myrmekophilen und Termitophilen Arthropoden*. Berlin: Verlag von Felix L. Dames.

Wassmann, E. 1912. The ants and their guests. *Annual Report of the Smithsonian Institute*, pp. 455–74.

Way, M.J. 1963. Mutulism between ants and honey-dew producing Homoptera. *Annual Review of Entomology*, **8**, 307–44.

Webberley, K.M., Buszko, J., Isham, V. and Hurst, G.D.D. 2006b. Sexually transmitted disease epidemics in a natural insect population. *Journal of Animal Ecology*, **75**, 33–43.

Webberley, K.M., Hurst, G.D.D., Buszko, J. and Majerus, M.E.N. 2002. Lack of parasite mediated sexual selection in a ladybird-sexually transmitted disease system. *Animal Behaviour*, **63**, 131–41.

Webberley, K.M., Hurst, G.D.D., Husband, R.W., et al. 2004. Host reproduction and an STD: causes and consequences of *Coccipolipus hippodamiae* distribution on coccinellids. *Journal of Animal Ecology*, **73**, 1–10.

Webberley, K.M., Tinsley, M., Sloggett, J.J., Majerus, M.E.N. and Hurst, G.D.D. 2006a. Spatial variation in the incidence of sexually transmitted parasites of the ladybird beetle *Adalia bipunctata*. *European Journal of Entomology*, **103**, 793–7.

Welch, V.L., Sloggett, J.J., Webberley, K.M. and Hurst, G.D.D. 2001. Short range clinal variation in the prevalence of a sexually transmitted fungus associated with urbanization. *Ecological Entomology*, **26**, 547–50.

Wellenstein, G. 1952. Zur Ernahrungsbiologie der Roten Waldameise (*Formica rufa* L.). *Z. Pflanzenkrankh. u. Pflanzenschutz*, **59**, 430–51.

Werren, J.H. 1987. The coevolution of autosomal and cytoplasmic sex ratio factors. *Journal of Theoretical Biology*, **124**, 317–34.

Werren, J.H., Hurst, G.D.D., Zhang, W., et al. 1994. Rickettsial relative associated with male killing in the ladybird beetle (*Adalia bipunctata*). *Journal of Bacteriology*, **176**, 388–94.

Wheeler, A.G. Jr. and Stoops, C.A. 1996. Status and spread of the Palearctic lady beetles *Hippodamia variegata* and *Propylea quatuordecimpunctata* (Coleoptera: Coccinellidae) in Pennsylvania, 1993–1995. *Entomological News*, **107**, 291–8.

Wheeler, W.M. 1911. An ant-nest coccinellid (*Brachyancantha quadripunctata* (Mels.)) *Journal of the New York Entomological Society*, **19**, 169–74.

Williams, G.C. 1966. *Adaptation and Natural Selection*. Princeton, NJ: Princeton University Press.

Williamson, M. 1996. *Biological Invasions*. London: Chapman and Hall.

Williamson, M. 1999. Invasions. *Ecography*, **22**, 5–12.

Wipperfürth, T., Hagen, K.S. and Mittler, T.E. 1987. Egg production by the coccinellid *Hippodamia convergens* fed on two morphs of the green peach aphid, *Myzus persicae*. *Entomologie Experimentalis et Applicata*, **44**, 195–8.

Witte, L., Ehmke, A. and Hartmann, T. 1990. Interspecific flow of pyrrolizidine alkaloids. *Naturwissenschaften*, **77**, 540–3.

Wootton, R.J. 1979. Function, homology and terminology in insect wings. *Systematic Entomology*, **4**, 81–93.

Wootton, R.J. 1981. Support and deformability in insect wings. *Journal of Zoology, London*, **193**, 447–68.

Wootton, R.J. 1992. Functional morphology of insect wings. *Annual Review of Entomology*, **37**, 113–40.

Wratten, S.D. 1976. Searching by *Adalia bipunctata* (L.) (Coleoptera:Coccinellidae) and escape behaviour of its aphid and ciccadellid prey on lime (*Tilia* x *vulgaris* Hayne). *Ecological Entomology*, **1**, 139–42.

Wratten, S.D., Gurr, G.M., Kehrli, P. and Scarratt, S. 2005. Cultural manipulations to enhance biological control in Australia and New Zealand: progress and prospects. *Proceedings of International Symposium on Biological Control of Aphids and Coccids*, Yamagata University; pp. 39–44.

Wright, S. 1982. The shifting balance theory and macroevolution. *Annual Review of Genetics*, **16**, 1–19.

Wynne-Edwards, V.C. 1962. *Animal Dispersal in Relation to Social Behaviour*. Edinburgh: Oliver and Boyd.

Wyss, E., Villiger, M., Hemptinne, J.-L. and Muller-Scharer, H. 1999. Effects of augmentative releases of eggs and larvae of the two-spot ladybird beetle, *Adalia bipunctata*, on the abundance of the rosy apple aphid, *Dysaphis plantaginea*, in organic apple orchards. *Entomologia Experimentalis et Applicata*, **90**, 167–73.

Yakhontov, V.V. 1960a. Seasonal migrations of lady-birds *Brumus octosignatus* Gebl and *Semiadalia undecimnotata* Schneid in Central Asia. *Proceedings of the 11th International Congress of Entomology*, **1**, 21–3.

Yakhontov, V.V. 1960b. *Utilisation of Coccinellids in the Control of Agricultural Pests*. Tashkent: Izdanii, Akademiia Nauk Uzbekskoi SSR.

Yarbrough, J.A., Armstrong, J.L., Blumberg, M.Z., et al. 1999. Allergic rhinoconjunctivitis caused by *Harmonia axyridis* (Asian lady beetle, Japanese lady beetle, or lady bug). *Journal of Allergy and Clinical Immunology*, **104**, 705.

Yasuda, H., Evans, E.W., Kajita, Y., Urakawa, K. and Takizawa, T. 2004. Asymmetric larval introductions between introduced and indigenous ladybirds in North America. *Oecologia*, **B141**, 722–31.

Yasuda, H., Kikuchi, T., Kindlmann, P. and Sato, S. 2001. Relationships between attacks and escape rates, cannibalism, and intraguild predation in larvae of two predatory ladybirds. *Journal of Insect Behavior*, **14**, 373–84.

Yasuda, H. and Ohnuma, N. 1999. Effect of cannibalism and predation on the larval performance of two ladybird beetles. *Entomologia Experimentalis et Applicata*, **93**, 63–7.

Yasuda, H. and Shinya, Y. 1997. Cannibalism and interspecific predation in two predatory ladybirds in relation to prey abundance in the field. *Entomophaga*, **42**, 153–63.

Yasui, Y. 1997. A "good-sperm" model can explain the evolution of costly multiple mating by females. *American Naturalist*, **149**, 573–84.

Yasui, Y. 1998. The 'genetic benefits' of female multiple mating reconsidered. *Trends in Ecology and Evolution*, **13**, 246–50.

Yinon, U. 1969. Food consumption of the armoured scale lady-beetle, *Chilocorus bipustulatus* (Coleoptera, Coccinellidae). *Entomologia Experimentalis et Applicata*, **12**, 139–46.

Young, O.P. 1989. Interactions between the predators *Phidippus audax* (Araneae: Salticidae) and *Hippodamia convergens* (Coleoptera: Coccinellidae) in cotton and in the laboratory. *Entomological News*, **100**, 43–7.

Zahavi, A. 1975. Mate selection – selection for a handicap. *Journal of Theoretical Biology*, **67**, 603–5.

Zaslavskii, V.A. and Bogdanova, T.P. 1965. Properties of imaginal diapause in two *Chilocorus* species (Coleoptera, Coccinellidae). *Trudy Zoologicheskii Institut, Leningrad*, **36**, 89–95.

Zeh, J.A. and Zeh, D.W. 1996. The evolution of polyandry. I: Intragenomic conflict and genetic incompatibility. *Proceedings of the Royal Society of London B*, **263**, 1711–17.

Zeh, J.A. and Zeh, D.W. 1997. The evolution of polyandry. II: Post-copulatory defences against genetic incompatibility. *Proceedings of the Royal Society of London B*, **264**, 69–75.

Zotov, V.A. 2008. Circadian rhythm of light sensitivity in beetles *Coccinella septempunctata* (Coleoptera, Coccinellidae): effects of illumination and temperature. *Zoologichesky Zhurnal*, **87**, 1472–5.

Index